Solar Heating and Cooling Systems

Solar Heating and Cooling Systems

Fundamentals, Experiments and Applications

Ioan Sarbu

Calin Sebarchievici

ELSEVIER

AMSTERDAM • BOSTON • HEIDELBERG • LONDON
NEW YORK • OXFORD • PARIS • SAN DIEGO
SAN FRANCISCO • SINGAPORE • SYDNEY • TOKYO
Academic Press is an imprint of Elsevier

Academic Press is an imprint of Elsevier
125 London Wall, London EC2Y 5AS, United Kingdom
525 B Street, Suite 1800, San Diego, CA 92101-4495, United States
50 Hampshire Street, 5th Floor, Cambridge, MA 02139, United States
The Boulevard, Langford Lane, Kidlington, Oxford OX5 1GB, United Kingdom

Notices
Knowledge and best practice in this field are constantly changing. As new research and experience broaden our understanding, changes in research methods, professional practices, or medical treatment may become necessary.

Practitioners and researchers must always rely on their own experience and knowledge in evaluating and using any information, methods, compounds, or experiments described herein. In using such information or methods they should be mindful of their own safety and the safety of others, including parties for whom they have a professional responsibility.

To the fullest extent of the law, neither the Publisher nor the authors, contributors, or editors, assume any liability for any injury and/or damage to persons or property as a matter of products liability, negligence or otherwise, or from any use or operation of any methods, products, instructions, or ideas contained in the material herein.

Library of Congress Cataloging-in-Publication Data
A catalog record for this book is available from the Library of Congress

British Library Cataloguing-in-Publication Data
A catalogue record for this book is available from the British Library

ISBN: 978-0-12-811662-3

For information on all Academic Press publications
visit our website at https://www.elsevier.com/

Working together
to grow libraries in
developing countries

www.elsevier.com • www.bookaid.org

Publisher: Joe Hayton
Acquisition Editor: Lisa Reading
Editorial Project Manager: Maria Convey
Production Project Manager: Mohana Natarajan
Cover Designer: Mark Rogers

Typeset by TNQ Books and Journals

Contents

Author Biographies

Ioan Sarbu is a professor and head of the Department of Building Services Engineering at the Polytechnic University of Timisoara, Romania. He obtained a diploma in civil engineering from the "Traian Vuia" Polytechnic Institute of Timisoara in 1975 and a PhD degree in civil engineering from the Timisoara Technical University in 1993. He is a European engineer, as designated by the European Federation of National Engineering Associations (Brussels) in 2001.

His main research interests are related to refrigeration systems, heat pumps, and solar energy conversion. He is also active in the field of thermal comfort and environmental quality, energy efficiency and energy savings, and numerical simulations and optimisations in building services. He is involved in preparation of regulations and standards related to energy and environment. Additionally, he is a member of the American Society of Heating, Refrigerating and Air-Conditioning Engineers (ASHRAE), International Association for Hydro-Environment Engineering and Research (IAHR), Romanian Association of Building Services Engineers (RABSE), and Society for Computer Aided Engineering (SCAE). He is a doctoral degree advisor in the civil engineering branch, an expert reviewer on the National Board of Scientific Research for Higher Education (Bucharest), vice president of the National Board of Certified Energetically Auditors Buildings (Bucharest), a member of the National Council for Validation of University Titles, Diplomas and Certificates, a member of the Technical Council for Civil Engineering from Ministry of Regional Development and Public Administration, and a reviewer of the *Journal of Thermal Science and Technology, Energy Conversion and Management, Applied Thermal Engineering, International Journal of Refrigeration, International Journal of Sustainable Energy, Energy and Buildings, Advances in Mechanical Engineering, Energy Efficiency, Applied Energy, Energy, and Thermal Science*. He was also supervisor of up to 20 PhD students, more than 40 MSc students, and of about 300 postgraduate students.

Honors (citation, awards, etc.): Award, Romanian General Association of Engineers, 1997; Diploma, Outstanding Scientist of the 21st Century, IBC,

2005; World Medal of Freedom, ABI, 2006; Plaque, Gold Medal for Romania, ABI, 2007; Lifetime Achievement Award, ABI, 2008; Ultimate Achiever Award Award Certificate for Engineering, IBC, 2009; Plaque, Hall of Fame for Distinguished Accomplishments in Science and Education, ABI, 2009; The Albert Einstein Award of Excellence, ABI, 2010; Diploma, Top 100 Engineers, IBC, 2012; Certificate for Outstanding Engineering Achievement, IBC, 2013; Medal of Achievement, IBC, 2014, Decoration for Achievement in Engineering, IBC, 2015; listed in several Who's Who publications (e.g., *Who's Who in the World, Who's Who in Science and Engineering*, and *Who's Who in America*) and other biographical dictionaries.

He has published 14 books, 7 chapters to books, more than 130 articles in indexed journals, and about 50 articles in proceedings of international conferences. He is also author of 5 patent certificates, 1 utility model, and of up to 20 computer programs.

Calin Sebarchievici is a lecturer in the Building Services Engineering Department at the Polytechnic University of Timisoara, Romania. He obtained a diploma in building services engineering from the Polytechnic University of Timisoara in 2003. He received his master's degrees from the same university and Bologna University, Italy, in 2004 and 2007, respectively, and his PhD degree in civil engineering from the Polytechnic University of Timisoara in 2013.

His research is focused on air conditioning, heat pump and refrigeration systems. He is also active in the field of thermal comfort, energy efficiency, and energy savings. Additionally, he is a member of the ASHRAE, RABSE, and GEO-EXCHANGE Romanian Society. He was also supervisor–mentor of more than 30 postgraduate students. He is coauthor of 3 books, 4 book chapters, 1 utility model, 17 journal articles, and more than 10 conference proceeding publications.

Preface

Energy is vital for progress and development of a nation's economy. The economic growth and technological advancement of every country depends on energy and the amount of available energy reflects that country's quality of life. Economy, population, and per capita energy consumption have increased the demand for energy during the last several decades. Fossil fuels continue to supply much of the energy used worldwide, and oil remains the primary energy source. However, fossil fuels are a major contributor to global warming. The awareness of global warming has been intensified in recent times and has reinvigorated the search for energy sources that are independent of fossil fuels and contribute less to global warming. Among the energy sources alternative to fossil fuels, renewable energy sources (RESs) such as solar and wind garner the public's attention, as they are available and have fewer adverse effects on the environment than do fossil fuels.

It is well known that there is a need to develop technologies to achieve thermal comfort in buildings lowering the cooling and heating demand.

Thermal energy obtained from the sun with a solar thermal or photovoltaic (PV) system can be used for domestic water heating, and heating or even cooling of buildings. Solar domestic hot-water (DHW) systems prevail because the hot-water requirement can be well covered by the solar energy offer. Air-conditioning systems are the dominating energy consumers in buildings in many countries, and their operation causes high electricity peak loads during the summer. The solar cooling technology can reduce the environmental impact and the energy consumption issues raised by conventional air-conditioning systems.

Solar energy can potentially contribute to 10% of the energy demand in Organisation for Economic Co-operation and Development (OECD) countries if all cooling and heating systems would be driven by solar energy. Data published by European Solar Thermal Industry Federation (ESTIF) indicates that even under such difficult economic conditions due to the financial and economic crisis, solar heating and cooling technologies has proved to be a driver in the European energy strategy, especially toward achieving the 2020 targets.

This book treats a modern issue of great current interest at a high scientific and technical level, based both on original research and achievements and on the synthesis of consistent bibliographic material to meet the increasing need for modernization and for greater energy efficiency of building services to significantly reduce CO_2 emissions. The text offers a comprehensive and consistent overview of all major solar energy technologies. The book is an impressive example of putting seamlessly together fundamentals, experiments, and applications of solar heating and cooling systems and covers the technical, economic, and energy savings aspects related to design, modeling, and operation of these systems. In addition, physical and mathematical concepts are introduced and developed sufficiently to make this publication a self-contained and up-to-date source of information for readers. The topics are presented in the logical and pedagogical order. Each of the chapters involves a learning aims, chapter summary, quick questions, and extensive bibliography. This book is structured as nine chapters. Chapter 1 summarizes a description of RESs covering some general aspects of regional policies and presents the necessity for using solar energy in heating/cooling of buildings and DHW production.

Chapter 2 presents the main characteristics of solar energy and exposes a methodology for calculating and predicting solar radiation including the main computation elements and the estimation of solar radiation on tilted surface likely to be available as input to a solar device or crop at a specific location, orientation, and time.

Chapter 3 presents a detailed description of energy balance for solar collector and of different types of solar thermal and PV collectors including the calculation of their efficiency and new materials for PV cells. The discussion considers both concentrating and nonconcentrating thermal technologies. Additionally, a brief overview on the research and development and application aspects for the hybrid PV/thermal (PV/T) collector systems is presented. Additionally, a simulation model of a PV/T collector with water heating in buildings closer to the actual system has been developed to analyze the PV and thermal performances of this system. The energetic performance of commercially available PV/T systems for electricity and DHW production is also evaluated for use in three European countries, each under entirely different climatic conditions.

Chapter 4 is focused on the analysis of thermal energy storage (TES) technologies that provide a way of valorizing solar heat and reducing the energy demand of buildings. The principles of several energy storage methods and calculation of storage capacities are described. Sensible heat storage technologies, including the use of water, underground and packed-bed are briefly reviewed. Latent heat storage systems associated with phase change materials (PCMs) for use in solar heating and cooling of buildings, solar water heating and heat pumps (HPs) systems, and thermochemical heat storage are also presented. Additionally, a three-dimensional heat transfer simulation model of latent heat TES is developed to investigate the quasisteady state and transient heat transfer of PCMs. The numerical simulation results using paraffin RT20 are compared with available experimental data for cooling and heating of buildings. Finally, outstanding information on the performance and costs of TES systems are included.

Chapter 5 provides a description of main types of solar space and water heating systems, concentrating on classifications, system components, and operation principles. It is also focused on active and combisystems. Important information on the space heating/cooling load calculation and the selection of the solar thermal systems are discussed. The f-chart method applicable to evaluate space and water heating in many climates and conditions and Transient System Simulation (TRNSYS) program is briefly described. Additionally, some installation, operation, and maintenance instructions for solar heating systems are analyzed and examples of DHW systems and combisystems application are presented. Finally, valuable information on the solar district heating and solar energy use for industrial applications is provided. A comprehensive discussion about pump control in district heating plants and the energy efficiency analysis of flow control methods is provided and also an optimal design model of hot-water branched heating networks based on the linear programming method is developed.

Chapter 6 present the heat distribution systems in buildings, including hot-water radiators, radiant panels (floor, wall, ceiling, and floor—ceiling), and room air heaters. First objective of this study is the analysis of the energy savings in central heating systems with reduced supply temperature, for different types of radiators taking into account the thermal insulation of the distribution pipes and the performance investigation of different types of low-temperature heating system with different methods. Additionally, a mathematical model for numerical modeling of the thermal emission at radiant floors is developed and experimentally validated, and a comparative analysis of the energy, environmental and economic performances of floor, wall, ceiling, and floor—ceiling heating using numerical simulation with TRNSYS software is performed. Finally, important information for control and efficiency of solar heating systems is included, an analytical model for energetically analysis of the solar heating systems is developed and some economic analysis indicators are presented to show the opportunity to implement these systems in buildings.

Chapter 7 provides a detailed review of different solar thermal-driven refrigeration and cooling systems. Theoretical basis and practical applications for cooling systems within various working fluids assisted by solar energy and their recent advances are presented. The first aim of this chapter is to give an overview of the state-of-the-art sorption and thermomechanical technologies that are available to deliver cooling from solar energy. The second aim is to compare the potential of these technologies in delivering competitive sustainable solutions. The topics approached in present chapter is similar to that of the other work, but it is focused on solar closed sorption refrigeration systems providing useful information updated and more extensive on their principles, development history, applications, and recent advances. The application areas of these technologies are categorized by their cooling temperature demands (air-conditioning, refrigeration, ice making). Additionally, the thermodynamic properties of most common working fluids and the use of ternary mixtures in solar-powered absorption systems are reviewed. A mini type solar absorption cooling system using both fan coils and the radiant ceiling as terminals was designed and installed. The system performance as well as the indoor thermal comfort is analyzed. Finally, some information on design, control, and operation of hybrid cooling and heating systems are included. The study also refers to a comparison of various solar thermal-powered cooling systems and to some use suggestions of these systems.

Chapter 8 covers solar electric cooling systems including the solar PV and thermoelectric systems. Thus the utilization of solar PV panels coupled with a vapor-compression air-conditioning system is described, and a good amount of information regarding ecological refrigerants trend is included. Additionally, the chapter presents the details referring to thermoelectric cooling parameters and formulations of the performance indicators and focuses on the development of thermoelectric cooling systems in recent decade with particular attention on advances in materials and modeling approaches, and applications.

Chapter 9 presents the operation principle of a HP, discusses the vapor compression—based HP systems, and describes the thermodynamic cycle and they calculation, as well as operation regimes of a vapor-compression HP with electro-compressor. The calculation of greenhouse gas emissions of HPs and energy and economic performance criteria that allow for implementing an IIP in a heating/cooling system is considered. A detailed description of the HP types and ground-source HP development is presented and important information on the selection of the heat source and HP systems are discussed. Additionally, other approach is to integrate the solar thermal system on the source side of the HP so that the solar thermal energy is either the sole heat source for the HP or provides supplementary heat. Additionally, the operation principle and calculation of the thermodynamic cycle for a solar-assisted absorption HP are also briefly analyzed. Finally, analytical and experimental studies are performed on a direct-expansion solar-assisted HP (DX-SAHP) water heating system. The effect of various parameters, including solar radiation, ambient air temperature, collector area, storage volume, and speed of compressor, are been investigated on the thermal performance of the DX-SAHP system. A novel heating, ventilating, and air-conditioning system consisting in a solar-assisted absorption ground-coupled HP is also described, and some of the influence parameters on its energy efficiency is analyzed. A model of the experimental installation is developed using the TRNSYS software and validated with experimental results obtained in the installation for its cooling mode operation.

Generous permissions have been provided by many publishers for the use of their figures, drawings, and tables to make this book more complete and useful.

This book provides a useful source of information and basis for extended research for all those involved in the field, whether as a graduate student, MSc student and also PhD student, academic, scientific researcher, industrialist, consultant, or government agency with responsibility in the area of solar energy.

Prof. Ioan Sarbu
Department of Building Services Engineering,
Polytechnic University of Timisoara, Romania

Lecturer Calin Sebarchievici
Department of Building Services Engineering,
Polytechnic University of Timisoara, Romania

Chapter 1

Introduction

1.1 GENERALITIES

Energy, similar to water, food and shelter, is an essential need of all human beings in the world. The technological advancement and economic growth of every country depends on it [1], and the quantity of available energy reflects that country's quality of life. The economy, population, and per capita energy consumption have caused an increase in demand for energy during the last few decades. Fossil fuels are the prominent source for generating utilizable forms of energy [2]. Therefore, fossil fuels are the major contributors to global warming and the greenhouse effect on the ozone.

The ever-increasing worldwide energy consumption has created an urgent need to find new ways to use the energy resources in a more efficient and rational way. It is estimated that the global energy consumption will increase by 71% from 2003 to 2030 [3,4].

European Union (EU) energy consumption patterns reveal that buildings are the greatest energy consumers using approximately 40% of the total energy demand followed by industry and transportation, which consume approximately 30% each (Fig. 1.1) [5]. This ratio means that 36% of greenhouse gas (GHG) emissions are released from buildings in 2013 [6]. The percentage of final energy consumption in buildings in 2013 is above the critical threshold value for future projection of the EU. That is why European Commission created future projections for 2020-energy-efficiency target of EU with respect

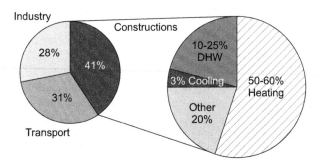

FIGURE 1.1   Primary energy consumption in EU.

Solar Heating and Cooling Systems. http://dx.doi.org/10.1016/B978-0-12-811662-3.00001-3
1

to energy efficiency investments as a financial action. Nearly same values were announced from the US Energy Information Administration (EIA), and 41% of the total US energy consumption was consumed in residential and commercial buildings in 2014 [7].

Buildings represent the largest and most cost-effective potential for energy savings. Studies have also shown that saving energy is the most cost-effective method for reducing GHG emissions.

Currently, heating is responsible for almost 80% of the energy demand in houses and utility buildings, used for space heating and hot-water generation, whereas the energy demand for cooling is growing yearly.

Furthermore, vapur compression-based refrigeration systems are generally employed in refrigeration, air-conditioning and heat pump (HP) units operating with synthetic refrigerants, such as chlorofluorocarbons (CFCs), hydro-chlorofluorocarbons (HCFCs) and hydro-fluorocarbons (HFCs). When released into the atmosphere, such refrigerants deplete the ozone layer and/or contribute to the greenhouse effect. In the late 1980s, it was estimated that the emissions of these compounds by refrigeration systems resulting from anomalies during operation accounted for 33.3% of the greenhouse effect [8]. As a result, several international protocols, such as the Montreal Protocol of 1987 [9] or the Kyoto Protocol of 1997 [10] were established to phase out, or at least to considerably reduce, the emissions of these refrigerants [11]. However, the situation continues, and there is still a need to develop alternative technologies operating with ecological substances, especially due to the increasing emissions of HFCs, although the emission of CFCs and HCFCs have been decreasing since the late 1980s [12]. Some directives have been approved to reduce the GHG emissions [13] and that another alternative can be used as natural [12] or synthetic refrigerants [14].

Usual vapor compression-based cycles are electrically powered, consuming large amounts of high-quality energy, which significantly increases the fossil fuel consumption. The *International Institute of Refrigeration* in Paris estimated that approximately 15% of all the electrical energy produced worldwide is employed for air-conditioning and refrigeration processes [15]. Moreover, electricity peak demands during summer are becoming more and more frequent due to the general increase in air-conditioning and refrigeration equipment usage.

The awareness of global warming has intensified in recent times and has reinvigorated the search for energy sources that are independent of fossil fuels and contribute less to global warming. The European strategy to decrease the energy dependence rests on two objectives: the diversification of the various sources of supply and policies to control consumption. The key to diversification is ecological and renewable energy sources (RES) because they have significant potential to contribute to sustainable development [16].

The use of surface-heating systems is increasing in Europe, but it is still much less than the use of hot water radiators. Also, low-temperature panel heating and cooling systems for residential buildings are increasingly used. According to some studies in 1994, use of these systems exceeds 50% [17]. Interest and growth in radiant heating and cooling systems have increased in recent years because they have been demonstrated to be energy efficient in comparison to all-air distribution systems [18].

Many theoretical works in the literature, for large-scale heating systems, compare thermal perception and energy efficiency of radiant heating systems with convection heating systems. Imanari et al. [19] concluded that radiant heating ceiling panel system is capable of creating a smaller vertical variation of air temperature than a conventional system. Additionally, there are many papers dealing with investigations of low-temperature radiant systems and their comparison with other heating systems regarding energy consumption and obtained thermal comfort.

Moreover, one solution for addressing worldwide energy-demand increase and climate changes will be using renewable energy to providing cooling instead of fossil fuel-consuming air-conditioning systems, making solar cooling technologies important for future. Solar energy is currently a subject of great interest, and refrigeration is a particularly attractive application due to the coincidence between the peak of cooling demand and the solar radiation availability.

Solar thermal energy is appropriate for both heating and cooling. Key applications for solar technologies are those that require low-temperature heat such as domestic water heating, space heating, pool heating, drying process, and certain industrial processes. Depending on the other uses of energy in buildings, domestic water heating can represent up to 30% of the energy consumed. Solar water heaters represent one of the most profitable applications of solar energy today. They constitute the bulk of the current market of solar heating and cooling, which itself produces almost four times more energy than all solar electric technologies combined (Fig. 1.2) [20].

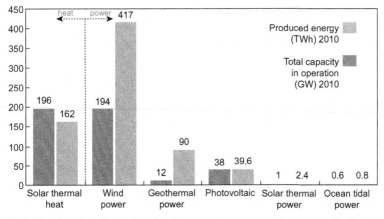

FIGURE 1.2 Capacities and produced energy of new renewable energy technologies.

Solar applications can also meet the cooling needs, with the advantage that the supply (sunny summer days) and the demand (desire for a cool indoor environment) are well matched. To generate synergy effects in climates with heating and cooling demand, combined systems should be used.

1.2 RENEWABLE ENERGY

Renewable energy refers to the form of energy that neither becomes depleted nor has the natural ability to renew itself. In its various forms, it derives directly from the sun, or from the heat generated deep within the earth. RES include wind, wave, solar, biomass, hydro, tidal, and geothermal energies, or even thermal waste from various processes [21,22]. Renewable energy replaces conventional fuels in four distinct areas: electricity generation, air and water heating/cooling, motor fuels, and rural (off-grid) energy services.

Renewable energy resources exist over wide geographical areas, in contrast to other energy sources, which are concentrated in a limited number of countries. Rapid deployment of renewable energy and energy efficiency is resulting in significant energy security, climate change mitigation, and economic benefits. It would also reduce environmental pollution such as air pollution caused by burning of fossil fuels and improve public health, reduce premature mortalities due to pollution, and save associated health costs that amount to several 100 billion $ annually only in the United States [23].

There have been numerous efforts undertaken by developed countries to implement different renewable energy technologies. The use of wind energy has increased over the last few years [24]. For example, the China, USA, Germany, Spain, and India are using wind turbines for producing electricity [25]. In northwestern Iran, mineral materials are used for the production of geothermal energy, and in Iceland, 70% of their factories utilize geothermal energy for industrial purposes [26].

On 23 April 2009, the European Parliament and the Council adopted the Renewable Energy Directive 2009/28/EC, which establishes a common framework for the promotion of RES. This directive has stated that by the year 2020, the average energy percentage coming out from renewable sources will be approximately 20% of the total energy consumption (Table 1.1).

The 2009/28/EC directive opens up a major opportunity for the further use of HPs for heating and cooling of new and existing buildings. HPs enable the use of ambient heat at useful temperature levels and need electricity or other form of energy to function. Some countries have much higher long-term policy targets of up to 100% renewable. Outside Europe, a diverse group of 20 or more other countries target renewable energy shares in the 2020−30 time frame that range from 10% to 50%.

Among the energy alternatives to fossil fuels, RES such as solar, wind, and hydropower are more available.

TABLE 1.1 Renewable Energy Percent out of the Total Energy Consumption for 2020

No.	Country	Code	Percentage (%)
1	Belgium	BE	13
2	Bulgaria	BG	16
3	Czech Republic	CZ	13
4	Denmark	DK	30
5	Germany	DE	18
6	Estonia	ES	25
7	Ireland	IE	16
8	Greece	GR	18
9	Spain	ES	20
10	France	FR	23
11	Italy	IT	17
12	Cyprus	CY	13
13	Latvia	LV	42
14	Lithuania	LT	23
15	Luxembourg	LU	11
16	Hungary	HU	13
17	Malta	MT	10
18	Netherlands	NL	14
19	Austria	AT	34
20	Poland	PL	15
21	Portugal	PT	31
22	Romania	RO	24
23	Slovenia	SI	25
24	Slovakia	SK	14
25	Finland	FI	38
26	Sweden	SE	49
27	United Kingdom	UK	15

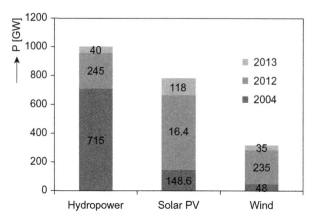

FIGURE 1.3 Renewable energy power capacity.

Renewable energy installation in recent years has seen further growth. This has hugely contributed to the awareness on the importance of renewable energy and government policies in revising energy priorities to ensure adoption and significant growth of renewable energy. The cost of renewable energy has also dramatically declined over the years, thus making them competitive compared to non-RES such as fossil powered power. The bar chart in Fig. 1.3 shows the total power generation capacity, P for RES on a global scale [27].

The amount of renewable powered capacity for the three main RES—hydro, solar, and wind are analyzed. The concentrating solar power (CSP) and solar hot-water capacity are also considered as entities for solar photovoltaic (PV) energy. It can be clearly observed that the hydropower constitutes as major percentage of RES. This is due to the constant availability and huge capacity of hydropower in many different parts of the world. Solar PV and wind power, which are intermittent in nature, have limited availability based on their geographical location. In 2013, the total installed capacity of hydropower is 1000 MW, and solar PV and wind power are at 783 MW and 318 MW, respectively [28].

Another important observation that can be viewed is that the total generation capacity in 2013 has decreased in comparison to 2012. This is in correlation to the global investment level in RES. Fig. 1.4 shows the global investment, I in renewable energy by region from 2004 till 2014 [29].

Majority of the regions in 2013 have experienced a drop in total investment in comparison to 2012. The decline in investment is mainly attributed to shifts on uncertainties in renewable energy policy as well as reduction on technology costs. Despite the decrease in investment, the ratio of installed capacity for solar PV in 2013 with respect to 2012 is at 22.85%. This ratio is greater than the hydropower and wind which, are at 16.22% and 14.89%, respectively.

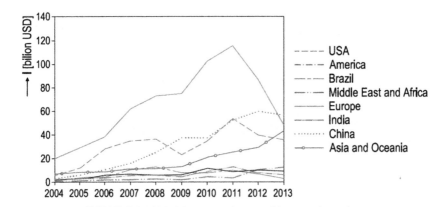

FIGURE 1.4   Global investment in renewable energy by region.

The global installed capacity of renewable energy has reached 480 GW, while the contribution of EU-27 was 210 GW. Global installed capacity of wind has reached to 318 GW in 2013 due to great installation capacity by China and Canada. Europe's installed wind capacity achieved to almost 121.5 GW in 2013 compared to 110 GW at the end of 2012 [30]. The world-installed capacity of solar energy (including PV, the CSP, and solar hot-water capacity) surpassed 783 GW at the end of 2013. In 2013, the contribution of European countries for solar energy was almost 114 GW.

Although Romania has a high potential of RES, in 2010 the RES share of the total energy consumption was 23.4%. Romania ranked second place in the European Union for the portion of energy from renewable sources out of gross final consumption from 2006 to 2010 [25].

For centuries, Romanians have been using wind and water to put in force mills, and wood and solar energy to heat water and buildings. The types of resources and the energy potential of each are summarized in Table 1.2 [31].

Romania is pursuing RES in three different directions:

- *Electricity:* The renewable energies used to produce electricity are wind, hydropower, solar PV, and biomass. In 2011, the electricity produced from renewable source achieved 20,673 GWh [31];
- *Heating/cooling*: The renewable energies most suited for heating and cooling are: biomass, geothermal, and solar resources;
- *Transportation*: Biofuels for transport are obtained by processing the rape, corn, sunflower, and soybean crops.

Undoubtedly, the different types of RES should be investigated further, but it is a reality that anyone of these energy sources could power the world without the use of any non-RES.

TABLE 1.2 Energy Potential of Renewable Energy Sources

No.	RES	Annual Energy Potential	Economic Energy Equivalent (ktoe)	Application
1	Solar energy: • Thermal	60×10^6 GJ	1433.0	Thermal energy
	• PV	1200 GWh	103.2	Electrical energy
2	Wind energy	23,000 GWh	1978.0	Electrical energy
3	Hydro-energy	40,000 GWh	3440.0	Electrical energy
4	Biomass	318×10^6 GJ	7594.0	Thermal energy
5	Geothermal energy	7×10^6 GJ	167.0	Thermal energy

Different ways could be considered to reduce primary energy consumption due to fossil fuel and equivalent CO_2 emissions related to space heating and cooling:

- Interventions on opaque and transparent building envelope reducing transmittance;
- Introducing high-efficiency energy conversion systems, such as condensing boiler, cogeneration system, gas engine-driven HP, and ground-source HP;
- Introducing renewable energy system based on solar thermal collectors considering different technologies;
- Solar electric driven system: solar PV collectors interacting with a space heating and cooling system based on an electric heat pump;
- Solar thermal-assisted electric heat pump: thermal energy available from solar collector used to operate an electric heat pump at lower evaporating temperature;
- A combination of renewable energy-based and high-energy conversion efficiency technologies.

If effective support policies are put in place in a wide number of countries during this decade, solar energy in its various forms (i.e., solar heat, solar PV, and solar thermal electricity) can make considerable contributions to solving some of the most urgent problems the world now faces: climate change, energy security, and universal access to modern energy services.

Solar energy offers a clean, climate-friendly, very abundant and inexhaustible energy resource to mankind, relatively well-spread over the globe. Its availability is greater in warm and sunny countries, and those countries that will experience most of the world's population and economic growth over the next decades. They will likely contain about seven billion inhabitants by 2050 versus 2 billion in cold and temperate countries (including most of Europe, Russia, parts of China, and the United States of America) [32].

The largest solar contribution to the world-energy needs is currently through solar heat technologies. The potential for solar water heating is considerable. Solar energy can provide a significant contribution to space-heating demand, both directly and through heat pumps. Direct solar cooling offers additional options but may face tough competition from standard cooling systems run by solar electricity.

Buildings are the largest energy consumers today. Positive-energy building combining excellent thermal insulation, smart design, and the exploitation of free solar resources can help change this. Ambient energy, that is, the low-temperature heat of the surrounding air and ground, transferred into build-ings with heat pumps, solar water heating, solar space heating, solar cooling and PV can combine to fulfill buildings' energy needs with minimal waste.

Industry requires large amounts of electricity and process heat at various temperature levels. Solar PV, solar thermal electricity, and solar heating and cooling can combine to address these needs in part, including those of agri-culture, craft industry, cooking, and desalination. Solar-process heat is currently untapped, but offers a significant potential in many sectors of the economy. Concentrating solar technologies can provide high-temperature process heat in clear-sky areas.

Support incentives reflect the willingness of an ever greater number of governments and policy makers to broaden the range of energy technology options with inexhaustible, clean RES. They drive early deployment, which, in turn, drives learning and cost reductions. Long-term benefits are expected to be considerable, from climate change mitigation to other reduced environmental impacts, reduced price volatility and increased energy security. Short-term costs, however, raise concerns among policy makers. In the last few years, drastic policy adjustments have affected PV in European countries and the financial sustainability of its unexpected rapid growth, driven by production incentives and unexpectedly rapid price declines.

This book presents a detailed theoretical study, numerical modelings, and experimental investigations on solar heating and cooling technologies.

REFERENCES

[1] Hassan HZ, Mohamad AA, Al-Ansary HA. Development of continuously operating solar-driven adsorption cooling system: thermodynamic analysis and parametric study. Applied Thermal Engineering 2012;48:332−41.

[2] Choudhury B, Saha BB, Chatterjee PK, Sarkar JP. An overview of developments in adsorption refrigeration systems towards a sustainable way of cooling. Applied Energy 2013;104:554−67.

[3] Sarbu I, Adam M. Applications of solar energy for domestic hot-water and buildings heating/cooling. International Journal of Energy 2011;5(2):34–42.

[4] Sarbu I, Sebarchievici C. Review of solar refrigeration and cooling systems. Energy and Buildings 2013;67(12):286–97.

[5] Anisimova N. The capability to reduce primary energy demand in EU housing. Energy and Buildings 2011;43:2747–51.

[6] European Commission COM. Financial support for energy efficiency in buildings. 2013. http://ec.europa.eu/energy/efficiency/buildings/buildings_en.htm.

[7] EIA, US. How much energy is consumed in residential and commercial buildings in the United States? http://www.eia.gov/tools/faqs.cfm?i86&t=1.

[8] Edminds JA, Wuebles DL, Scott MJ. Energy and radiative precursor emissions. In: Proceedings of the 8th international conference on alternative energy sources, Miami; 1987. p. 14–6.

[9] UNEP. Montreal protocol on substances that deplete the ozone layer. New York (USA): United Nations Environment Program; 1987.

[10] GECR. Kyoto protocol to the United Nations framework conservation on climate change. New York (USA): Global Environmental Change Report; 1997.

[11] Anisur MR, Mahfuz MH, Kibria MA, Saidur R, Metselaar IHSC, Mahlia TMI. Curbing global warming with phase change materials for energy storage. Renewable and Sustainable Energy Reviews 2013;18:23–30.

[12] Sarbu I. A review on substitution strategy of non-ecological refrigerants from vapour compression-based refrigeration, air-conditioning and heat pump systems. International Journal of Refrigeration 2014;46(10):123–41.

[13] Regulation (EU) No 517/2014 of the European Parliament and the Council of 16 April 2014 on fluorinated greenhouse gases and repealing Regulation (EC) No 842/2006. Official Journal of European Union; 2014.

[14] Mota-Babiloni A, Navarro-Esbri J, Barragan-Cervera A, Moles F, Peris B. Analysis based on EU Regulation No 517/2014 of new HFC/HFO mixtures as alternatives of high GWP refrigerants in refrigeration and HVAC systems. International Journal of Refrigeration 2015;52:21–31.

[15] Santamouris M, Argiriou A. Renewable energies and energy conservation technologies for buildings in southern Europe. International Journal of Solar Energy 1994;15:69–79.

[16] Al Z. Management of renewable energy and regional development: European experiences and steps forward. Theoretical and Empirical Researches in Urban Management 2011;6(3):35–42.

[17] Kilkis I, Sager S, Uludag M. A simplified model for radiant heating and cooling panels. Simulation Practice and Theory 1994;2:61–76.

[18] Henze GP, Felsmann C, Kalz D, Herkel S. Primary energy and comfort performance of ventilation assisted thermo-active building system in continental climates. Energy and Buildings 2008;40(2):99–111.

[19] Imanari T, Omori T, Bogaki K. Thermal comfort and energy consumption of the radiant ceiling panel system comparison with the conventional all-air system. Energy and Buildings 1999;1999(30):167–75.

[20] Weiss W, Mauthner F. Solar heat worldwide: markets and contributions to the energy supply 2009, IEA Solar Heating and Cooling Programme. Gleisdorf (Austria): AEE Intec; 2011.

[21] Hassan HZ, Mohamad AA. A review on solar-powered closed physisorption cooling systems. Renewable and Sustainable Energy Reviews 2012;16:2516–38.

[22] Fernandes MS, Brites GJVN, Costa JJ, Gaspar AR, Costa VAF. Review and future trends of solar adsorption refrigeration systems. Renewable and Sustainable Energy Reviews 2014;39:102−23.

[23] Jacobson MZ, Delucchi MA, Bazouin G, Baner ZAF, Heavey CC, Fisher E. 100% clean and renewable wind, water and sunlight (WWS) all-sector energy roadmaps for the 50 United States. Energy and Environmental Science 2015;8:2093−117.

[24] Ullah KR, Saidur R, Ping HW, Akikur RK, Shuvo NH. A review of solar thermal refrigeration and cooling methods. Renewable and Sustainable Energy Reviews 2013;24: 490−513.

[25] Colesca SE, Ciocoiu CN. An overview of the Romanian renewable energy sector. Renewable and Sustainable Energy Reviews 2013;24:149−58.

[26] Abu-Hamdeh NH, Al-Muhtaseb MtA. Optimization of solar adsorption refrigeration system using experimental and statistical techniques. Energy Conversion and Management 2010;51:1610−5.

[27] Renewables 2014 Global Status Report, REN21 (renewable energy policy network for the 21st century), http://www.ren21.ne.

[28] The First Decade: 2004−2014, http://www.ren21.net/Portals/0/documents/activities/Topical%20Reports/REN21_10yr.pdf.

[29] Karimi M, Mokhlis H, Naidu K, Uddin S, Bakar AHA. Photovoltaic penetration issues and impact in distribution network − a review. Renewable and Sustainable Energy Reviews 2016;53(1):594−605.

[30] Rehman S, Al-Hadhrami LM, Alam MM. Pumped hydro energy storage system: a technological review. Renewable and Sustainable Energy Reviews 2015;44:586−98.

[31] ANRE, National Authority of Energy Settlement 2012, http://www.anre.ro.

[32] IEA. Renewable energy technologies: solar energy perspective. Paris (France): International Energy Agency Publications; 2011. http://www.iea.org/publications/freepublications/publication/solar_energy_perspectives2011.pdf.

Chapter 2

Solar Radiation

2.1 GENERALITIES

Among the renewable resources, hydro is the predominant source used for electricity generation in several countries. The other significant growth in the renewable sector is solar energy, particularly if the cost of the construction of new solar power systems continues to decrease. In fact, experts predict that the price of producing solar power can be comparable to other sources of energy within the next 10 years.

Hermann [1] illustrates the availability of energy in the terrestrial environment. Most of the solar radiation reaching the terrestrial environment is dissipated and only a very small amount is converted into solar energy (959 PJ/year, while 1,356,048,000 PJ/year is dissipated only into surface heating).

International Energy Agency (IEA) [2] reports the worldwide solar conversion in 2011 of 711 PJ (74%) thermal, 234 PJ (24%) photovoltaic (PV), and 14 PJ (2%) concentrating solar power (CSP), in total 959 PJ/year (1 Peta-Joule $= 10^{15}$ Joule). The potential of solar energy use is thus huge.

The absolute size of the solar electricity market is still very tiny, generating around 0.1% of electricity globally. Not surprisingly, there are significant differences between regions. As a result, electricity demand in non−*Organization for Economic Co-operation and Development* (non-OECD) countries will surpass the cumulative electricity demand in OECD countries before 2015. Over the past 20 years, solar electricity-generation technologies have grown by leaps and bounds, registering annual growth rates between 25% and 41%. Global solar electric generation technologies contribute roughly 2000 MW of electricity today.

In recent years, scientists have paid increasing attention to solar energy. There is a sudden demand in the utilization of solar energy for various applications such as water heating, building heating/cooling, cooking, power generation, and refrigeration [3].

Fig. 2.1 shows the distribution of the energy consumption in the residential sector of OECD member countries in 2011 [2]. The total yearly energy consumption of 25,000 PJ is a small fraction of the solar energy dissipated into surface heating. Space heating consumed almost 50% of the total (12,350 PJ) while space cooling consumed 6% (1610 PJ). The sector "services" that

Solar Heating and Cooling Systems. http://dx.doi.org/10.1016/B978-0-12-811662-3.00002-5
13

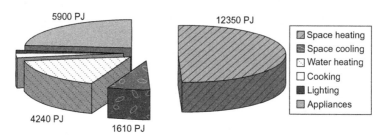

FIGURE 2.1   Energy use in the residential sector of OECD countries during 2011.

include utility buildings are not included in these numbers but also consume large amounts of energy for heating and cooling purposes (estimated around 7000 PJ). This indicates that about 10% of the energy use of OECD countries (225,752 PJ) has the potential to be served by solar-driven refrigeration/heat pump cycles. Although solar cooling is mainly considered for removing cooling loads from the applications, the same cycles used as heat pumps will also have an advantage in comparison to the direct use of solar heat for heating purposes.

Most countries are now accepting that solar energy has enormous potential because of its cleanliness, low price, and natural availability. For example, it is being used commercially in solar power plants. Sweden has been operating a solar power plant since 2001. In the Middle East, solar energy is used for desalination and absorption air-conditioning. In China and, to a lesser extent, Australasia, solar energy is widely used, particularly for water heating. In Europe, government incentives have fostered the use of PV and thermal systems for both domestic hot water and space heating/cooling. Romania's experience in solar energy represents a competitive advantage for the future development of this area, the country being a pioneer in this field. Between 1970 and 1980, around 800,000 m^2 of solar collectors were installed that placed the country third worldwide in the total surface of photovoltaic panels. The peak of solar installations was achieved between 1984 and 1985, but after 1990 unfavorable macroeconomic developments led to the abandonment of the production and investments in the solar energy field. Today about 10% of the former installed collector area is still in operation [4].

For solar energy use, its conversion into other forms of energy is needed. Solar technologies are broadly characterized as either passive solar or active solar depending on the way they capture, convert, and distribute solar energy. Passive solar techniques include orienting a building to the sun, selecting materials with favorable thermal mass or light dispersing properties, and designing spaces that naturally circulate air. Active solar technologies encompass solar thermal energy, using solar collectors for heating, and solar power by converting sunlight into electricity either directly using photovoltaic or indirectly using concentrated solar power.

Because of the many benefits that solar systems can provide, governments and the energy industry should be encouraged to find any new approach to develop a more cost-effective system. However, the majority of disadvantages about the solar power are more of economics in nature. Even after lots of technological development, the solar panels used to produce electricity are still quite expensive. A single solar panel can generate only a small amount of power. This means a larger number of solar panels are required to generate enough amount of electricity to power buildings and industries.

This chapter presents the main characteristics of solar energy and exposes a methodology for calculating and predicting solar radiation including the main computation elements and the estimation of solar radiation on tilted surface likely to be available as input to a solar device or crop at a specific location, orientation, and time.

2.2 CALCULATION OF SOLAR RADIATION

Before installing a solar energy system, it is necessary to predict both the demand and the likely solar energy available, together with their variability. Knowing this and the projected pattern of energy usage from the device, it is possible to calculate the size of collector and storage. Ideally, the data required to predict the solar input are several years of measurements of irradiance on the proposed collector plane. These are very rarely available, so the required (statistical) measures have to be estimated from meteorological data available (1) from the site, (2) (more likely) from some "nearby" site having similar irradiance, or (3) (most likely) from database.

2.2.1 Characteristics of Solar Radiation

The Sun is a sphere of intensely hot gaseous matter with a diameter of 1.39×10^9 m and is, on the average, 1.5×10^{11} m from the earth. As seen from the earth, the sun rotates on its axis about once in every 4 weeks. However, it does not rotate as a solid body; the equator takes about 27 days and the polar regions take about 30 days for each rotation. The sun has an effective blackbody temperature of 5777 K.

Solar energy is the result of electromagnetic radiation released from the sun by the thermonuclear reactions occurring inside its core. All of the energy resources on earth originate from the sun (directly or indirectly), except for nuclear, tidal, and geothermal energy.

The sun radiates considerable energy onto the earth. Solar radiation intensity, rarely over 950 W/m^2 has led to the creation of many types of devices to convert this energy into useful forms, mainly heat and electricity. Radiant light and heat from the sun is harnessed using a range of ever-evolving technologies such as solar heating, PV, CSP, solar architecture, and artificial photosynthesis.

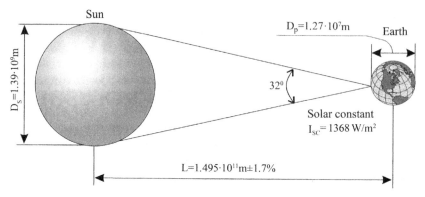

FIGURE 2.2   Geometry of the sun–earth system.

Fig. 2.2 shows schematically the geometry of the sun–earth relationships. The eccentricity of the earth's orbit is such that the distance between the sun and the earth varies by 1.7%. At a mean earth–sun distance $L = 1.495 \times 10^{11}$ m, the sun subtends an angle of 32 degree. The radiation emitted by the sun and its spatial relationship to the earth result in a nearly fixed intensity of solar radiation outside of the earth's atmosphere.

The *solar constant* I_{SC} signifies the energy from the sun per unit time received on a unit area of surface perpendicular to the direction of propagation of the radiation at mean earth–sun distance outside the atmosphere. The mean value of solar constant is equal to 1368 W/m^2. Therefore, considering a global plane area of 1.275×10^{14} m^2 and the mean radius of the earth being approximately 6371 km, the total solar radiation transmitted to the earth is 1.74×10^{17} W, whereas the overall energy consumption of the world is approximately 1.84×10^{13} W [5].

2.2.1.1 Solar Angles

The axis about which the earth rotates is tilted at an angle of 23.45 degrees to the plane of the earth's orbital plane and the sun's equator.

The earth's axis results in a day-by-day variation of the angle between the earth–sun line and the earth's equatorial plane called the solar declination δ. This angle may be estimated by the following equation [6]:

$$\delta = 23.45 \sin\left[\frac{360}{365}(284 + N)\right], \tag{2.1}$$

where $N =$ year day, with January 1 + 1.

The position of the sun can be defined in terms of its altitude β above the horizon and its azimuth ϕ measured in horizontal plane (Fig. 2.3).

To determine the angle of incidence θ between a direct solar beam and the normal to the surface, the surface azimuth ψ and the surface-solar

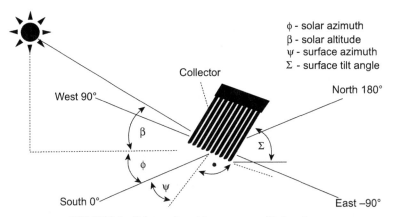

FIGURE 2.3   Solar angles with respect to a tilted surface.

azimuth γ must be known. The surface-solar azimuth is designated by γ and is the angular difference between the solar azimuth ϕ and the surface azimuth ψ. For a surface facing the east of south, $\gamma = \phi - \psi$ in the morning, and $\gamma = \phi + \psi$ in the afternoon. For surfaces facing the west of south, $\gamma = \phi + \psi$ in the morning and $\gamma = \phi - \psi$ in the afternoon. For south-facing surfaces, $\psi = 0$ degree, so $\gamma = \phi$ for all conditions. The angles δ, β, and ϕ are always positive.

For a surface with tilt angle Σ (measured from the horizontal), the angle of incidence θ is given by [6]

$$\cos\theta = \cos\beta \cos\gamma \sin \Sigma + \sin\beta \cos \Sigma. \qquad (2.2)$$

For vertical surfaces, $\Sigma = 90$ degrees, $\cos\Sigma = 0$, and $\sin\Sigma = 1.0$, so Eq. (2.2) becomes

$$\cos\theta = \cos\beta \cos\gamma \qquad (2.3)$$

For horizontal surfaces, $\Sigma = 0$ degree, $\sin\Sigma = 0$, and $\cos\Sigma = 1.0$, so Eq. (2.2) leads to

$$\theta = 90° - \beta \qquad (2.4)$$

Latitude ϕ is the angular location north or south of the equator, north positive; -90 degrees $\leq \phi \leq 90$ degrees.

Zenith angle θ_z, the angle between the vertical and the line to the sun, is the angle of incidence of direct (beam) radiation on a horizontal surface ($\theta_z = \theta$).

Hour angle ω is the angular displacement of the sun east or west of the local meridian due to rotation of the earth on its axis at 15 degrees per hour;

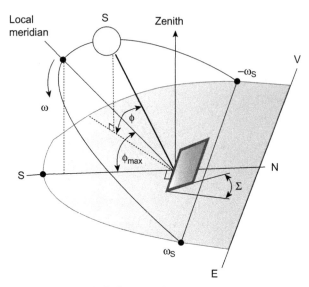

FIGURE 2.4   Hour angle ω.

morning negative $(-\omega_s)$ and afternoon positive $(+\omega_s)$ (Fig. 2.4). The sun position at any hour τ can be expressed as follows:

$$\omega = 15(12 - \tau) \tag{2.5}$$

If the angles δ, φ, and ω are known, then the sun position in the interest point can be easily determined for any hour and day using following expressions [7]:

$$\sin\beta = \sin\delta \sin\varphi + \cos\delta \cos\varphi \cos\omega = \cos\theta_z \tag{2.6}$$

$$\cos\phi = \frac{\sin\beta \sin\varphi - \sin\delta}{\cos\beta \cos\varphi}. \tag{2.7}$$

For any day of a year, solar declination δ can be determined in Eq. (2.1) and for the hour τ, hour angle ω can be calculated in Eq. (2.5). Latitude φ is also known and thus solar altitude β can be determined.

2.2.1.2 Design Value of Total Solar Radiation

Solar radiation reaches earth's surface as (1) direct (beam) solar radiation, (2) diffuse solar radiation, and (3) reflected radiation, which can be neglected. The total radiation received from the sun, of a horizontal surface at the level of the ground for a serene day, is the sum of the direct and diffuse radiations. Direct radiation depends on the orientation of receiving surface. Diffuse radiation can be considered the same, irrespective of the receiving surface orientation, although in reality there are small differences. Fig. 2.5 represents the proportion of the diffuse radiation in total radiation I_T [8].

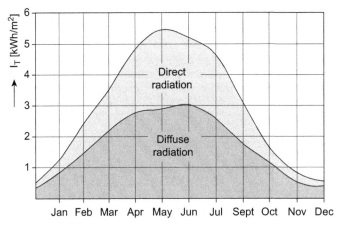

FIGURE 2.5   Total and diffuse solar radiation.

Attenuation of the solar rays is determined by the composition of the atmosphere and the length of the atmospheric path through which the rays travel. The path length is expressed in terms of the air mass m, which is the ratio of the mass of atmosphere in the actual earth–sun path to the mass that would exist if the sun was directly overhead at sea level ($m = 1$). Beyond the earth's atmosphere, $m = 0$. For zenith angles $\theta_z \in$ [0 degree, 70 degrees] at the sea level, to a close approximation, $m = 1/\cos\theta_z$.

The total solar radiation I_T of a terrestrial surface of any orientation and tilt with an incident angle θ is the sum of the direct component I_D plus the diffuse component I_d:

$$I_T = I_D + I_d \tag{2.8}$$

in which

$$I_D = I_{DN}\cos\theta \tag{2.9}$$

$$I_{DN} = A_0 e^{-B/\sin\beta}, \tag{2.10}$$

where I_{DN} is the direct normal radiation estimated by Stephenson [9] using Eq. (2.10); A_0 is the apparent extraterrestrial radiation at the air mass $m = 0$; B is the atmospheric extinction coefficient; β is the sun's altitude above horizon, in degrees [9].

Irradiance, expressed in W/m^2, is the rate at which radiant energy is incident on a surface per unit area of surface. *Irradiation*, expressed in J/m^2 or kWh/m^2, is the incident energy per unit area on a surface, found by integration of irradiance over a specified time, usually an hour or a day. *Insolation* is a term applying specifically to solar energy irradiation.

A simpler method for estimating clear-sky radiation by hours is to use data for the American Society of Heating, Refrigerating, and Air-Conditioning

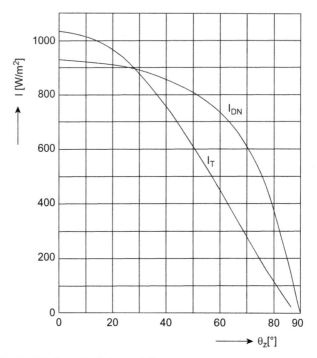

FIGURE 2.6   Total horizontal radiation and direct normal radiation for the ASHRAE standard atmosphere.

Engineers (ASHRAE) standard atmosphere. Farber and Morrison [10] provide tables of direct normal radiation and total radiation on a horizontal surface as a function of zenith angle θ_z. These are plotted in Fig. 2.6. For a given day, hour-by-hour estimates of radiation I (W/m^2) can be made based on midpoints of the hours.

The most common measurements of solar radiation are total radiation on a horizontal surface. The commonest instruments used for measuring solar radiation are of two basic types: a *pyroheliometer*, which measures the direct radiation I_D, and a pyranometer or *solarimeter*, which measures total radiation I_T.

Only the active cavity radiometer (ACR) gives an absolute reading. In this instrument, the solar beam falls on an absorbing surface of area A, whose temperature increase is measured and compared with the temperature increase in an identical (shaded) absorber heated electrically. In principle, then

$$\alpha A I_D = P_{el}. \tag{2.11}$$

The geometry of the ACR is designed so that, effectively, $\alpha = 0.999$ [11].

Unitary thermal energy received from the sun, measured at the level of earth's surface, perpendicularly on the direction of solar rays, for the conditions where the sky is perfectly clear and without pollution, in the areas of Western, Central, and Eastern Europe, around noon, can provide maximum 1000 W/m^2. This value represents the sum between direct and diffuse radiations. Atmosphere modifies the intensity, spectral distribution, and spatial distribution of the solar radiation by two mechanisms: absorption and diffusion. The absorbed radiation is generally transformed into heat, while the diffuse radiation is resent in all the directions into the atmosphere.

Meteorological factors that have a big influence on the solar radiation at the earth's surface are atmosphere transparency, nebulosity, and clouds' nature and their position.

Romania disposes an important potential of solar energy due to the favorable geographical position and climatic conditions. On 1 m^2 plate of horizontal surface, perpendicular on the incidence direction of the sun's rays, energy of 900 to 1450 kWh/year can be received, depending on the season, altitude, and geographical position. The daily mean solar radiation can be up to five times more intense in the summer than in the winter. There are situations when, in the winter, under favorable conditions (clear sky, low altitude, etc.), values of solar energy received can reach approx. 4—5 kWh/(m^2·day), the solar radiation being practically independent of the environment air temperature. Quantifying this value related to Romania's annual energy requirement situated around the value of 260,900,000 MWh, around the year 2011, energy of approx. 285,000,000,000 MWh/year radiated by the sun in the country's territory is obtained. This represents Romania's total energy consumption for a period of 1092 years!

2.2.2 Solar Radiation on a Tilted Surface

For purposes of solar process design and performance calculations, it is often necessary to calculate the hourly radiation on a tilted surface of a collector from measurements or estimates of solar radiation on a horizontal surface. The most commonly available data are total radiation for hours or days on the horizontal surface, whereas the need is for direct and diffuse radiation on the plane of a collector.

Fig. 2.7 indicates the angle of incidence of direct radiation on the horizontal and tilted surfaces. The geometric factor R_D, the ratio of direct radiation on the tilted surface to that on a horizontal surface at any time, of $I_{D,\Sigma}/I_D$ is given by:

$$R_D = \frac{I_{D,\Sigma}}{I_D} = \frac{I_{DN} \cos \theta}{I_{DN} \cos \theta_z} = \frac{\cos \theta}{\cos \theta_z}, \qquad (2.12)$$

where $\cos \theta$ and $\cos \theta_z$ are both determined from Eq. (2.2).

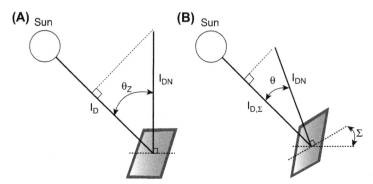

FIGURE 2.7   Direct radiation on horizontal and tilted surfaces.

Equations relating the angle of incidence θ of direct radiation on a surface to the other angles [12]:

$$\cos\theta = \sin\delta\,\sin\varphi\,\cos\Sigma - \sin\delta\,\cos\varphi\,\sin\Sigma\,\cos\psi$$
$$+ \cos\delta\,\cos\varphi\,\cos\Sigma\,\cos\omega + \cos\delta\,\sin\varphi\,\sin\Sigma\,\cos\psi\,\cos\omega \qquad (2.13)$$
$$+ \cos\delta\,\sin\Sigma\,\sin\psi\,\sin\omega$$

and

$$\cos\theta = \cos\theta_z\,\cos\Sigma + \sin\theta_z\,\sin\Sigma\,\cos(\phi - \psi). \qquad (2.14)$$

There are several commonly occurring cases for which Eq. (2.13) is simplified. For fixed surfaces sloped toward the south or north, with a surface azimuth angle ψ of 0 or 180 degrees (a very common situation for fixed flat-plate collectors), the last term drops out.

For vertical surfaces, $\Sigma = 90$ degrees, the equation becomes

$$\cos\theta = -\sin\delta\,\cos\varphi\,\cos\psi + \cos\delta\,\sin\varphi\,\cos\psi\,\cos\omega + \cos\delta\,\sin\psi\,\sin\omega. \qquad (2.15)$$

For horizontal surfaces, the angle of incidence is the zenith angle of the sun, θ_z. Its value must be between 0 degree and 90 degrees when the sun is above the horizon. For this situation $\Sigma = 0$ degree and Eq. (2.13) becomes

$$\cos\theta_z = \cos\varphi\,\cos\delta\,\cos\omega + \sin\varphi\,\sin\delta. \qquad (2.16)$$

Useful relationships for the angle of incidence of surfaces sloped due north or due south can be derived from the fact that surfaces with slope Σ to the north or south have the same angular relationship to beam radiation as a horizontal surface at an artificial latitude of $\varphi - \Sigma$. Modifying Eq. (2.16) yields

$$\cos\theta = \cos(\varphi - \Sigma)\,\cos\delta\,\cos\omega + \sin(\varphi - \Sigma)\,\sin\delta. \qquad (2.17)$$

For the special case of solar noon, for the south-facing sloped surface in the northern hemisphere,

$$\theta_{noon} = |\varphi - \delta - \Sigma| \tag{2.18}$$

and in the southern hemisphere,

$$\theta_{noon} = |-\varphi + \delta - \Sigma|, \tag{2.19}$$

where $\Sigma = 0$ degree, the angle of incidence is the zenith angle, which for the northern hemisphere is

$$\theta_{z,noon} = |\varphi - \delta| \tag{2.20}$$

and for the southern hemisphere,

$$\theta_{z,noon} = |-\varphi + \delta|. \tag{2.21}$$

For a plane rotated about a horizontal east–west axis with continuous adjustment to minimize the angle of incidence,

$$\cos\theta = \left(1 - \cos^2\delta \ \sin^2\omega\right)^{1/2}. \tag{2.22}$$

The optimum azimuth angle for flat-plate collectors is usually 0 degree in the northern hemisphere (or 180 degrees in the southern hemisphere). Thus, it is a common situation that $\psi = 0$ degree (or 180 degrees). In this case, Eqs. (2.16) and (2.17) can be used to determine $\cos\theta_z$ and $\cos\theta$, respectively, leading in the northern hemisphere, for $\psi = 0$ degree, to

$$R_D = \frac{\cos(\varphi - \Sigma) \ \cos\delta \ \cos\omega + \sin(\varphi - \Sigma) \ \sin\delta}{\cos\varphi \ \cos\delta \ \cos\omega + \sin\varphi \ \sin\delta} \tag{2.23}$$

In the southern hemisphere, $\psi = 180$ degrees and the equation is

$$R_D = \frac{\cos(\varphi + \Sigma) \ \cos\delta \ \cos\omega + \sin(\varphi + \Sigma) \ \sin\delta}{\cos\varphi \ \cos\delta \ \cos\omega + \sin\varphi \ \sin\delta}. \tag{2.24}$$

Eq. (2.13) can also be applied to other-than fixed flat-plate collectors. For example, for a plane rotated continuously about a horizontal east–west axis to maximize the direct radiation on the plane, from Eq. (2.22), the ratio of direct radiation on the plane to that on a horizontal surface at any time is

$$R_D = \frac{\left(1 - \cos^2\delta \ \sin^2\omega\right)^{1/2}}{\cos\varphi \ \cos\delta \ \cos\omega + \sin\varphi \ \sin\delta}. \tag{2.25}$$

Eqs. (2.23)–(2.25) are used to determine optimum tilt angle Σ for duration of 1 h or 1 day. For use in solar process design procedures, also it needs the monthly average daily radiation on the tilted surface. Thus, for surfaces that are sloped toward the equator in the northern hemisphere, that is, for surfaces with $\psi = 0$ degree,

$$\bar{R}_D = \frac{\cos(\varphi - \Sigma) \cos \delta \sin \omega'_s + (\pi/180)\omega'_s \sin(\varphi - \Sigma)}{\cos \varphi \cos \delta \sin \omega_s + (\pi/180)\omega_s \sin \varphi \sin \delta} \qquad (2.26)$$

in which ω'_s is the sunset hour angle for the tilted surface for the mean day of the month, which is given by,

$$\omega'_s = \min \begin{bmatrix} \cos^{-1}(-\tan \varphi \ \tan \delta) \\ \cos^{-1}(-\tan(\varphi - \Sigma) \ \tan \delta) \end{bmatrix} \qquad (2.27)$$

and

$$\omega_s = \cos^{-1}(\tan \varphi \ \tan \delta), \qquad (2.28)$$

where "min" means the smaller of the two items in the brackets.

For surfaces in the southern hemisphere sloped toward the equator, with $\psi = 180$ degrees, the equations are

$$\bar{R}_D = \frac{\cos(\varphi + \Sigma) \cos \delta \sin \omega'_s + (\pi/180)\omega'_s \sin(\varphi + \Sigma) \sin \delta}{\cos \varphi \cos \delta \sin \omega_s + (\pi/180)\omega_s \sin \varphi \sin \delta} \qquad (2.29)$$

and

$$\omega'_s = \min \begin{bmatrix} \cos^{-1}(-\tan \varphi \ \tan \delta) \\ \cos^{-1}(-\tan(\varphi + \Sigma) \ \tan \delta) \end{bmatrix} \qquad (2.30)$$

$$\omega_s = \cos^{-1}(\tan \varphi \ \tan \delta). \qquad (2.31)$$

The numerator of Eqs. (2.26) and (2.29) is the extraterrestrial radiation on the tilted surface, and the denominator is that on the horizontal surface. Each of them is obtained by integration of Eq. (2.13) over the appropriate time period, from true sunrise to sunset for the horizontal surface, and from apparent sunrise to apparent sunset on the tilted surface.

The R_D was calculated by Duffie and Beckman depending on angle difference $\varphi - \Sigma$ and latitude φ, and its values were included in [12].

For a known value of ratio R_D can be determined direct solar radiation on tilted surface from Eq. (2.12):

$$I_{D,\Sigma} = R_D I_D. \qquad (2.32)$$

The procedure for calculating total solar radiation $I_{T,\Sigma}$ is by summing the contributions of the direct radiation, the components of the diffuse radiation, and the radiation reflected from the ground.

The diffuse radiation and the radiation reflected from the ground can be evaluated using the isotropic diffuse model of the sky proposed by Liu and Jordan [13] and extended by Klein [14]. The diffuse radiation on the tilted surface $I_{d,\Sigma}$ is defined by

$$I_{d,\Sigma} = \frac{1}{2}(1 + \cos \Sigma)I_d, \qquad (2.33)$$

TABLE 2.1 Typical Values of Ground Diffuse Reflectance

Ground Characteristic	ρ_g
Plowing	0.2
Verdure ground	0.3
Sandy desert	0.4
Snow	0.7

where I_d is the diffuse radiation on the horizontal surface.

The radiation reflected from the ground on the tilted surface $I_{r,\Sigma}$ is given by formula

$$I_{r,\Sigma} = \frac{1}{2}(1 - \cos \Sigma)\rho_g I_T, \qquad (2.34)$$

where ρ_g is the diffuse reflectance of the ground (Table 2.1) [15] and I_T is the total solar radiation on the horizontal surface.

Consequently, total solar radiation on the tilted surface is the sum of the three components:

$$I_{T,\Sigma} = R_D I_D + \frac{1}{2}(1 + \cos \Sigma)I_d + \frac{1}{2}(1 - \cos \Sigma)\rho_g I_T. \qquad (2.35)$$

The third term of Eq. (2.35) can be neglected as suggested by Hottel and Woertz [16] because the combination of diffuse and ground-reflected radiation is assumed isotropic.

The prediction of collector performance requires information on the solar energy absorbed by the collector absorber plate. Eq. (2.35) can be modified to give the absorbed radiation S by multiplying each term by the appropriate transmittanceabsorptance product $(\tau\alpha)$ as

$$S = I_D R_D(\tau\alpha)_D + I_d(\tau\alpha)_d \left(\frac{1 + \cos \Sigma}{2}\right) + \rho_g I_T(\tau\alpha)_g \left(\frac{1 - \cos \Sigma}{2}\right), \qquad (2.36)$$

where $(1 + \cos \Sigma)/2$ and $(1 - \cos \Sigma)/2$ are the view factors from the collector to the sky and from the collector to the ground, respectively.

In addition to the regular variations, there are also substantial irregular variations. Of them, perhaps the most significant for engineering purposes are the day-to-day fluctuations because they affect the amount of energy storage required within a solar energy system. Thus, even a complete record of past radiation can be used to predict future radiation only in a statistical sense. Therefore design methods usually rely on approximate averages, such as monthly means of daily insolation. To estimate these cruder data from other measurements are easier than to predict a shorter-term pattern of radiation.

2.3 PREDICTION OF SOLAR RADIATION USING IMPROVED BRISTOW–CAMPBELL MODEL

Bristow and Campbell [17] proposed a method for estimating the daily I_T from diurnal temperature range ($t_M - t_m$) and the atmospheric transmittance ρ. The ratio between I_T and I_0 is calculated as a function of ($t_M - t_m$), which can be expressed as,

$$\frac{I_T}{I_0} = \rho = A\left[1 - \exp\left(-B(t_M - t_m)^C\right)\right], \qquad (2.37)$$

where I_T is the total daily solar radiation; I_0 is the extraterrestrial radiation; ρ is the daily total atmospheric transmittance; A is the maximum radiation expected on a clear day, being distinctive for each location and depending on air quality and altitude; coefficients B and C control the rates at which A is approached as the temperature difference increases; t_M is the maximum air temperature; and t_m is the minimum air temperature.

This simple model neglects other factors that affect the amount of solar radiation that reaches the earth's surface, such as relative humidity, cloud cover, etc. Improved Bristow–Campbell (IBC) model considers the influencing factors of surface-solar radiation. So, more meteorological variables (e.g., t_M, t_m, R_H, and OP) should be employed to represent the actual atmospheric transmittance. The IBC model is proposed as follows [18]:

$$I_0 = 37.54 \times \left(\frac{L_m}{L}\right)^2 (\omega_s \sin\varphi \sin\delta + \cos\varphi \cos\delta \sin\omega_s), \qquad (2.38)$$

where I_0 is the daily extraterrestrial insolation incident on a horizontal surface in MJ/(m^2·day); ω_s is the solar angle at sunset (half-day length), in rad, computed using Eq. (2.31); δ is the solar declination, in rad; φ is the latitude of the location of interest, in rad; L_m is the mean value of the distance from the sun to earth, in km; and L is the distance from the sun to earth, in km.

The solar radiation at the top of the atmosphere can be calculated as

$$\frac{I_T}{I_0} = (b_0 + b_1 \sin w + b_2 \cos w + b_3 R_H + b_4 OP) \times \left[1 - \exp\left(-b_5(t_M - t_m)^{b_6}\right)\right]$$

$$(2.39)$$

where R_H is the relative humidity; OP is the occurrence of precipitation; $w = 2\pi j/365$; and j is the Julian day.

The ratio L_m/L is known as the correction factor for the sun–earth distance, which can be obtained from equation [19]

$$\frac{L_m}{L} = \sqrt{1.00011 + 0.034221\cos\xi + 0.034221\sin\xi + 0.000719\cos 2\xi + 0.000077\sin 2\xi}.$$

$$(2.40)$$

The daily angle ξ, in rad, is calculated as a function of the Julian day ($\xi = 2\pi(j - 1)/365$). Then, the coefficients of the IBC model (b_0, b_1, b_2, b_3, b_4,

b_5, and b_6) can be obtained by minimizing the sums of the squares of deviations between observed and expected values. These derivatives are numerically computed using finite differences and the Newton–Raphson algorithm can be used for multivariate nonlinear optimization. The procedure can be easily implemented using Solver application of the Excel software. The detailed procedures for calculating the model parameters for IBC model can be seen in Meza et al. [20] and Pan et al. [21].

One additional factor that can be considered to improve the estimations of solar global radiation is the altitude of the station. Solar radiation that reaches the earth's surface is influenced by altitude above sea level due to the diminishing of the layer of air upon it. As a result, with the same meteorological conditions, higher locations receive more global solar radiation than at sea level.

The IBC model is easy to use in any location where measurements of temperature, precipitation, and relative humidity are available and present a simple solution that can be used as proxy for relative humidity in case that variable is not been measured.

Other solar radiation prediction models use the artificial intelligence methods including multilayer perception neural network, radial basis neural network, and generalized regression neural network, which have been recently discussed and tested by Wang et al. [18].

REFERENCES

[1] Hermann WA. Quantifying global exergy resources. Energy 2006;31:1685–702.
[2] IEA. Tracking clean energy Progress 2013. Paris, France: International Energy Agency Publications; 2013.
[3] Li ZF, Sumathy K. Technological development in the solar absorption air-conditioning systems. Renewable and Sustainable Energy Reviews 2000;4:267–93.
[4] EBRD. Renewable energy resource assessment. Bucharest. Romania: European Bank for Reconstruction and Development; 2010.
[5] Sarbu I, Sebarchievici C. General review of solar-powered closed sorption refrigeration systems. Energy Conversion and Management 2015;105(11):403–22.
[6] ASHRAE handbook, HVAC applications. Atlanta, GA: American Society of Heating, Refrigerating and Air Conditioning Engineers; 2015.
[7] Bougard J. Conversion d'energie: Machines solaires. Mons, France: Faculte Politechnique de Mons, AGADIR; 1995.
[8] Sarbu I, Adam M. Applications of solar energy for domestic hot-water and buildings heating/cooling. International Journal of Energy 2011;5(2):34–42.
[9] Stephenson DG. Tables of solar altitude and azimuth; Intensity and solar heat gain tables, Technical Paper 243. Otawa: National Research Council of Canada; 1967.
[10] Farber EA, Morrison CA. Clear-day design values. In: Jordan RC, Liu BYH, editors. Applications of solar energy for heating and cooling of buildings. New York: ASHRAE GRP-170; 1977.
[11] Iqbal M. An introduction to solar radiation. Toronto: Toronto University Press; 2004.
[12] Duffie JA, Beckman WA. Solar engineering of thermal processes. Hoboken, NJ: Wiley & Sons, Inc.; 2013.

[13]   Liu BYH, Jordan RC. The interrelationship and characteristic distribution of direct, diffuse and total solar radiation. Solar Energy 1960;4(3):1−19.

[14]   Klein SA. Calculation of monthly average insolation on tilted surfaces. Solar Energy 1977;19:325−9.

[15]   Bostan I, Dulgheru V, Sobor I, Bostan V, Sochireanu A. Conversion systems of renewable energies. Chisinau: Tehnica-Info Publishing House; 2007 [in Romanian].

[16]   Hottel HC, Woertz BB. Performance of flat-plate solar heat collectors. ASME Transactions 1942;64:91.

[17]   Bristow KL, Campbell GS. On the relationship between incoming solar radiation and daily maximum and minimum temperature. Agricultural and Forest Meteorology 1984;31(2):159−66.

[18]   Wang L, Kisi O, Zounemat-Kermani M, Salazar GA, Zhu Z, Gong W. Solar radiation prediction using different techniques: model evaluation and comparison. Renewable and Sustainable Energy Reviews 2016;61:384−97.

[19]   Spencer JW. Fourier series representation of the position of the Sun. Search 1971;2(5):172−3.

[20]   Meza FJ, Yebra ML. Estimation of daily global solar radiation as a function of routine meteorological data in Mediterranean areas. Theoretical and Applied Climatology 2015;6:1−10. http://dx.doi.org/10.1007/s00704-015-1519-6.

[21]   Pan T, Wu S, Dai E, Liu Y, et al. Estimating the daily global solar radiation spatial distribution from diurnal temperature ranges over the Tibetan Plateau in China. Applied Energy 2013;107:384−93.

Chapter 3

Solar Collectors

3.1 GENERALITIES

Solar energy can be converted to chemical, electrical, and thermal processes. Photosynthesis is a chemical process that produces food and converts CO_2 to O_2. PV cells convert solar energy to electricity. Traditionally, devices intended for using solar energy fall into two main classes depending on the method of its conversion: either heat or electricity, like thermal collectors and PV modules, respectively.

Solar thermal collectors are a special kind of heat exchangers that convert solar radiation into thermal energy through a transport medium or a heat transfer fluid (HTF). Classification of various solar collectors is illustrated in Fig. 3.1. The major component of any solar system is the solar collector. This is a device that absorbs the incoming solar radiation, converts it into heat energy, and transfers it through a fluid (usually water, air, or oil) for useful purpose/applications.

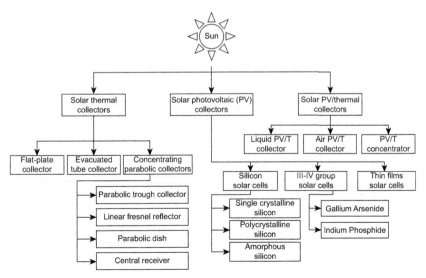

FIGURE 3.1 Classification of various solar collectors.

Solar Heating and Cooling Systems. http://dx.doi.org/10.1016/B978-0-12-811662-3.00003-7

The thermal conversion process provides thermal energy for space heating and cooling, domestic water heating, power generation, distillation, and process heating.

Photovoltaic (PV) system is the most useful way of utilizing solar energy by directly converting it into electricity. Energy conversion device that is used to convert sunlight to electricity by the use of the photoelectric effect is called solar cell (module). A PV system consists of the solar cells and ancillary components. It converts the solar radiation directly into electricity.

The solar energy conversion into electricity and heat with a single device is called hybrid photovoltaic/thermal (PV/T) collector.

This chapter presents a detailed description of energy balance for solar collector and of different types of solar thermal and PV collectors including the calculation of their efficiency and new materials for PV cells. The discussion considers both concentrating and nonconcentrating thermal technologies. Additionally, a brief overview on the research and development and application aspects for the hybrid PV/T collector systems is presented. Additionally, a simulation model of a PV/T collector with water heating in buildings closer to the actual system has been developed to analyze the PV and thermal performances of this system. The energetic performance of commercially available PV/T systems for electricity and domestic hot water (DHW) production is also evaluated for use in three European countries, each under entirely different climatic conditions.

3.2 SOLAR THERMAL COLLECTORS

A system for converting solar energy into thermal energy is generally provided with the following equipments (Fig. 3.2): solar collectors, heat storage devices, circulating pumps, heat transport and distribution network, automation, control and safety devices. This system can be supplied heat for a house heating system or/and a DHW system. If insufficient solar energy is available, then an auxiliary energy source is used.

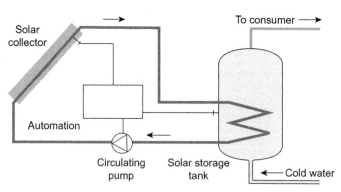

FIGURE 3.2   Solar energy conversion in thermal energy.

Solar thermal collector collects the solar radiation energy, transforms it into thermal energy, and transports heat toward an HTF (water, antifreeze-water mix, or air). A collector heat exchanger can be isolated with the antifreeze solution in the collector loop from the water storage tank loop. The solar storage tank is an insulated water storage unit that is sized in proportion to the collector area. The tank can be fully mixed or stratified. The solar DHW subsystem consists of a heating coil located inside the main storage tank. The solar space heating subsystem withdraws water from the top of the tank and circulates it through a water-to-air load heat exchanger and returns it to the tank.

There are basically two types of solar thermal collectors: nonconcentrating or stationary and concentrating. A nonconcentrating collector has the same area for intercepting and for absorbing solar radiation, whereas a Sun-tracking concentrating solar collector usually has concave reflecting surfaces to intercept and focus the sun's beam radiation to a smaller receiving area, thereby increasing the radiation flux. A large number of solar collectors are available in the market. A comprehensive list is shown in Table 3.1 [1]. In this section, a

TABLE 3.1 Types of Solar Thermal Collectors

Motion	Collector Type	Absorber Type	Concentration Ratio	Indicative Temperature Range (°C)
Stationary	Flat-plate collector (FPC)	Flat	1	30–80
	Evacuated tube collector (ETC)	Flat	1	50–200
	Compound parabolic collector (CPC)	Tubular	1–5	60–240
Single-axis tracking	Linear Fresnel reflector (LFR)	Tubular	10–40	60–250
	Parabolic trough collector (PTC)	Tubular	15–45	60–300
	Cylindrical through collector (CTC)	Tubular	10–50	60–300
Two-axes tracking	Parabolic dish reflector (PDR)	Point	100–1000	100–500
	Heliostat field collector (HFC)	Point	100–1500	150–2000

review of the various types of collectors currently available will be presented. This includes flat-plate collector (FPC), evacuated tube collector (ETC), and concentrating collectors.

Temperatures needed for space heating and cooling do not exceed 90°C, even for absorption refrigeration, and they can be attained with carefully designed FPCs.

3.2.1 Flat-Plate Collectors

The FPC is the heart of any solar energy collection system designed for operation in the low-temperature range (less than 60°C) or in the medium temperature range (less than 100°C). It is used to absorbed solar energy, convert it into heat, and then to transfer that heat to stream of liquid or gases. They use both direct and diffuse solar radiations, do not require tracking of the Sun, and require little maintenance. They are mechanically simpler than concentrating collectors. The major applications of these units are in solar water heating, building heating, air conditioning, and industrial process heat. An FPC generally consists of the following components (Fig. 3.3):

- *Glazing* consists of one or more sheets of glass or other radiation-transmitting material.
- *Tubes, fins, or passages* to conduct or direct the HTF from the inlet to the outlet.
- *Absorber plates* are flat, corrugated, or grooved plates, to which the tubes, fins, or passages are attached. The plate may be integral with the tubes.
- *Headers or manifolds* to admit and discharge the HTF.
- *Insulation* to minimize heat loss from the back and sides of the collector.

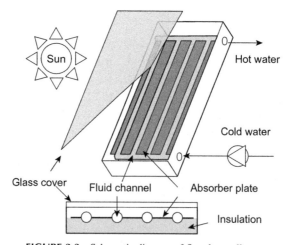

FIGURE 3.3   Schematic diagram of flat-plate collector.

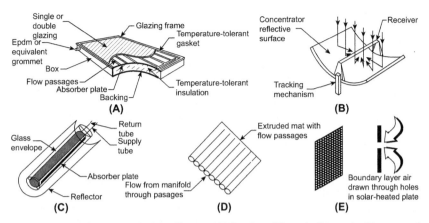

FIGURE 3.4 Various types of solar collectors: (A) flat-plate, (B) parabolic trough, (C) evacuated tubes, (D) unglazed EPDM collector, and (E) perforated plate.

- *Container* or *casing* to surround the other components and protect them from dust or moisture.

FPCs have been built in a wide variety of designs from many different materials (Fig. 3.4). Their major purpose is to collect as much solar energy as possible at lowest possible total cost. The collector should also have a long effective life, despite: the adverse effects of the Sun's ultraviolet radiation; corrosion or clogging because of acidity, alkalinity, or hardness of the HTF; freezing or air binding in the case of water, or deposition of dust or moisture in the case of air.

Glass has been widely used to glaze FPCs because it can transmit as much as 90% of the incoming shortwave solar radiation while transmitting very little of the longwave radiation emitted outward from the absorber plate. Glass with low iron content has a relatively high transmittance for solar radiation (0.85−0.90 at normal incidence).

Plastic films and sheets also have high shortwave transmittance, but because most usable varieties also have transmission bands in the middle of the thermal radiation spectrum, their longwave transmittance may be as high as 0.40.

The glass generally used in solar collectors may be either of single-strength (2.2−2.5 mm) or double-strength (2.92−3.38 mm). For direct radiation, the transmittance varies markedly with the angle of incidence θ, as shown in Table 3.2 [2]. Antireflective coating and surface texture can improve transmission significantly.

The absorber plate absorbs as much of the radiation as possible through the glazing, while losing as little heat as possible up to the atmosphere and down through the back of the casing. The absorber plate transfers retained heat to the

TABLE 3.2 Variation With Incident Angle of Transmittance and Absorptance

Incident Angle θ (°)	Transmittance τ		Absorptance α (for Flat-Black Paint)
	Single Glazing	Double Glazing	
0	0.87	0.77	0.96
10	0.87	0.77	0.96
20	0.87	0.77	0.96
30	0.87	0.76	0.95
40	0.86	0.75	0.94
50	0.84	0.73	0.92
60	0.79	0.67	0.88
70	0.68	0.53	0.82
80	0.42	0.25	0.67
90	0.00	0.00	0.00

transport fluid. The absorptance of the collector surface for shortwave solar radiation depends on the nature and color of the coating and on the incident angle, as shown in Table 3.2 for a typical flat-black paint.

By suitable electrolytic or chemical treatments, selective surfaces can be produced with high values of solar radiation absorptance α and low values of longwave emittance ε. Selective surfaces are particularly important when the collector surface temperature is much higher than the ambient air temperature.

Materials most frequently used for absorber plates are copper, aluminum, and stainless steel. UV-resistant plastic extrusions are used for low-temperature applications. Potential corrosion problems should be considered for any metals.

Materials most used for insulation are mineral wool, glass wool, and foam glass. Polyurethane foam is used for low-temperature application and styropor is used very rarely. The main properties of insulation materials are summarized in Table 3.3.

FPC is usually permanently fixed in position and requires no tracking of the Sun. The collectors should be oriented directly toward the equator, facing south in the northern hemisphere and north in the southern. The optimum tilt angle of the collector is equal to the latitude of the location with angle variations of 10–15°C, more or less depending on the application.

Solar air collectors have the operating principles similar to FPCs (Fig. 3.5). The difference is that instead of liquid fluid, an electric fan pumps air through

TABLE 3.3 Main Properties of Insulation Materials

No.	Materials	Accepted Temperature (°C)	Density (kg/m³)	Thermal Conductivity (W/(m K))
1	Mineral wool	>200	60–200	0.040
2	Glass wool	>200	30–100	0.040
3	Foam glass	>200	130–150	0.048
4	Polyurethane foam	<130	30–80	0.030
5	Styropor	<80	30–50	0.034

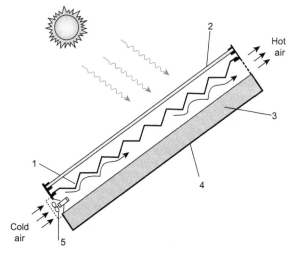

FIGURE 3.5   Schematic of solar air collector.

the collector. The main components are: absorber plate (1), glass cover (2), insulation (3), casing (4), and exhaust fan (5).

This collector type is not very common in Europe (covers only 1−2% of the solar liquid collector market). Reasons for this might be that on the one hand, the lack of experience and the lack of knowledge of the end users, and on the other hand, this collector type cannot be used directly for DHW production, which dominates the market today [3].

In Fig. 3.6 is illustrated energy balance of a standard FPC. From total solar radiation I_T incident on glass cover, a part ($\tau \cdot I_T$) determined of transmittance τ reaches the absorber plate where it is transformed in heat. The glass cover

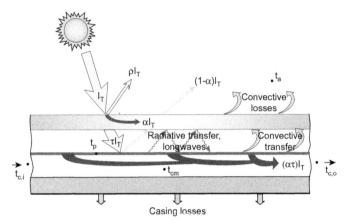

FIGURE 3.6   Energy balance of flat-plate collector.

reflects in space radiation ($\rho \cdot I_T$) and absorbs radiation ($\alpha \cdot I_T$), where ρ and α are the glass cover reflectance and absorptance, respectively. One part of radiation ($\tau \cdot I_T$) incident on absorber plate is reflected and the most part of this radiation is transformed in heat. The sum of the coefficients τ, ρ, and α is

$$\tau + \rho + \alpha = 1 \qquad (3.1)$$

3.2.2 Evacuated Tube Collectors

Evacuated tube collector (ETC) consists of single tubes that are connected to a header pipe. To reduce heat losses of the water-bearing pipes to the ambient air, each single tube is evacuated. Besides different geometrical configurations, it has to be considered that the collector must always be mounted with a certain tilt angle in order to allow the condensed internal fluid of the heat pipe to return to the hot absorber.

The ETCs provide the combined effects of a highly selective surface coating and vacuum insulation of the absorber element so that they can have high heat extraction efficiency compared with FPCs in the temperature range above 80°C [4]. At present, the glass-evacuated tube has become the key component in solar thermal utilization, and they are proved to be very useful especially in residential applications for higher temperatures. So, ETCs are widely used to supply the DHW or heating, including heat pipe-evacuated solar collectors and U-tube glass ETCs [5−10]. An ETC uses liquid−vapor phase-change materials (PCMs) to transfer heat at high efficiency. A schematic diagram of an ETC is given in Fig. 3.7 [11].

These collectors feature a heat pipe (a highly efficient thermal conductor) placed inside a vacuum-sealed tube. The pipe, which is a sealed copper pipe,

(A)

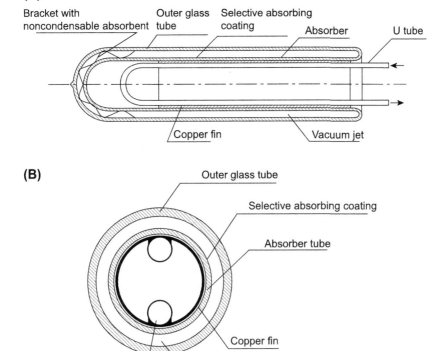

FIGURE 3.7 Glass ETC with U-tube (A) illustration of the glass evacuated tube; (B) cross-section.

attached to a black copper fin that fills the tube (absorber plate) is a metal tip attached to the sealed pipe (condenser). The heat pipe contains a small amount of fluid that undergoes an evaporating–condensing cycle. In this cycle, solar heat evaporates the liquid, and the vapor travels to the heat sink region where it condenses and releases its latent heat. The condensed fluid return back to the solar collector and the process is repeated. When these tubes are mounted, the metal tips up into a heat exchanger (manifold).

All these collectors are installed at a fixed tilt that optimizes the performance for a specified period; for a summer use, a tilt equal to the latitude φ minus 10 degrees can be considered as optimal.

Thermal tube collector (TTC) consists of the same components as the previous collector except that instead of a U-tube, there exists a thermal tube tightly loaded with a substance that vaporises under the influence of solar radiation. The produced vapors rise to the top of the tube called condenser, which yields the latent heat of condensation to HTF, whereby the substance

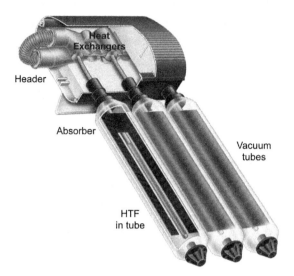

FIGURE 3.8   Thermal tube collector.

from the thermal tube condenses and the process restarts. Thus, the HTF is heated and flows through the collecting pipe. The operation of the thermal tube collector is illustrated in Fig. 3.8.

3.2.3 Concentrating Collectors

A concentrating collector comprises a *receiver*, where the radiation is absorbed and converted to some other energy form, and a *concentrator*, which is the optical system that directs beam radiation onto the receiver. The aperture A_a of the system is the projected area of the concentrator facing the beam. These collectors can be viewed as FPCs with augmented radiation. Analyses of these concentrators have been presented by Hollands [12], Selcuk [13], and others.

Concentrating collectors provide energy at temperatures higher than those of FPCs and ETCs. They redirect solar radiation passing through an aperture into an absorber and usually require tracking of the Sun. In concentrating collectors, solar energy is optically concentrated before being transferred into heat.

Concentration can be obtained by reflection or refraction of solar radiation by the use of mirrors or lens. A concentrating collector exhibits certain advantages and disadvantages as compared with the conventional flat-plate-type collector [14] some of them are given in Table 3.4.

Concentrating collectors can also be classified into non-imaging and imaging depending on whether the image of the Sun is focused at the receiver or not. The concentrator belonging in the first category is the compound parabolic

TABLE 3.4 Advantage and Disadvantage of Concentrating Collectors

No.	Advantages	Disadvantages
1	The working fluid can achieve higher temperatures in a concentrator system when compared to a flat-plate system of the same solar energy collecting surface. This means that a higher thermodynamic efficiency can be achieved.	Concentrator systems collect little diffuse radiation depending on the concentration ratio.
2	It is possible with a concentrator system to achieve a thermodynamic match between temperature level and task. The task may be to operate thermo ionic, thermodynamic, or other higher temperature devices.	Some form of tracking system is required so as to enable the collector to follow the Sun.
3	The thermal efficiency is greater because of the small heat-loss area relative to the receiver area.	Solar reflecting surfaces may lose their reflectance with time and may require periodic cleaning and refurbishing.
4	Reflecting surfaces require less material and are structurally simpler than FPC. For a concentrating collector the cost per unit area of the solar collecting surface is therefore less than that of an FPC.	
5	Owing to the relatively small area of receiver per unit of collected solar energy, selective surface treatment and vacuum insulation to reduce heat losses and improve the collector efficiency are economically viable.	

collector (CPC) whereas all the other types of concentrators belong to the imaging type. The collectors falling in this category are

- parabolic trough collector (PTC);
- linear Fresnel reflector (LFR); and
- parabolic dish reflector (PDR).

Non-imaging concentrators, as the name implies, do not produce clearly defined images of the Sun on the absorber but rather distribute radiation from all parts of the solar disk onto all parts of the absorber. The concentration ratios of linear non-imaging collectors are in the low range and are generally below 10. Imaging concentrators, in contrast, are analogous to camera lenses in that they form images (usually of very low quality by ordinary optical standards) on the absorber.

In general, concentrators with receivers much smaller than the apertures are effective only on the beam radiation. It is evident also that the angle of incidence of the beam radiation on the concentrator is important and that Sun tracking will be required for these collectors. A variety of orienting mechanisms have been designed to move focusing collectors so that the incident direct radiation will be reflected to the receiver. The motions required to accomplish tracking vary with the design of the optical system, and a particular resultant motion may be accomplished by more than one system of component motions.

The basis parameters of a concentrating collector are the concentration ratios. The geometric (area) concentration ratio C_A is the ratio of the area of aperture A_a to the area of the receiver A_r as

$$C_A = \frac{A_a}{A_r}. \tag{3.2}$$

The optical concentration ratio C_o is the ratio of the direct radiation intensity (power density) at the receiver $I_{D,r}$ to that at the aperture $I_{D,a}$ as

$$C_o = \frac{I_{D,r}}{I_{D,a}}. \tag{3.3}$$

For an ideal concentrating collector, $C_A = C_o$, but in practice, the solar intensity varies greatly across the receiver. The temperature of the receiver cannot be increased indefinitely by simply increasing C_A, since by Kirchhoff's laws the receiver temperature t_r cannot exceed the equivalent temperature t_s of the Sun. Moreover, the Sun (radius R_s, distance L) subtends a finite angle Φ_s at the Earth, which limits the achievable geometric concentration ratio to a value given by Eqs. (3.4) or (3.5):

- for circular concentrators:

$$C_A < (L/R_s)^2 = 45000; \tag{3.4}$$

- for linear concentrators:

$$C_A < L/R_s = 215 \tag{3.5}$$

The non-imaging concentrators have the capability of reflecting to the receiver all of the incident radiation on the aperture over ranges of incidence angles within wide limits. The limits define the acceptance angle of the concentrator. As all incident radiation within the acceptance angle is reflected to the receiver, the diffuse radiation within these angles is also useful input to the collector. The basic concept of the CPC is shown in Fig. 3.9 [15]. These concentrators are potentially most useful as linear or trough-type concentrators.

Each side of the CPC is a parabola; the focus and axis of only the right-hand parabola are indicated. Each parabola extends until its surface is parallel with the CPC axis. The angle between the axis of the CPC and the line connecting the focus of one of the parabolas with the opposite edge of the aperture is the acceptance half-angle θ_c. If the reflector is perfect, any radiation entering the aperture at angles between $\pm\theta_c$ will be reflected to a receiver at the base of the concentrator by specularly reflecting the parabolic reflectors.

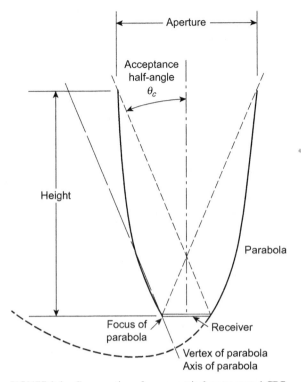

FIGURE 3.9 Cross-section of a symmetrical no-truncated CPC.

CPCs have area concentration ratios depending on the acceptance half-angle θ_c as

$$C_A = \frac{1}{\sin\theta_c}. \tag{3.6}$$

Parabolic Trough Collector. The first practical experience with PTC goes back to 1870, when a successful engineer, John Ericsson, a Swedish immigrant to the United States, designed and built a 3.25 m^2-aperture collector that drives a small 373 W engine [16]. Nowadays, PTCs target application in which a solar field can be successfully integrated for supplying thermal energy at temperatures up to 250°C (Fig. 3.10) [11]. Nevertheless, there are other applications, such as heat-driven refrigeration and cooling, low-temperature heat demand with high consumption rates, irrigation water pumping, desalination, and detoxification. On the one hand, these temperature requirements cannot be achieved by conventional low-temperature collectors.

Horizontal reflective parabolic trough, oriented east and west, requires continuous adjustment to compensate for changes in the Sun's declination. There is inevitably some morning and afternoon shading of the reflector if the

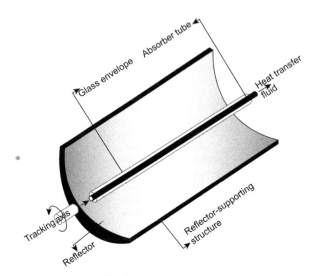

FIGURE 3.10   Parabolic trough collector.

concentrator has opaque end panels. The necessity of moving the concentrator to accommodate the changing solar declination can be reduced by moving the absorber or by using a trough with two sections of a parabola facing each other, known as a CPC. By using multiple internal reflections, any radiation that is accepted finds its way to the absorber surface located at the bottom of the apparatus.

In thermal steady state, maximum temperature of the receiver t_R, in K, can be calculated using the following equation [17]

$$t_R = \sqrt[4]{\frac{\rho \alpha I_D}{\varepsilon \sigma \Phi_s}} \tag{3.7}$$

in which

$$\Phi_s = \frac{R_s}{L}, \tag{3.8}$$

where ρ is the reflector reflectance; α is the reflector absorptance; I_D is the direct solar radiation, in W/m^2; ε is the reflector emittance; σ is the Stefan–Boltzmann constant (5.67×10^{-8} $W/(m^2 \cdot K^4)$); R_s is the solar disk radius (0.695×10^9 m); and L is the mean distance between the Sun and the Earth (1.495×10^{11} m).

For $I_D = 600$ W/m^2, $\alpha/\varepsilon = 1$, $\rho = 0.8$, $\Phi_s = 0.00465$ rad   is   obtained maximum temperature of 1162K or 889°C.

Linear Fresnel Reflector (LFR) differs from that of the PTCs in that the absorber is fixed in space above the mirror field (Fig. 3.11). Also, the reflector is composed of many low-row segments, which focus collectively on an

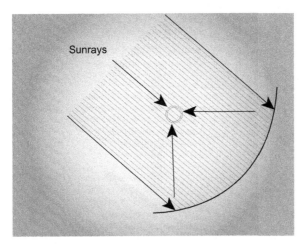

FIGURE 3.11   Linear Fresnel reflector collector.

elevated long tower receiver running parallel to the reflector rotational axis [18]. This type of concentrator has been used to attain temperatures well above those that can be reached with FPCs.

This system offers a lower cost solution as the absorber row is shared among several rows of mirrors. However, one fundamental difficulty with the LFR technology is the avoidance of shading of the incoming solar radiation and blocking of the reflected solar radiation by adjacent reflectors. Blocking and shading can be reduced by using absorber towers elevated higher and/or by increasing the absorber size, which allows increased spacing between reflectors remote from the absorber. Both these solutions increase costs, besides a larger ground usage is required. The compact linear Fresnel reflector (CLFR) offers an alternate solution to the LFR problem. The classic LFR has only one linear absorber on a single linear tower. This prohibits any option of the direction of orientation of a given reflector. Since this technology would be introduced in a large field, one can assume that there will be many linear absorbers in the system. Therefore, if the linear absorbers are close enough, individual reflectors will have the option of directing the reflected solar radiation onto at least two absorbers. This additional factor gives potential for more densely packed arrays, since patterns of alternative reflector inclination can be set up such that the closely packed reflectors can be positioned without shading and blocking solar radiation [19].

Parabolic-Dish Reflector. A PDR, shown schematically in Fig. 3.12, is a point-focusing collector [11]. Concentrating solar energy onto a receiver located at the focal point of the dish, it tracks the Sun in two axes. The dish structure must track fully the Sun to reflect the beam into the thermal receiver. Because the receivers are distributed throughout a collector field, like parabolic troughs, parabolic dishes are often called distributed-receiver systems.

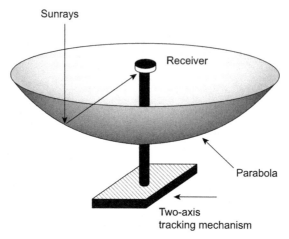

FIGURE 3.12   Schematic of a parabolic dish collector.

Parabolic-dish systems that generate electricity from a central power converter collect the absorbed sunlight from individual receivers and deliver it via a HTF to the power conversion systems.

The temperature of the receiver of parabolic-dish collector t_R, in K, can be calculated using following equation [17]

$$t_R = \sqrt[4]{\frac{3\rho\alpha I_D}{8\varepsilon\sigma\Phi_s^2}}. \tag{3.9}$$

For the same value of direct solar radiation is obtained the maximum theoretical temperature $t_R = 3480K$ (3208°C). The real parabolic-dish systems can achieve temperatures in excess of 1500°C.

3.2.4 Solar Thermal Power Plants With Central-Receiver

The central-receiver systems are considered to have a large potential for midterm cost reduction of electricity compared to parabolic trough technology since they allow many intermediate steps between the integration in a conventional Rankin cycle up to the higher energy cycles using gas turbines at temperatures above 1000°C, and this subsequently leads to higher efficiencies and huge outputs. Another alternative is to use Brayton cycle, which requires higher temperature than the ones employed in Rankin cycle.

Solar thermal power plants with heliostats and central-receiver (solar tower) uses thousands of mirrors (heliostats) that reorientate the concentrated solar flux toward a receiver mounted on a tower. For most receivers, a salt solution heated in the receiver is used to generate vapors direct in a receiver, which are used by a steam turbine generator to produce electricity. The melted

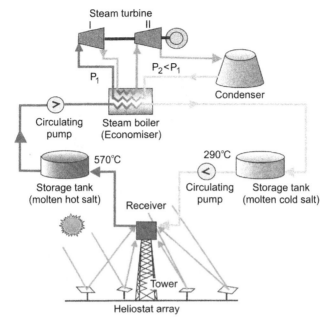

FIGURE 3.13   Schematic of a solar thermal power plant.

nitrate salt has superior thermal proprieties and energy storage capacities. The commercial solar towers can be designed to produce 50–200 MW electricity. The solar tower technology with high-temperature solar receiver is illustrated in Fig. 3.13.

The central-receiver concept for generation of electrical energy from solar energy is based on the use of very large concentrating collectors. The optical system consists of a field of a large number of heliostats, each reflecting direct radiation onto a central receiver. The result is a Fresnel-type concentrator, a parabolic reflector broken up into small segments. Several additional optical phenomena must be taken into account. Shading and blocking can occur (shading of incident direct radiation from a heliostat by another heliostat and blocking of reflected radiation from a heliostat by another that prevents that radiation from reaching the receiver). As a result of these considerations, the heliostats are spaced apart, and only a fraction of the ground area ψ is covered by mirrors. A ψ of approximately 0.3–0.5 has been suggested as a practical value [15].

The maximum concentration ratio C_{Amax} for a three-dimensional concentrator system with radiation incident at an angle θ_i on the plane of the heliostat array ($\theta_i = \theta_z$ for a horizontal array), a rim angle of ϕ_r, and a dispersion angle of δ, if all reflected direct radiation is to be intercepted by a spherical receiver, is

$$C_{Amax} = \frac{\psi \sin^2 \phi_r}{4 \sin^2(0.267 + \delta/2)} - 1. \tag{3.10}$$

For a flat receiver, the concentration ratio is

$$C_{Amax} = \psi \left[\frac{\sin\phi_r \cos(\phi_r + 0.267 + \phi/2)}{\sin(0.267 + \delta/2)} \right]^2 - 1. \tag{3.11}$$

As with linear concentrators, the optimum performance may be obtained with intercept factors less than unity. The heliostat field need not be symmetrical, the ground cover ψ does not have to be uniform, and the heliostat array is not necessarily all in one plane. These collectors operate with high concentration ratios and at relatively higher receiver temperatures.

In the early 1990s of the previous century, some pilot-power plants with heliostats and central tower were constructed in Russia, Italy, Spain, and France. The biggest solar power plant in the world called *Solar One* was constructed in the year 1982 in Barstow, California, and has a power of 10 MW [20]. An interesting European project, similarly with the former one, is the system Solair 3000 installed in the region Almeria, Spain.

3.2.5 Thermal Collector Efficiency

For a certain solar collector, the performance depends on fluid inlet/outlet temperatures of collector, $t_{c,i}/t_{c,o}$ (°C) or absorber plate temperature t_p (°C), ambient air temperature t_a (°C), and solar radiation intensity I_T (kW/m²).

Balance energy equation of a solar collector can be written as

$$Q_u = Q_a - Q_{hl} = mc_p(t_{c,o} - t_{c,i}) \tag{3.12}$$

where Q_u is the usable energy collected by solar collector, in kW; Q_a is the total absorbed energy by collector, in kW; Q_{hl} is the heat loss of collector, in kW; m is mass flow rate of water-antifreeze solution, in kg/s; and c_p is the specific heat of water-antifreeze solution, in kJ/(kg K).

The instantaneous efficiency η_c of a collector operating under steady conditions is defined mathematically as

$$\eta_c = \frac{Q_u}{A_c I_T} = \frac{mc_p(t_{c,o} - t_{c,i})}{A_c I_T}, \tag{3.13}$$

where Q_u is the usable energy of collector; in kW; A_c is the collector surface area, in m²; and I_T is the solar radiation intensity, in (kW/m²).

Taking into account the Q_a and Q_{hl} expressions, the first equality of Eq. (3.12) receives the form Hottel-Whillier-Bliss [2,15]:

$$Q_u = A_c[I_T(\alpha\tau) - U_L(t_p - t_a)], \tag{3.14}$$

where α is the absorptance; τ is the transmittance; U_L is the overall heat loss coefficient, in kW/(m²K); t_p is the absorber plate temperature, in °C; and t_a is the ambient air temperature, in °C.

Taking into account the first equality of Eq. (3.13), the expression for collector efficiency η_c becomes

$$\eta_c = (\alpha\tau) - \frac{U_L(t_p - t_a)}{I_T}. \tag{3.15}$$

The product $(\alpha\tau)$ characterizes the optical properties of the collector.

In 1977 Jordan and Liu [21] suggested introducing an additional term, the collector heat removal factor F_R, to allow use of the fluid inlet temperature in Eqs. (3.14) and (3.15) as

$$Q_u = F_R A_c \left[I_T(\alpha\tau) - U_L\left(t_{c,i} - t_a\right) \right] \tag{3.16}$$

$$\eta_c = F_R(\alpha\tau) - \frac{F_R U_L\left(t_{c,i} - t_a\right)}{I_T} \tag{3.17}$$

in which [15]

$$F_R = \frac{mc_p\left(t_{c,o} - t_{c,i}\right)}{A_c\left[S - U_L\left(t_{c,i} - t_a\right)\right]}, \tag{3.18}$$

where F_R equals the ratio of the heat actually delivered by the collector to the heat that would be delivered if the absorber were at $t_{c,i}$ and S is the absorbed radiation, in W/m^2. F_R is found from the result of a test performed in accordance with American standard ASHRAE 93 [22] or European norms EN 12,975-1 [23], and EN 12,975-2 [24].

Eq. (3.17) provides the basis for simulation models of a collector operating under steady conditions.

Other equations also used the collector efficiency factor and the arithmetic average of the fluid inlet and outlet temperatures [25] as

$$Q_u = FA_c[I_T(\alpha\tau) - U_L(t_{cm} - t_a)] \tag{3.19}$$

$$\eta_c = F(\alpha\tau) - \frac{FU_L(t_{cm} - t_a)}{I_T}, \tag{3.20}$$

where F is the collector efficiency factor equal with ratio between the heat transfer resistance from the fluid to the ambient air and the heat transfer resistance from the absorber plate to the ambient air and t_{cm} is the average of the fluid inlet and outlet temperatures $((t_{c,i} + t_{c,o})/2)$ for solar collector, in °C.

In Fig. 3.14 is illustrated the efficiency for a solar collector with selective absorber plate depending ratio $(t_{cm} - t_a)/I_T$ computed using Eq. (3.20), in which $F = 0.8$, and mean efficiency obtained experimentally according American standard ASHRAE 93. For the water temperatures between 30°C and 60°C, it is noticed as a good agreement, but a large deviation for temperatures up to 60°C.

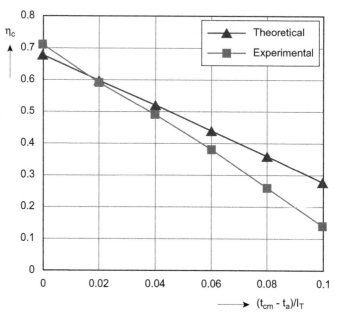

FIGURE 3.14   Collector efficiency variation.

The instantaneous efficiency η_c of a collector can be also written in a general form [15,26] as

$$\eta_c = \eta_0 - a_1 \frac{t_{cm} - t_a}{I_T} - a_2 \frac{(t_{cm} - t_a)^2}{I_T}, \qquad (3.21)$$

where η_0, a_1, and a_2 are constants relative to the considered solar collector, t_{cm} is the average of the fluid inlet and outlet temperatures, and I_T is the total solar radiation for nonconcentrating collectors and the direct radiation for concentrating collectors. The second-order coefficient is usually negligible and is not always considered.

Then the losses are negligible and η_0 (optical efficiency or zero-loss collector efficiency) is the fraction of solar insolation transmitted through the shield and absorbed by the plate; a_1 and a_2 are heat loss coefficients temperature-dependent. In Fig. 3.15 is illustrated the η_c efficiency variation of different collector types with total radiation I_T, for a temperature difference $\Delta t = t_{cm} - t_a = 60°C$ between the fluid and ambient air.

Usual thermal efficiency of solar collectors varies between 40% and 55%.

The three coefficients account for different collector types and realization, and they can be experimentally evaluated by the solar collector efficiency test. When constants a_1 and a_2 are large, the collector efficiency is more sensible to high operative temperatures. Even if tracking collectors are not so widespread, parabolic concentrating collectors (PTCs) should be considered for operating

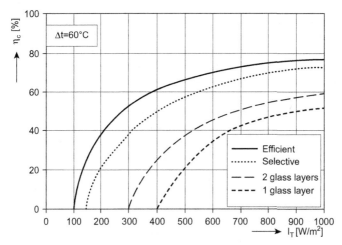

FIGURE 3.15 Variation of collector efficiency with irradiation.

temperatures as high as 160°C, suitable for driving double-effect sorption chillers. In a PTC, a reflector focuses the direct solar radiation parallel to the collector axis onto the receiver placed on the focal line (Fig. 3.16) [27].

The general equation describing a PTC behavior is quite similar to Eq. (3.21); the main difference is that the optical collector efficiency must be multiplied by the ratio between direct solar and global radiation intensity as only a negligible fraction of diffuse radiation can be concentrated onto the focus [28,29].

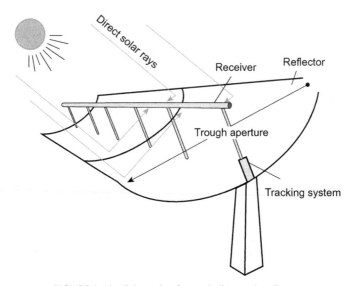

FIGURE 3.16 Schematic of a parabolic trough collector.

TABLE 3.5 Parameters of Efficiency Curve of the Four-Considered Solar Collectors

Parameter	FPC	ETC	TTC	PTC	Unit
η_0	0.791	0.718	0.825	0.520	(–)
a_1	3.940	0.974	1.190	0.475	$(W/(m^2\ K))$
a_2	0.012	0.005	0.009	0.0031	$(W/(m^2\ K^2))$

The three constants of the general Eq. (3.21) differ greatly from one collector type to another, but even within the same typology of collector, the differences can be significant.

For the following evaluations, reasonable average parameters are listed on Table 3.5 for the four solar collector typologies considered. The parameters of the ETC are of a very widespread model sold in Italy [30]. As for the concerns, the PTC Cabrera et al. [27] consider three different models with not-so-different performances of one from another. The selected one presents an optical (zero loss) coefficient of 0.6931, mainly referred to a direct radiation. The comparison of the efficiency curves requires an estimate of the diffuse radiation fraction. Fig. 3.17 represents the efficiency curves of the three collectors, assuming a 25% diffuse radiation fraction. The slope is quite different for the three collectors. The intercept with the ordinate axis is low for the PTC just as it can use only the direct solar radiation.

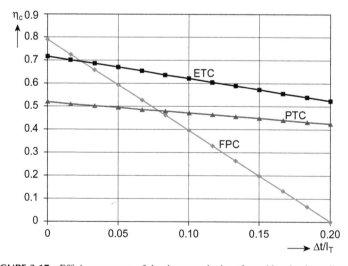

FIGURE 3.17 Efficiency curves of the three typologies of considered solar collectors.

Fluid mass flow rate m, in kg/s, can be determined by second equality of Eq. (3.15) as

$$m = \frac{Q_c}{c_p(t_{c,o} - t_{c,i})}.$$ (3.22)

Maximum fluid temperature t_{max} in collector is the temperature for mass flow rate equal to zero, that is, $t_{max} = I_{Tmax}\eta_0/U_L$. This temperature imposes conditions on the materials that are used to realize the collector, but also in the choice of fluid and the protection against superpressure for the fluid circuit.

Solar collector area A_c, in m^2, can be expressed as

$$A_c = \frac{\rho_w V c_p(t_{ST} - t_{cw})}{I_m \eta_c \eta_s},$$ (3.23)

where ρ_w is the water density, in kg/m^3; V is the volume of heated water, in m^3; c_p is the specific heat of water, in J/(kg K); t_{ST} is the water temperature in storage tank, in °C; t_{cw} is the cold water temperature, in °C; I_m is the daily mean solar radiation, in J/m^2; η_c is the solar collector efficiency; and η_s is the solar circuit efficiency.

Utilizability Φ has been used to describe the fraction of solar flux absorbed by collector, which is delivered to the HTF. On a monthly timescale

$$\Phi = \frac{Q_u}{F_R \eta_0 I_T} < 1.0,$$ (3.24)

where Φ is the fraction of the absorbed solar flux, which is delivered to the fluid in a collector operating at a fixed temperature t_{cm}; Q_u is the monthly averages daily total useful energy delivery; F_R is the heat removal factor; η_0 is the optical efficiency; and I_T is the total solar radiation. The fixed-temperature mode will occur if the collector provides only a minor fraction of the thermal demand. When t_{cm} is not constant in time as in the case of a collector coupled to storage, the Φ concept cannot be applied directly.

An important implication of the performance of solar thermal collectors is the cut in irradiance. The cut in irradiance is the radiation for which a collector actually starts producing heat at the desired temperature level. Van Leeuwen [31] used the average solar radiation for a specific location in South Spain in combination with the corresponding ambient temperatures and derived the heat production of collectors for three different heat-delivery temperature levels. His results are illustrated in Fig. 3.18. ETCs produce daily more heat than FPCs. However, at low heat-delivery temperatures, the difference is only about 25% while at high temperature differences, the difference becomes quite large, as more than 60%. Since FPCs cost approximately 50% of ETCs, the heat generated by FPCs is always cheaper (for the specific location considered).

There are a variety of solar collectors produced by different manufacturers as Viesmann, Wagner Co, Solvis, Conergy, Rehau, Thermomax, Schott, etc.

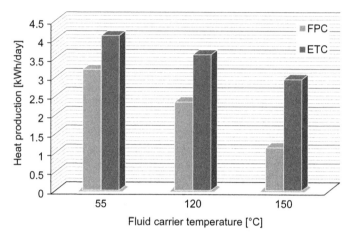

FIGURE 3.18 Daily solar collector heat production for different temperature levels.

Table 3.6 summarizes the technical data of some FPCs and Table 3.7 presents the technical data of some ETCs [32].

Price of the nonconcentrating solar thermal collectors varies widely. Gebreslassie et al. [33] report prices of 271 €/m² for a specific FPC. Cabrera et al. [27] report prices of 650 €/m² for a specific ETC. These prices show that ETCs will be approximately 2.5 times more expensive than FPCs. To make these values comparable to the installed values for the PV systems, these values have been added with 100 €/m² to account for installation and distribution costs—in total 270 €/m² for PDC, 540 €/m² for PTC, 750 €/m² for ETC and 371 €/m² for FPC.

3.3 SOLAR PV COLLECTORS

One of the methods used to generate significant electric power is the PV generation using solar cells. These devices produce electricity directly from electromagnetic radiation, especially light, without any moving parts. The physical process in which a PV cell converts sunlight into electricity is known as the PV effect. The PV effect was discovered by Becquerel in 1839 but not developed as a power source until 1954 by Chapin, Fuller, and Pearson using doped semiconductor silicon [34].

In recent years, PV technology has been developed quickly and made this technology viable even for small-scale power generation in distribution systems. Solar PV capacity for grid-connected system around the world was 10 GW in 2007, 16 GW in 2008, 24 GW in 2009, and 40 GW in 2010 [35]. At the end of 2014, worldwide PV capacity reached at least 177 GW. PVs grew fastest in China, followed by Japan and the USA, while Germany remains the world's largest overall producer of PV power, contributing approximately 7%

TABLE 3.6 Technical Data for Some Models of Flat-Plate Collectors

Manufacturer	Model	Dimensions L × l × H (m)	Absorber Plate (m²/m²)	Weight (kg)	Temp. (°C)	Collector Parameters		
						η_{l0} (−)	a_1 (W/(m² K))	a_2 (W/(m² K²))
Viesmann	Vitosol 100	2.38 × 1.06 × 0.09	2.3/2.51	52	221	0.810	3.48	0.0164
BBT solar Diamante	Logasol SKS 4.0	2.07 × 1.145 × 0.09	2.1/2.37	46	204	0.851	4.04	0.108
Wagner Co	Euro C20 AR	2.151 × 1.215 × 0.11	2.39/2.61	48	232	0.854	3.37	0.0104
Solvis	Fera F 552	3.793 × 1.48 × 0.105	5.25/5.61	109	201	0.832	3.62	0.0152
Conergy	F 2000	2.01 × 1.07 × 0.09	1.91/2.15	39	220	0.765	3.44	0.0183
Rehau	Solect WK	2.356 × 1.081 × 0.1	2.2/2.55	46	218	0.770	3.49	0.0170

TABLE 3.7 Technical Data for Some Models of Evacuated Tube Collectors

Manufacturer	Model	Dimensions L × l × H (m)	Absorber Plate (m²/m²)	Weight (kg)	Temp. (°C)	η_0 (−)	a_1 (W/(m² K))	a_2 (W/(m² K²))
							Collector parameters	
Paradigma	CPC 14 star azurro	1.61 × 1.62 × 0.12	2.33/2.61	42	–	0.691	0.929	0.0020
Thermomax	Mazdon 20	2.021 × 1.5 × 0.161	2.03/3.03	53	150	0.804	1.150	0.0064
Schott	ETC 16	1.684 × 0.765 × 0.1	0.808/1.29	20	–	0.773	1.090	0.0094

to the overall electricity generation. By 2018, worldwide capacity is projected to reach as much as 430 GW. This corresponds to a tripling within 5 years.

In terms of globally installed capacity, PV is the third most important RES after hydro and wind power [36]. For instance in European Union, PV represents approximately 37% from all new capacity of energy sources installed in 2012.

3.3.1 PV Converters

PV uses solar cells assembled into solar panels (modules) to convert sunlight directly into electricity. The term "photo" means light and "voltaic," electricity. A PV cell is a semiconductor device that generates electricity when light falls on it.

Incident solar radiation can be considered as discrete "energy units" called photons. The product of the frequency ν and wavelength λ is the speed of light as

$$c = \lambda\nu. \tag{3.25}$$

The energy E of a photon is a function of the frequency of the radiation (and thus also of the wavelength) and is given in terms of Planck's constant h by

$$E = h\nu. \tag{3.26}$$

Thus the most energetic photons are those of high frequency and short wavelength.

The most common PV cells are made of single-crystal silicon. An atom of silicon in the crystal lattice absorbs a photon of the incident solar radiation, and if the energy of the photon is high enough, an electron from the outer shell of the atom is freed. This process thus results in the formation of a hole— electron pair, a hole where there is a lack of an electron and an electron out in the crystal structure. These normally disappear spontaneously as electrons recombine with holes. The recombination process can be reduced by building into the cells a potential barrier, a thin layer or junction across which a static charge exists. This barrier is created by doping the silicon on one side of the barrier with very small amounts (of the order of one part in 10^6) of boron to form p-silicon, which has a deficiency of electrons in its outer shell, and that on the other side with phosphorus to form n-silicon, which has an excess of electrons in its outer shell. The barrier inhibits the free migration of electrons, leading to a build-up of electrons in the n-silicon layer and a deficiency of electrons in the p-silicon. If these layers are connected by an external circuit, electrons (i.e., a current) will flow through that circuit. Thus free electrons created by absorption of photons are in excess in the n-silicon and flow through the external circuit to the p-silicon. Electrical contacts are made by metal bases on the bottom of the cell and by metal grids or meshes on the top layer. Schematic of a silicon solar cell in a circuit is shown in Fig. 3.19, where

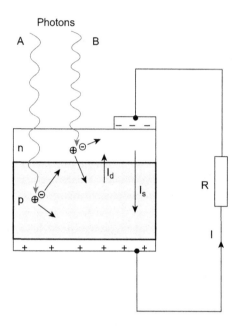

FIGURE 3.19 Schematic of a silicon solar cell.

pn junction diode current I_d has the opposite sense to PV current I_s. The Shockley equation for the junction diode is usually written as [34]

$$I_d = I_0 \left[\exp\left(\frac{eU}{kt}\right) - 1 \right], \tag{3.27}$$

where I_0 is the diode reverse saturation current (leakage or diffusion current), in A; e is the electron charge ($e = 1.602 \times 10^{-19}$ C); U is the voltage, in V; k is the Boltzmann's constant (1.381×10^{-23} J/K); and t is the cell temperature, in K.

There are many variations on cell material, design, and methods of manufacture. Amorphous or polycrystalline silicon (Si), cadmium sulfide (CdS), gallium arsenide (GaAs), and other semiconductors are used for cells [15].

The predominant PV technology is crystalline silicon, while thin-film solar cell technology accounts for approximately 10% of the global PV deployment. In Fig. 3.20 are illustrated two technologies: conventional c-Si solar cells and heterostructure with intrinsic thin layer (HIT) cells. However, the relatively high cost of manufacturing these silicon cells has prevented their widespread use. Another disadvantage of silicon cells is the use of toxic chemicals in manufacture. These aspects prompted the search for environmentally friendly and low-cost solar cell alternatives.

Several different semiconductor materials have been used to make the layers in different types of solar cells, and each material has its own quantities and drawbacks. The first requirement of a material to be suitable for solar cell

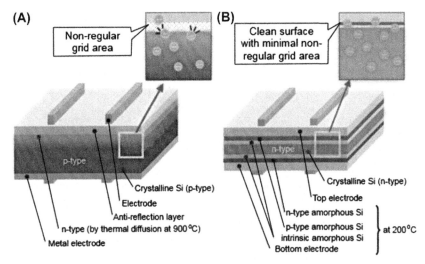

FIGURE 3.20   Solar cell types (A) Conventional c-Si solar cell; (B) HIT solar cell.

application is a band-gap matching to the solar spectrum. The band gap should be between 1.1 and 1.7 eV. The material must also have high motilities and lifetime of charge carriers.

Other requirements are (1) direct band structure, (2) consisting of readily availability, (3) nontoxicity, (4) easy reproducible deposition technique, (5) suitable for large area production, (6) good PV conversion efficiency, and (7) long-term stability. Highest efficiencies achieved using the different semiconductor materials are given in Table 3.8 [37]. Depending on the material used, solar cells can be categorized into three main groups: (1) silicon solar cells, (2) III-V group solar cells, and (3) thin-film solar cells.

Solar cells are assembled into solar panels (modules) that constitute a significant part of a PV system. Solar cell arrays are often assembled from a combination of individual modules usually connected in series and parallel. Each module is itself a combination of cells in series. Each cell is a set of surface elements connected in parallel.

Cell modules can be purchased on the market that are over 15% efficient and have design lifetimes of over 10 years. Experimental single crystalline silicon cells have been produced with efficiencies of 25% and cells with multiple junctions (i.e., two or more layers of materials with varying spectral response) have been constructed that have efficiencies of more than 30%.

3.3.2 PV Generator Characteristics

An assessment of the operation of solar cells and the design of power systems based on solar cells must be based on the electrical characteristics, that is, the

TABLE 3.8 Best Efficiencies Reported for the Different Types of Solar Cells

Cell Type	Highest Reported Efficiency for Small Area Produced in the Laboratory	Highest Reported Module Efficiency
c-Si (crystalline Si)	24.7% (UNSW, PERL)	22.7% (UNSW/Gochermann)
Multi-c-Si	20.3% (FhG-ISE)	15.3% (Sandia/HEM)
αSi:H, amorphous Si	10.1% (Kaneka), N.B. single junction	Triple junction. Stabilized efficiency = 10.4%
μc-Si/αSi:H (micro-morph cell)	11.7% (Kaneka), N.B. minimodule	11.7% (Kaneka), N.B. minimodule
Heterostructure with intrinsic thin layer (HIT) cell	21% (Sanyo)	18.4% (Sanyo)
Gallium arsenide (GaAs) cell	25.1% (Kopin)	Not relevant
Indium phosphide (InP) cell	21.9% (Spire)	Not relevant
GaInP/GaAs/Ge multijunction	32% (Spectolab), N.B. 37.3% under concentration	Not relevant
Cadmium telluride (CdTe)	16.5% (NREL)	10.7% (BP Solarex)
Cu(In,Ga)Se$_2$ (CIGS)	19.5% (NREL)	13.4% (Showa shell), N.B. for copper gallium indium sulfur selenide
Dye-sensitized cell (DSC)	8.2% (ECN)	4.7% submodule (INAP)

voltage–current relationships of the cells under various levels of radiation and at various cell temperatures.

The *pn* junction with photon absorption is a DC source of current and power, with positive polarity at the p-type material. For a simplified equivalent circuit (Fig. 3.21), the solar cell current I is determined by subtracting the diode current I_d, expressed by Eq. (3.27), from the photon-generated current I_s as

$$I = I_s - I_0 \left[\exp\left(\frac{eU}{kt}\right) - 1 \right]. \tag{3.28}$$

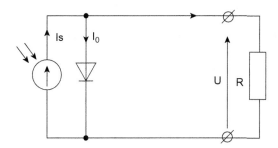

FIGURE 3.21   Equivalent circuit of a PV cell.

Shunt resistance is caused by structural defects across and at the edge of the cell. Present technology has reduced these to a negligible effect, so shunt resistance may be considered infinite and series resistance equal to zero in single-crystal Si cells, and that simplified equivalent circuit is satisfactory.

Currentvoltage IV characteristics of a typical PV module are shown in Fig. 3.22. The current axis ($V = 0$) is short-circuit current I_{sc}, and the intersection with the voltage axis ($I = 0$) is the open-circuit voltage U_o. For this module, the current decreases slowly to about U_M and then decreases rapidly to the open-circuit conditions at about U_o.

The power as a function of voltage is also shown in Fig. 3.22. The maximum power that can be obtained corresponds to the rectangle of maximum area under the *IV* curve. At the maximum power point M, the power is P_M, the current is I_M, and the voltage is U_M. Ideally, cells would always operate at the maximum power point, but practically cells operate at a point on the *IV* curve that matches the *IV* characteristic of the load.

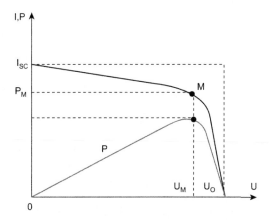

FIGURE 3.22   Typical *IV* and *PV* curves for a PV module.

Electrical power of a PV cell is

$$P = UI = U\left\{ I_s - I_0 \left[\exp\left(\frac{eU}{kt} \right) \right] - 1 \right\}. \qquad (3.29)$$

The current I_M, the voltage U_M, and the power P_M are expressed as

$$I_M = I_s \left(\frac{I_0}{I_s} + 1 \right) \frac{U_M}{U_M + U_t} \qquad (3.30)$$

$$U_M = U_0 - U_t \ln\left(\frac{U_M}{U_t} + 1 \right) \qquad (3.31)$$

$$P_M = U_M I_M, \qquad (3.32)$$

where $U_t = kt/e$ is the thermal voltage.

The fill factor (FF) of a cell can be determined using following equation as

$$FF = \frac{U_M I_M}{U_0 I_{sc}} \qquad (3.33)$$

so that

$$P_M = FF \cdot U_0 \cdot I_{sc}. \qquad (3.34)$$

Cell efficiency η is defined as the ratio between maximum power generated by cell for a specified temperature and solar radiation power as

$$\eta = \frac{P_M}{A I_T}, \qquad (3.35)$$

where P_M is the delivered power, in W; A is the cell area, in m^2; and I_T is the total incident radiation on the cell surface, in W/m^2.

Cell temperature: The temperature of operation of a PV module is determined by an energy balance. The solar energy that is absorbed by a module is converted partly into thermal energy and partly into electrical energy, which is removed from the cell through the external circuit. For best operation of the cells, they should be at the minimum possible temperature, which limits the possible applications of combined systems to situations where thermal energy is needed at low temperatures. The nominal operating cell temperature (NOCT) is defined as the cell temperature that is reached when the cells are mounted in their normal way at a solar radiation level of 800 W/m^2, a wind speed of 1 m/s, an ambient temperature of 20°C, and no-load operation (i.e., with $\eta = 0$). In actual operating conditions, the cell temperature t, in °C, can be calculated in function of NOCT [38] as

$$t = t_a + \frac{NOCT - 20}{0.8} \times I_T, \qquad (3.36)$$

where t_a is the ambient air temperature, in °C and I_T is the total solar radiation, in W/m^2.

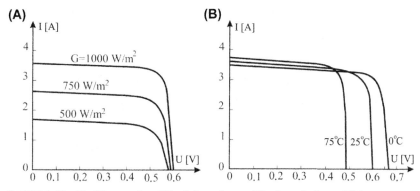

FIGURE 3.23   The *IV* curves for a PV cell depending on (A) solar radiation and (B) temperature.

Influence of solar radiation and temperature on cells characteristics: Currentvoltage curves are shown in Fig. 3.23A for a module operating at a fixed temperature and at several radiation levels. The locus of maximum power points is also shown. For this module, short-circuit current increases in proportion to the solar radiation while the open-circuit voltage increases logarithmically with solar radiation. As long as the curved portion of the *IV* characteristic does not intersect the current axis ($V = 0$), the short-circuit current is nearly proportional to the incident radiation. Cell temperature influences significantly on open-circuit voltage U_o and less on short-circuit current (Fig. 3.23B). The U_o parameter can be calculated using following equation

$$U_o = U_{o,25} - 0.0023(t - 25), \tag{3.37}$$

where $U_{o,25}$ is the open-circuit voltage of cell, in V, for standard temperature (25°C) and t is the cell temperature, in °C.

3.3.3 PV Collector Efficiency

The reference efficiency for the most recent single crystal silicon products is $\eta_R = 0.17$ [39]. The instantaneous efficiency η_c of a PV collector is mainly dependent on the cell temperature and the following general equation can be proposed [40]:

$$\eta_c = \eta_R \left[1 - \beta_t(t_a - t_R) - \frac{\beta_t \tau \alpha I_T}{U_L} \right], \tag{3.38}$$

where β_t is the temperature coefficient of the cell (it can be $0.004K^{-1}$ for single-crystal silicon); $\tau \alpha$ is the product between the glass cover transmittance and the cell absorptance (possible value 0.95); t_a is the outdoor air temperature, in °C; t_R is the reference temperature at which the reference efficiency η_R is

TABLE 3.9 Constants in Eq. (3.38) When Applied for Crystalline Silicon Cells

k_1 (–)	k_2 (–)	k_3 (°C^{-1})	k_4 (°C^{-1})	k_5 (°C^{-1})	k_6 (°C^{-1})
−0.017162	−0.040289	−0.004681	0.000148	0.000169	0.000005

evaluated (usually 25°C); and U_L is the overall heat loss coefficient of the PV module (possible value 20 W/(m^2 K).

Huld et al. [41] proposed an approach to predict the solar radiation-dependent performance and efficiency of wafer-based crystalline silicon cells. The efficiency η_c can be calculated using Eq. (3.39) as

$$
\eta_c = \left\{ 1 + k_1 \ln I_T + k_2 (\ln I_T)^2 + t_{\text{mod}} \left[k_3 + k_4 \ln I_T + k_5 (\ln I_T)^2 \right] + k_6 t_{\text{mod}}^2 \right\} \eta_{\text{sys}},
\tag{3.39}
$$

where t_{mod} is the panel temperature, in °C and η_{sys} is the system losses that cover the inverter, connections, and dirt accumulation on the surface of cells and is generally taken as 0.86. Table 3.9 gives the values of coefficients k_1-k_6, which apply for crystalline silicon cells [42].

Efficiency of a PV array η_a depends of the average module temperature t as

$$
\eta_a = \eta_R [1 - \beta_t (t - t_R)].
\tag{3.40}
$$

3.3.4 Control and Delivered Energy Estimation for a PV System

PV modules, connected into a PV system, are very reliable, have no moving parts and require no maintenance or external inputs such as fuel but only a flux of solar energy.

PV systems can be divided in two categories: stand-alone systems and on-grid systems. Most stand-alone PV systems use a battery to store and help regulate the power. Storage battery is needed, since the solar input does not coincide with use; for example, for lighting at night or for peak power when signals are transmitted. Regulation is needed, usually with the addition of electronic controllers, since otherwise there would be no voltage reference.

Systems that consist of PV generators, storage batteries, and loads need controls to protect the battery from overcharge or deep discharge. Overcharge will damage the storage batteries used in these systems, and high-voltage cut-off or power-shunting devices are used to interrupt the current to batteries after a full charge has been achieved. Stand-alone systems requiring very high degrees of reliability have been studied by Klein and Beckman [43] and Gordon [44].

If AC power is needed, DC/AC inverters will be required. This may be the case if AC machinery is to be operated or if the PV system is to be tied into a utility grid. A stand-alone inverter uses an internal frequency generator and switching circuitry to transform the low-voltage DC power to higher voltage AC power.

Grid-connected systems may have batteries and be controlled so as to maximize the contribution of the PV system to meeting the load and thus minimize the energy purchased; or, they may "float" on the grid without batteries, with the system adding to or drawing from the grid depending on the loads and available radiation. For practical purposes, this type of system functions as one with infinite storage capacity if the capacity of the grid is large compared to the capacity of the PV system. Grid-connected systems are sometimes connected in such a manner that the power meter runs backward whenever excess power is produced by the PV system and, of course, forward when power is needed from the grid.

Monthly average daily amount of energy E_a, in kWh, generated by a PV array can be calculated as follows:

$$E_a = A_c I_m \eta_a, \tag{3.41}$$

where A_c is the array area, in m^2 and I_m is the monthly average radiation, in $kWh/(m^2h)$.

For a grid-connected system, available grid energy is equal to the generated energy by PV array minus power losses in inverter.

Energy E_L delivered direct to a DC consumer (energy to the load) can be estimated using the utilizability method by equation [45] as

$$E_L = E_a(1 - \Phi), \tag{3.42}$$

where energy usability factor Φ is defined by Eq. (3.43) [40] as

$$\begin{aligned} \Phi &= 0 \quad \text{if } X_c \geq X_m \\ \Phi &= 1 - \frac{X_c}{X_m} \quad \text{if } X_m = 2 \end{aligned} \tag{3.43}$$

in which

$$X_c = \frac{I_c}{I_m} \tag{3.44}$$

$$X_m = 185 + 0.169 \frac{R_\Sigma}{K_t^2} - 0.0696 \frac{\cos \Sigma}{K_t^2} - 0.981 \frac{K_t}{(\cos \delta)^2}, \tag{3.45}$$

where Φ is the energy usability factor (dimensionless); I_c is the critical radiation level in which the output of the array is equal to the monthly average hourly load for that hour, in kWh/(m²h); I_m is monthly average radiation, in kWh/(m²h); R_Σ is the ratio of monthly average radiation on the tilted surface of array to that on a horizontal surface; K_t is the monthly clarity index; Σ is the tilted angle, in degree; and δ is the declination angle, in degree.

Energy E_B that exceeds the load and is delivered to storage battery can be written as

$$E_B = E_a - E_L. \tag{3.46}$$

3.3.5 Design of a PV System

The basic general principle for the design of a PV system is that the balance between the electrical energy produced by PV generator and the energy consumed by the electricity user must be perpetually respected. This balance is achieved for a defined period (1 day or 1 month).

Calculation of solar radiation on PV module surface can be performed using the methodology described in Chapter 2.

Daily electricity consumption E_c, in Wh, is expressed in Eq. (3.47) as

$$E_c = \sum_{i=1}^{n} \frac{P_i^{DC} \tau_i}{\eta_B} + \sum_{j=1}^{m} \frac{P_j^{AC} \tau_j}{\eta_{FC}}, \tag{3.47}$$

where n is the number of DC consumers; m is the number of AC consumers; P_i is the nominal power of DC consumers, in W; P_j is the nominal power of AC consumers, in W; τ_i, τ_j is the operating time for respective consumers, in h; and η_B, η_{FC} are the efficiencies of battery controller and frequency converter, respectively ($\eta_B = 0.80-0.90$, $\eta_{FC} = 0.85-0.95$). The values of the nominal power for each consumer are usually provided in the manufacturer's data sheets.

Electrical energy that must be produced by PV module E_p is

$$E_p = \frac{E_c}{K}, \tag{3.48}$$

where K is a factor that takes into account the uncertainty of weather data and the cable losses. The K value for PV system including storage batteries ranges between 0.75 and 0.85.

The PV module power is selected depending on the critical power of the module and the *number of PV modules connected in series* N_s is determined as

$$N_s = \frac{U_{DC}}{U_m}, \tag{3.49}$$

where U_{DC} is the nominal voltage of DC consumers and U_m is the nominal voltage of a PV module considered usually of 12 V.

Number of PV modules connected in parallel N_p is calculated as follows:

$$N_p = \frac{I_{PV}}{I_{sc}}$$

(3.50)

in which

$$I_{PV} = \frac{24I_m}{HSR}$$

(3.51)

$$I_m = \frac{E_p}{24U_{DC}},$$

(3.52)

where I_{PV} is the current of PV module; I_{sc} is the short-circuit current of a PV module considered approximately equal to the current in point M (Fig. 3.22); I_m is the mean current of the load for 1 day duration; and HSR is the number of hours per day of standard radiation, equal to 1000 W/m^2.

Calculation of storage batteries' capacity C_B, in Ah, can be performed using the equation as

$$C_B = \frac{NE_c}{K_d U_{DC}},$$

(3.53)

where N is the number of the cloudy days and K_d is the discharge coefficient of the battery (0.5–0.6, for Pb and 1.0, for NiCd).

The number of storage batteries connected in series is

$$N_{Bs} = \frac{U_{DC}}{U_B},$$

(3.54)

where U_B is the nominal voltage of the battery, usually equal to 12 V.

3.4 SOLAR PV/THERMAL HYBRID COLLECTORS

A PV/T collector is a module in which the PV is not only producing electricity but also serves as a thermal absorber. In this way, both heat and power are produced simultaneously. Since the demand for solar heat and solar electricity are often supplementary, it seems to be a logical idea to develop a device that can comply with both demands. PV cells utilize a fraction of the incident solar radiation to produce electricity and the remainder is turned mainly into waste heat in the cells and substrate raising the temperature of PV as a result, the efficiency of the module decreased. The PV/T technology recovers part of this heat and uses it for practical applications. The simultaneous cooling of the PV module maintains electrical efficiency at satisfactory level and thus the PV/T collector offers a better way of using solar energy with higher overall efficiency. The attractive features of the PV/T system are as follows [46]:

- it is dual-purpose: the same system can be used to produce electricity and heat output;

- it is efficient and flexible: the combined efficiency is always higher than using two independent systems and is especially attractive in building integrated PV (BIPV) when roof-panel spacing is limited;
- it has a wide application: the heat output can be used both for heating and cooling (desiccant cooling) applications depending on the season and practically being suitable for domestic applications;
- it is cheap and practical: it can be easily retrofitted/integrated to building without any major modification and replacing the roofing material with the PV/T system can reduce the payback period.

Different types of PV/T collectors are being used presently such as, PV/T-air, PV/T-water, and PV/T-concentrated collector [47]. The worldwide markets for both solar thermal and solar PV technologies are growing rapidly and have reached a very substantial size.

3.4.1 PV/T Liquid Collector

Similar to FPC water heating system, liquid PV/T collectors are used to heat up the water and simultaneously produce electricity for various domestic and industrial applications. The domestic water heater generally uses FPCs in parallel connection and run automatically with the thermosiphon action, whereas in the industrial water heating system, a number of FPCs in series are used and hence, it uses a photovoltaic-driven water pump to maintain a flow of water inside the collector. A schematic diagram of a PV/T-water collector is shown in Fig. 3.24 [11].

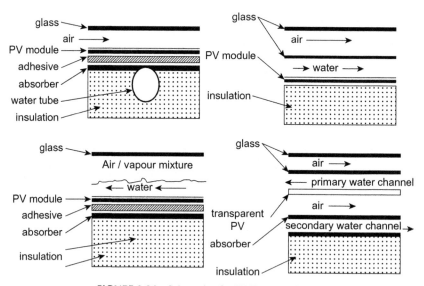

FIGURE 3.24 Schematic of a PV/T-water collector.

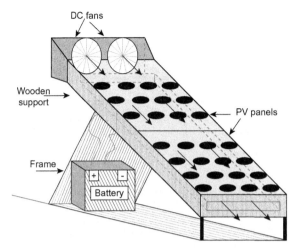

FIGURE 3.25   Schematic of the prototype of PV/T-air collector.

3.4.2 PV/T-Air Collector

Air and water both have been used as HTFs in practical PV/T-solar collectors, yielding PV/T-air and PV/T-water heating systems, respectively. PV/T-water systems are more efficient than those of PV/T-air systems [48] due to the high thermophysical properties of water as compared to air. However, PV/T-air systems (Fig. 3.25) [11] are used in many practical applications due to low-construction cost (minimal use of material) and operating cost among others.

Sarhaddi et al. [49] investigated the thermal and electrical performances of a solar PV/T-air collector. The thermal and electrical parameters of a PV/T-air collector include solar cell temperature, back-surface temperature, outlet air temperature, open-circuit voltage, short-circuit current, maximum power point voltage, maximum power-point current, etc. The electrical model presented can estimate the electrical parameters of a PV/T-air collector such as open-circuit voltage, short-circuit current, maximum power point voltage, and maximum power point current, etc. Furthermore, an analytical expression for the overall energy efficiency of a PV/T-air collector is derived in terms of thermal, electrical, design, and climatic parameters. The results of numerical simulation show good agreement with the experimental measurements. They found that the thermal efficiency, electrical efficiency, and overall energy efficiency of PV/T-air collector are about 17.18%, 10.01%, and 45%, respectively, for a typical climatic, operating, and design parameters.

3.4.3 PV/T Concentrator

Concentrating photovoltaic (CPV) systems can operate at higher temperatures than those of the FPCs. Collecting the rejected heat from a CPV system leads to

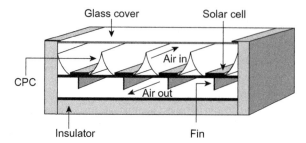

FIGURE 3.26   Schematic of a double-pass PV/T-solar collector with CPC and fins.

a CPV/thermal (CPV/T) system, providing both electricity and heat at medium temperatures. The use of CPV/T in combination with concentrating reflectors has a significant potential to increase the power production from a given solar-cell area. Presently, research is going to develop CPV/T collector for more electricity as well as heat generation. Few authors have worked in this direction enabling the multipurpose hybrid systems to fulfill the increasing demand of both electrical and thermal energies, while protecting the environment [50,51].

Hjothman et al. [52] designed and fabricated a double-pass PV thermal solar air collector with CPC and studied the performance over a range of operating conditions. The absorber of the hybrid PV/T collector consists of an array of solar cells for generating electricity, CPC to increase the radiation intensity falling on the solar cells, and fins attached to the back side of the absorber plate to improve heat transfer rate to the flowing air. The basic components of the experimental setup are as follows (Fig. 3.26): double-pass PV/T collector; the air flow measurement system; the temperature measurement system; the wind speed measurement system; the current and voltage measurement system; the solar radiation measurement system, and the data acquisition system. The results showed that electricity production in a PV/T hybrid module decreases with increasing the temperature of the airflow. This implies that the air temperature should be kept as low as possible.

Kostic et al. [51] designed the optimal orientation of PV/T collector with reflectors. In order to get more thermal and electrical energies, flat reflectors for solar radiation have been mounted on the PV/T collector. To obtain higher solar radiation intensity on PV/T collector, they designed reflectors with the movable PV/T collector system (Fig. 3.27). The thermal and electrical efficiencies of PV/T collector without reflectors and with reflectors in optimal position have been calculated. Using the experimental results, the total efficiency and energy-saving efficiency of PV/T collector have been determined. Energy-saving efficiency for PV/T collector without reflectors is found to be 60.1%, which is significantly higher than for the conventional solar thermal collector. On the other hand, the energy-saving efficiency for PV/T collector with reflectors in optimal position is found to be 46.7%, which is almost equal to the thermal efficiency of a conventional solar thermal collector. The energy-

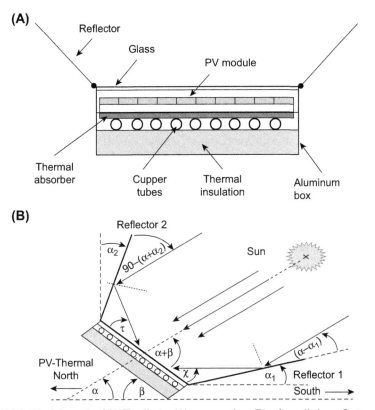

FIGURE 3.27 Schematic of PV/T collector (A) cross-section; (B) solar radiation reflectors.

saving efficiency of PV/T collector decreases slightly with the solar radiation intensity concentration factor.

The simultaneous use of hybrid PV/T, CPC, and fins have a potential to significantly increase the power production and reduce the cost of PV electricity.

3.4.4 Novel Applications of PV/T Collectors

PV/T collector has good potential for some other applications such as solar cooling, water desalination, solar greenhouse, solar still, PV/T-solar heat pump/air-conditioning system, and building integrated PV/T (BIPVT) solar collector.

Mittelman et al. [53] designed the desalination with the concentrating PV/T systems. The combined system produces solar electricity and simultaneously exploits the waste heat of the PV cells to desalinate water. The results indicate that this approach can be competitive relative to other solar-driven desalination systems and even relative to conventional reverse osmosis desalination. The result depends on the economic conditions, and is valid when the prevailing price of electricity is high enough, and the installed cost of the solar collectors is low enough.

Fang et al. [54] performed an experimental study on the operation performance of a PV/T-solar heat pump/air-conditioning system. The performance parameters such as the evaporation pressure, the condensation pressure and the coefficient of performance (COP) of heat pump/air-conditioning system, the water temperature and receiving heat capacity in water heater, the PV module temperature, and the PV efficiency have been investigated. The experimental results showed that the mean PV efficiency of a PV/T-solar heat pump/air-conditioning system reaches 10.4%, and can enhance to 23.8% in comparison to that of a conventional PV module. Additionally, the mean COP of heat pump/air-conditioning system may attain 2.88 and the water temperature in water heater can increase to 42°C. These results indicate that the PV/T-solar heat pump/air-conditioning system has better performances, can work with stability for such useful and potential applications, and hereby reducing the greenhouse gases, enabling a clean environment.

A novel BIPVT solar collector has been theoretically analyzed by Anderson et al. [55] using a modified Hottel-Whillier model. This model was validated with the experimental data from the testing facility on a prototype BIPVT collector. The results showed that the key design parameters such as the fin efficiency, the thermal conductivity between the PV cells and their supporting structure, and the lamination method had a significant influence on both the electrical and thermal efficiencies of the BIPVT. Finally, there appears to be a significant potential to utilize the low natural-convection heat transfer in the attic at the rear of the BIPVT system to act as an insulating layer rather than using additional insulation material. The use of such air would reduce the material cost of such a system to be significantly.

3.4.5 Energy Indicators

The most important energy performance indicator of a solar thermal installation is the *useful solar heat output* per *unit collector area* q_u, in kWh/m^2, defined as

$$q_u = \frac{Q_u}{A_c}, \tag{3.55}$$

where Q_u is the usable energy collected by solar collectors, in kWh and A_c is the collector surface area, in m^2.

The *thermal efficiency* η_c of a solar system is defined as the ratio of solar thermal energy used to solar energy received on the collectors according to the first equality of Eq. (3.13).

The *solar fraction* (*f*) is defined as the percentage of the total thermal load satisfied by solar energy and is given by

$$f = \frac{Q_{load} - Q_{aux}}{Q_{load}}, \tag{3.56}$$

where Q_{load} is the energy required by the load, in kJ and Q_{aux} is the auxiliary energy, in kJ.

The *PV-system yield Y*, in kWh/kW$_p$, is expressed by the following equation

$$Y = \frac{E_{el}}{P_{PV}}, \tag{3.57}$$

where E_{el} is the generated electrical energy, in kWh and P_{PV} is the installed PV power, in kW$_p$.

The *performance ratio* (PR) describes the percentage of energy generated by a PV installation with respect to its ideal performance.

The *PV conversion efficiency* η_{PV} is defined as follows:

$$\eta_{PV} = \frac{P_{el}}{A_c I_T}, \tag{3.58}$$

where P_{el} is the generated electrical power, in kW; A_c is the collector array surface area, in m^2; and I_T is the total solar radiation on the collector array surface, in kW/m^2;

3.5 PERFORMANCES OF A PV/T COLLECTOR WITH WATER HEATING IN BUILDINGS

In the solar thermal system, conventional electrical energy is used to circulate HTF through a collector. If both types of collectors (thermal and PV) are combined in a hybrid unit (PV/T collector), the use of conventional electrical energy can be avoided for circulating HTF through a collector. The PV/T collector in building can produce thermal and electrical energy simultaneously.

3.5.1 Description of the System

The schematic diagram of the PV/T collector with water heating is shown in Fig. 3.28 [56]. A piece of high-quality low-iron tempered glass with ideal transmission coefficient covers the solar cell for protection. What distinguishes this hybrid PV/T system from a conventional PV module is that a duct with rectangular cross-section is attached below the solar cell for water flowing. Three sides of the duct are adiabatic, but the remaining one as the backplane next

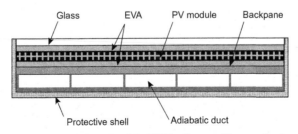

FIGURE 3.28   Schematic of the PV/T collector with water heating.

to the back surface of solar cell has excellent thermal conductivity. For the purpose of flowing uniformly, the whole duct is divided into several narrow ones in the flow direction. Both the glass and the backplane are glued to the solar cell with ethylene-vinyl acetate (EVA) copolymer. The entire system with layer structure is emplaced into a protective shell. A kind of antirust aluminum alloy with prominent characteristics of high strength/density ratio, easy machining, and high thermal conductivity is used as the backplane material.

3.5.2 Simulation Model

In actual working conditions, the PV/T collector with water heating is easily influenced by many complicated factors. To develop a simulation model for this system [57], the following assumptions have been made:

- The internal longitudinal temperature distribution of each component layer is neglected due to its small thickness.
- The heat capacity of the PV module is neglected in comparison with the heat capacity of water.
- The heat flux received by the backplane is equally distributed.
- The influence of temperature on the physical characteristics of each component layer is neglected.

According to the law of conservation of energy, the energy balance equations for each component layer of the PV/T collector with water heating can be written as follows [57]:

- for the glass plate:

$$\alpha_g I_T(t) = k_g(T_g - T_a) + U_{g1}(T_g - T_1) \qquad (3.59)$$

where: α_g is the absorptance of glass; $I_T(t)$ is the solar radiation intensity, in W/m^2; k_g is the convective heat transfer coefficient, in W/(m^2 °C); U_{g1} is the heat transfer coefficient (glass−EVA-1), in W/(m^2 °C); T_a is the ambient air temperature, in °C; T_g is the glass temperature, in °C; T_1 is the EVA-1 temperature, in °C; and t is the time.

- for the EVA-1:

$$\alpha_1 \tau_g I_T(t) + U_{g1}(T_g - T_1) = U_{1c}(T_1 - T_c), \qquad (3.60)$$

where α_1 is the absorptance of EVA-1; τ_g is the transmittance of glass; U_{1c} is the heat transfer coefficient (EVA-1−solar cell), in W/(m^2 °C); and T_c is the solar cell temperature, in °C.

- for the solar cell:

$$\alpha_c \tau_1 \tau_g I_T(t) + U_{1c}(T_1 - T_c) = \eta_c \alpha_c \tau_1 \tau_g I_T(t) + U_{c2}(T_c - T_2), \qquad (3.61)$$

where α_c is the absorptance of solar cell; τ_1 is the transmittance of EVA-1; U_{c2} is the heat transfer coefficient (solar cell−EVA-2), in W/(m^2 °C);

T_2 is the EVA-2 temperature, in °C; and η_c is the PV efficiency of solar cell.

- for the EVA-2:

$$\beta_c \alpha_2 \tau_c \tau_1 \tau_g I_T(t) + (1 - \beta_c)\alpha_2 \tau_1 \tau_g I_T(t) + \beta_c U_{c2}(T_c - T_2) = U_{2b}(T_2 - T_b),$$
(3.62)

where β_c packing factor of solar cell; α_2 is the absorptance of EVA-2; τ_c is the transmittance of solar cell; and U_{2b} is the heat transfer coefficient (EVA-2–backplane), in W/(m^2 °C).

- for the backplane:

$$\beta_c \alpha_b \tau_2 \tau_c \tau_1 \tau_g I_T(t) + (1 - \beta_c)\alpha_b \tau_2 \tau_1 \tau_g I_T(t) + U_{2b}(T_2 - T_b) = U_{bw}(T_b - T_w)$$
(3.63)

where α_b is the absorptance of backplane; τ_2 is the transmittance of EVA-2; U_{bw} is the heat transfer coefficient (backplane-water), in W/(m^2 °C); and T_w is the water temperature, in °C.

- for the water in the flow direction:

$$U_{bw}(T_b - T_w)b \times dx = c_w m_w \left(\frac{dT_w}{dx}\right) \times dx$$
(3.64)

where b is the width of solar cell, in m; c_w is the specific heat of water, in J/(kg °C); m_w is the mass flow rate of water, in kg/s; and x is the distance in flowing direction, in m.

From Eqs. (3.59)–(3.64), the temperature of water $T_w(x, t)$ and the temperature of solar cell $T_c(x, t)$ can be obtained with the initial condition $T_{w,in} = T_w|_{x=0}$.

Suppose the flow distance $x = S$, the outlet temperature of water can be obtained as $T_{w,out} = T_w|_{x=S}$.

In order to simplify the calculation, the average temperature of water and cell for a flow distance S is obtained as

$$\overline{T}_w(t) = \frac{1}{S} \int_0^S T_w(x,t)dx$$
(3.65)

$$\overline{T}_c(t) = \frac{1}{S} \int_0^S T_c(x,t)dx.$$
(3.66)

In practical applications, an empirical relationship between the PV efficiency and the temperature of solar cell is given by [58]

$$\eta_c = \eta_0[1 - \beta_0(T_c - T_a)],$$
(3.67)

where η_0 is the optical efficiency of solar cell.

From Eqs. (3.66) and (3.67), the PV efficiency $\eta_c(t)$ can be obtained. Consequently, the overall PV and thermal power of the PV/T collector with water heating can be obtained, respectively, as follows:

$$Q_p = \eta_c \alpha_c \tau_1 \tau_g (\beta_c A_c) I_T(t) \tag{3.68}$$

$$Q_t = c_w m_w \left(T_{w,out} - T_{w,in} \right), \tag{3.69}$$

where A_c is the area of solar cell. For neighboring component layers a and b, the heat transfer coefficient can be estimated as

$$U_{ab} = \frac{1}{(\delta_a/\lambda_a) + (\delta_b/\lambda_b)}, \tag{3.70}$$

where δ_a, δ_b are the thicknesses of layers a and b, respectively, in m; λ_a, λ_b are the thermal conductivities of layers a and b, respectively, in W/(m °C).

For turbulent flow in the duct, the forced convective heat transfer coefficients can be estimated as

$$U_{ab} = \mathrm{Nu}\frac{\lambda}{d_e} \tag{3.71}$$

in which Nu can be obtained using Dittus-Boelter equation [59] as

$$\mathrm{Nu} = 0.023 \mathrm{Re}^{0.8} \mathrm{Pr}^{0.4}, \tag{3.72}$$

where d_e is the equivalent diameter of duct, in m; Nu is the Nusselt number; Re is the Reynolds number; and Pr is the Prandtl number.

3.5.3 Model Validation

All the parameter values of the PV/T collector with water heating are given in Tables 3.10 and 3.11. Consequently, the PV and thermal performances of this system can be numerically calculated [57]. All calculations are realized in MATLAB software.

TABLE 3.10 The Physical Parameters of the PV Module

Material	λ (W/(m °C))	τ (–)	α (–)	δ (m)
Glass	0.70	0.91	0.05	0.0050
EVA	0.35	0.90	0.08	0.0005
Solar cell	148	0.09	0.80	0.0003
Backplane	144	0.00039	0.40	0.0005

TABLE 3.11 The Design Parameters of the Hybrid PV/T-Water Heating System

Parameters	Values
L	1.5 m
b	1.0 m
c_w	4.18×10^3 J/(kg °C)
β_c	0.80
η_0	0.12
β_0	0.0045
k_g	7.44 W/(m^2 °C)
k_{bw}	427.57 W/(m^2 °C)
U_{ga}	116.67 W/(m^2 °C)
U_{1c}, U_{c2}	699.01 W/(m^2 °C)
U_{2b}	698.30 W/(m^2 °C)

In order to validate the developed model, the numerical simulation has be conducted again under the same experimental conditions (Table 3.12 and Fig. 3.29) given by Tiwari et al. [60] for the climate of India to compare experimental data and simulation results. The variations of backplane temperatures of numerical and experimental results are shown in Fig. 3.30.

The coefficient of multiple determinations R^2 is calculated as follows [61]:

$$R^2 = 1 - \frac{\sum_{i=1}^{n} \left(y_{\text{sim},i} - y_{\text{mea},i} \right)^2}{\sum_{i=1}^{n} y_{\text{mea},i}^2}, \tag{3.73}$$

where n is the number of measured data in the independent data set; $y_{\text{mea},i}$ is the measured value of one data point i; $y_{\text{sim},i}$ is the simulated value; and $\bar{y}_{\text{mea},i}$ is the mean value of all measured data points.

It can be observed from Fig. 3.30 that the simulation result is in good agreement with the experimental data. In detail, the correlation coefficient between the experimental curve and the simulation curve is $R^2 = 0.994$.

3.5.4 Results and Discussion

The influences of the PV module number, the inlet temperature of water and the mass flow rate of water on the performance parameters of the system, including the PV efficiency, the temperature of solar cell, the outlet temperature of water, and PV power were analyzed in detail [57].

TABLE 3.12 The Design Parameters of the PV/T-Solar Air Collector System

Parameters	Values
L	1.0 m
b	0.605 m
c_w	1005 J/(kg °C)
β_c	0.83
η_0	0.12
β_0	0.0045
k_g	7.44 W/(m² °C)
k_{bw}	37.4 W/(m² °C)
U_{ga}	116.67 W/(m² °C)
U_{1c}, U_{c2}	699.01 W/(m² °C)
U_{2b}	60.60 W/(m² °C)

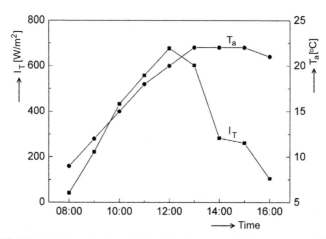

FIGURE 3.29   Variation of solar radiation intensity and ambient temperature with time.

3.5.4.1 Influence of the PV Module Number

The hourly variations of solar cell temperature T_c, photovoltaic efficiency η_c, outlet temperature of water $T_{w\text{-}out}$, and photovoltaic power Q_P are evaluated and analyzed by varying the tandem PV module number ($n = 1, 2$ and 4) at the

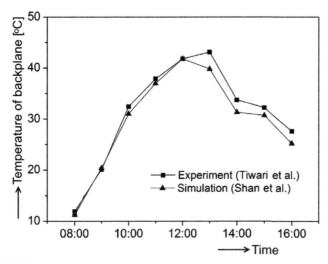

FIGURE 3.30 Comparisons of numerical simulation with experimental results.

constant mass flow rate of water ($m_w = 0.50$ kg/s) and inlet temperature of water ($T_{w\text{-}in} = 20°C$).

Eq. (3.65) has been used for evaluating the outlet temperature of water ($T_{w\text{-}out}$). The variation of outlet water temperature is shown in Fig. 3.31. The lowest curve ($n = 1$) indicates that the outlet water temperature $T_{w\text{-}out}$ rises from 20.4°C to 24.0°C with the increase in solar radiation intensity from 07:00

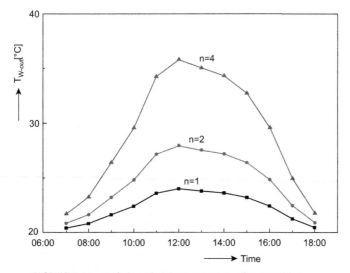

FIGURE 3.31 Variation of outlet temperature of water with time.

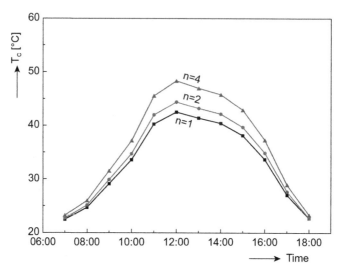

FIGURE 3.32   Variation of temperature of solar cell with time.

to 12:00, and then drops to 20.4°C with the decrease in solar radiation intensity from 12:00 to 18:00. Meanwhile, it can be observed clearly that the other curves ($n = 2$ and 4) move up with the increase of n, which means that the increase in tandem PV module number leads to a rise in outlet temperature of water correspondingly.

Eq. (3.66) has been used for evaluating the temperature of solar cell (T_c). The variation of temperature of solar cell is presented in Fig. 3.32. The lowest curve ($n = 1$) shows that the T_c rises from 22.5°C to 42.5°C with the increase in solar radiation intensity from 07:00 to 12:00, and then drops to 22.6°C with the decrease in solar radiation intensity from 12:00 to 18:00. It can be also known that the other curves ($n = 2$ and 4) move up with the increase of n, which indicates that the increase in tandem PV module number leads to a rise in temperature of solar cell.

Eq. (3.67) has been used for evaluating the PV efficiency η_c. The variation of PV efficiency is illustrated in Fig. 3.33. The highest curve ($n = 1$) indicates that the η_c drops from 12.35% to 11.54% with the increase in solar radiation intensity from 07:00 to 12:00, and then rises to 12.56% with the decrease in solar radiation intensity from 12:00 to 18:00. It can be found that the other curves ($n = 2$ and 4) move down with the increase of n, which means that the increase in tandem PV module number leads to a drop in PV efficiency correspondingly.

Eq. (3.68) has been used for evaluating the PV power Q_P. The variation of the PV power is shown in Fig. 3.34. The lowest curve ($n = 1$) indicates that the Q_P increases from 0.019 to 0.091 kW with the increase in solar radiation intensity from 07:00 to 12:00, and then decreases to 0.010 kW with the decrease

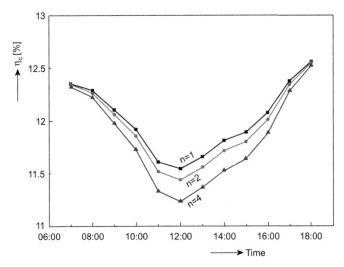

FIGURE 3.33   Variation of PV efficiency with time.

in solar radiation intensity from 12:00 to 18:00. It can be observed clearly that the other curves ($n = 2$ and 4) move up with the increase of n, which indicates that the increase in tandem PV module number results in an increase in PV power.

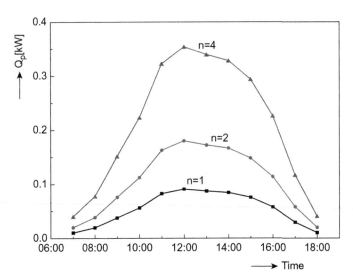

FIGURE 3.34   Variation of PV power with time.

3.5.4.2 Influence of Inlet Temperature of Water

The variation of PV efficiency, temperature of solar cell, outlet temperature of water, and PV power are evaluated and analyzed by varying the inlet temperature of water ($T_{w\text{-}in} = 20$, 25 and 30°C) at the constant tandem PV module number ($n = 4$) and mass flow rate of water ($m_w = 0.50$ kg/s). The variation of outlet temperature of water $T_{w\text{-}out}$ is presented in Fig. 3.35. The lowest curve ($T_{w\text{-}in} = 20$°C) indicates that the $T_{w\text{-}out}$ increases from 21.7°C to 35.8°C with the increase in solar radiation intensity from 07:00 to 12:00, and then decreases to 21.8°C with the decrease in solar radiation intensity from 12:00 to 18:00. It can be also observed that the other curves ($T_{w\text{-}in} = 25$°C and 30°C) move up with the increase of $T_{w\text{-}in}$, which means that the increase in inlet temperature of water leads to a rise in outlet temperature of water correspondingly.

The variation of solar cell temperature T_c is illustrated in Fig. 3.36. The lowest curve ($T_{w\text{-}in} = 20$°C) indicates that the T_c increases from 23.1°C to 48.3°C with the increase in solar radiation intensity from 07:00 to 12:00, and then decreases to 23.3°C with the decrease in solar radiation intensity from 12:00 to 18:00. It can be found that the other curves ($T_{w\text{-}in} = 25$°C and 30°C) move up with the increase of $T_{w\text{-}in}$, which means that the increase in inlet temperature of water results in a rise in temperature of solar cell.

The variation of PV efficiency η_c is presented in Fig. 3.37. The highest curve ($T_{w\text{-}in} = 20$°C) shows that the η_c drops from 12.32% to 11.23% with the increase in solar radiation intensity from 07:00 to 12:00, and then rises to 12.53% with the decrease in solar radiation intensity from 12:00 to 18:00. It

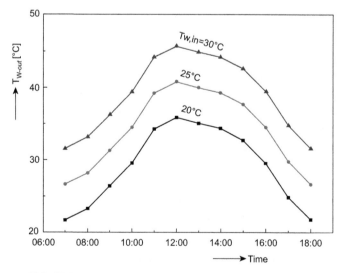

FIGURE 3.35 Variation of outlet temperature of water with time.

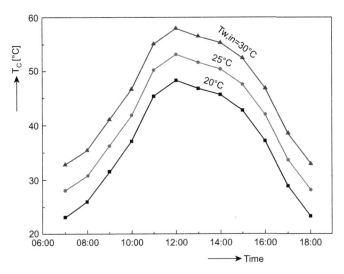

FIGURE 3.36   Variation of temperature of solar cell with time.

can be also observed clearly that the other curves ($T_{w\text{-}in} = 25°\text{C}$ and $30°\text{C}$) move down with the increase of $T_{w\text{-}in}$, which indicates that the increase in inlet temperature of water leads to a drop in PV efficiency.

The variation of PV power Q_P is shown in Fig. 3.38. The highest curve ($T_{w\text{-}in} = 20°\text{C}$) indicates that the Q_P increases from 0.039 to 0.354 kW with the increase in solar radiation intensity from 07:00 to 12:00, and then decreases to

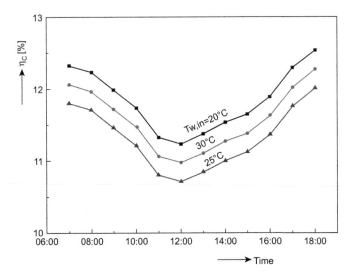

FIGURE 3.37   Variation of PV efficiency with time.

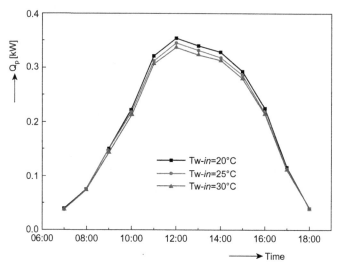

FIGURE 3.38 Variation of PV power with time.

0.040 kW with the decrease in solar radiation intensity from 12:00 to 18:00. It can be also observed clearly that the other curves ($T_{w\text{-}in} = 25°C$ and $30°C$) move down with the increase of $T_{w\text{-}in}$, which means that the increase in inlet temperature of water results in a decrease in PV power.

3.5.4.3 Influence of Mass Flow Rate of Water

The variation of PV efficiency, temperature of solar cell, outlet temperature of water, and PV power are evaluated and analyzed by varying the mass flow rate of water ($m_w = 0.25$, 0.50 and 0.75 kg/s) at the constant tandem PV module number ($n = 4$) and inlet temperature of water ($T_{w\text{-}in} = 25°C$).

The variation of outlet temperature of water $T_{w\text{-}out}$ is shown in Fig. 3.39. The highest curve ($m_w = 0.25$ kg/s) indicates that the $T_{w\text{-}out}$ rises from 28.2°C to 56.3°C with the increase in solar radiation intensity from 07:00 to 12:00, and then drops to 28.4°C with the decrease in solar radiation intensity from 12:00 to 18:00. It can be observed clearly that the other curves ($m_w = 0.50$ and 0.75 kg/s) move down with the increase of m_w, which means that the increase in mass flow rate of water leads to a drop in outlet temperature of water.

The variation of temperature of solar cell T_c is illustrated in Fig. 3.40. The highest curve ($m_w = 0.25$ kg/s) shows that T_c rises from 28.7°C to 60.7°C with the increase in solar radiation intensity from 07:00 to 12:00, and then drops to 28.9°C with the decrease in solar radiation intensity from 12:00 to 18:00. It can be found that the other curves ($m_w = 0.50$ and 0.75 kg/s) move down with the increase of m_w, which indicates that the increase in mass flow rate of water results in a drop in temperature of solar cell.

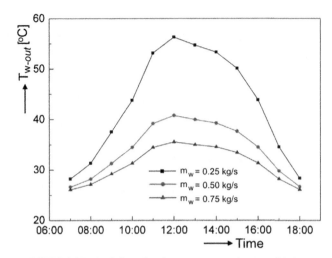

FIGURE 3.39 Variation of outlet temperature of water with time.

The variation of PV efficiency η_c is presented in Fig. 3.41. The lowest curve ($m_w = 0.25$ kg/s) indicates that η_c drops from 12.02% to 10.56% with the increase in solar radiation intensity from 07:00 to 12:00, and then rises to 12.22% with the decrease in solar radiation intensity from 12:00 to 18:00. It can be also known that the other curves ($m_w = 0.50$ and 0.75 kg/s) move up with the increase of m_w, which means that the increase in mass flow rate of water leads to a rise in PV efficiency.

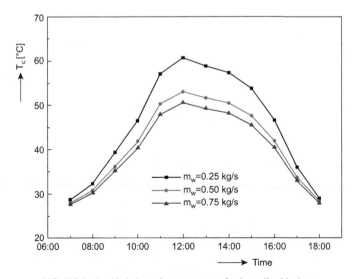

FIGURE 3.40 Variation of temperature of solar cell with time.

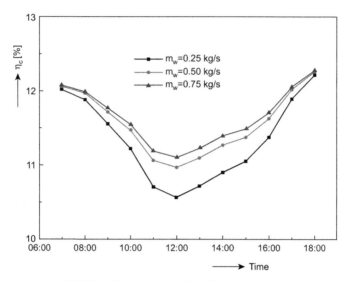

FIGURE 3.41   Variation of PV efficiency with time.

The variation of PV power Q_P is shown in Fig. 3.42. The lowest curve ($m_w = 0.25$ kg/s) indicates that Q_P increases from 0.038 to 0.333 kW with the increase in solar radiation intensity from 07:00 to 12:00, and then decreases to 0.039 kW with the decrease in solar radiation intensity from 12:00 to 18:00. It can be observed clearly that the other curves ($m_w = 0.50$ and 0.75 kg/s) move up with the increase of m_w, which means that the increase in mass flow rate of water results in an increase in PV power.

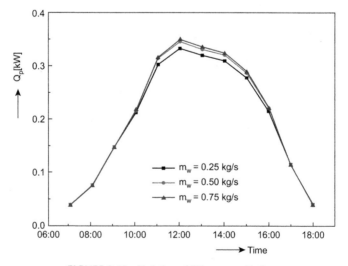

FIGURE 3.42   Variation of PV power with time.

3.5.5 Conclusions

The main conclusions that can be drawn from this study are listed below:

1. With the increase in solar radiation intensity at the constant series-connected PV module number, inlet temperature and mass flow rate of water, the outlet temperature of water rises, while the PV efficiency drops. The PV power of the PV/T collector with water heating increases.
2. With the increase in series-connected PV module number at the constant inlet temperature and mass flow rate of water in the same climate conditions, the outlet temperature of water rises, while the PV efficiency drops. The PV power of the PV/T collector with water heating increases.
3. With the increase in inlet temperature of water at the constant series-connected PV module number and mass flow rate of water in the same climate conditions, the outlet temperature of water rises, while the PV efficiency drops. The PV power of the PV/T collector with water heating decreases.
4. With the increase in mass flow rate of water at the constant series-connected PV module number and inlet temperature of water in the same climate conditions, the outlet temperature of water drops, while the PV efficiency rises. The PV power of the PV/T collector with water heating increases.

These conclusions can put forward corresponding approaches to improve the PV and thermal performances of the PV/T collector with water heating in buildings.

3.6 PERFORMANCES OF A HYBRID PV/T-SOLAR SYSTEM FOR RESIDENTIAL APPLICATIONS

The main aim of this study is the assessment of energy efficiency for a residential system comprising commercially available PV/T panels, therefore capable of electricity generation and DHW production simultaneously. The study is performed for three locations in Europe with entirely different climatic characteristics [62].

3.6.1 System Configuration

The hybrid solar system (Fig. 3.43) is designed to produce DHW for a single family household and to supply the electricity generated by its PV component to the grid. The hybrid solar system consists of liquid glazed flat-plate PV/T collectors, an inverter, a single cylindrical water storage tank with a capacity of 300 L and an immersed heat exchanger from the solar collector circuit located at the bottom of the tank, a pump, tee-piece, flow diverter, controllers, and insulated piping. Each PV/T collector has an aperture area of 1.42 m^2 and consists of 72 monocrystalline solar cells [62].

The absorber of the PV/T collectors is constructed of copper tubes, ultrasonic welded on a copper sheet in contact with the rear panel of the PV

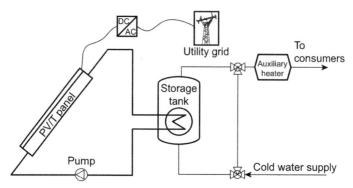

FIGURE 3.43   Schematic of the PV/T system configuration.

module. The collector is suitably insulated and enclosed in an aluminum box supporting the absorber plate, the PV module, and the transparent cover. The transparent cover is a low-iron tempered glass with a thickness of 4 mm.

The pump is controlled by a differential thermostat that automatically starts the pump when the temperature difference between the output of the solar collector and the water at the bottom of the storage tank exceeds 5°C. The HTF used in the closed loop of the collectors is a waterglycol mixture of 40% by volume to prevent freezing and its flow rate is the same as the recommended by the manufacturer (45 L/(h m^2)).

An auxiliary heating unit is present as part of the system. It simply is a heater placed between the storage tank and the consumption in order to supply the auxiliary energy required to heat the water, if necessary. As all of the installation sites examined in this study are in Europe, the orientation of the solar collectors is facing toward the true south, while its tilt angle is taken equal to the latitude of each city.

The PV/T collector used for the means of this study is a commercially available product that has been experimentally tested by an approved center for the solar collector testing [62]. Most of the system's components are identical at all three of the locations in this study, with the exception of two—the number of the PV/T collectors and the PV system inverter. The collector area varies in accordance to the local climate, in order to deliver the same annual hot water energy at the outlet of the solar PV/T collectors. However, due to the different PV/T collector area, the nominal electric power of the PV system changes as well and thus a different inverter is required at each location. The resulting design parameters at each city are shown in Table 3.13.

3.6.2 Simulation Model

The simulation software TRNSYS was used for the study of the aforementioned hybrid PV/T system. At each time-step of the simulation, the simulation

TABLE 3.13 PV/T System Design Parameters

Parameter	Athens	Munich	Dundee
Collector aperture area (m²)	5.68	12.78	24.14
Nominal electric power of system (W)	676.4	1521.9	2874.7
Array tilt angle (degree)	38.0	48.3	56.5
Piping length (m)	17.0	25.0	35.0

software calculates the incident solar radiation on the PV/T collectors and, subsequently, its electrical and thermal production by using the appropriate system parameters and the meteorological data for the three European locations, which were generated using Meteonorm [63].

3.6.2.1 Solar Thermal System

The model used for the thermal collector, which is part of the hybrid collector, is based on the performance data that have been determined by test procedures. Data from solar collector testing can be statistically analyzed to obtain the instantaneous efficiency η_c of the solar collector, which can be described by Eq. (3.21) written as

$$\eta_c = F(\alpha\tau)K_\theta - a_1\frac{t_{cm} - t_a}{I_T} - a_2\frac{(t_{cm} - t_a)^2}{I_T}, \tag{3.74}$$

where F is the collector efficiency factor; $(\alpha\tau)$ is the absorptancetransmittance product at normal incidence; K_θ is the incidence angle modifier; t_{cm} is the average of the fluid inlet and outlet temperatures, in °C; t_a is the ambient air temperature, in °C; I_T is the total solar radiation at collector's aperture, in W/m²; and a_1, in W/(m²K) and a_2, in W/(m²K²) are the collector performance parameters defined by the experimental collector testing [64].

The performance parameters of the tested collector were found to be: $F(\alpha\tau)$ $K_\theta = 0.486$, $a_1 = 4.028$ W/(m²K), and $a_2 = 0.067$ W/(m²K²) [62].

For the PV/T collector used, the experimental values of the incidence angle modifier K_θ for different angles of incidence θ are shown in Table 3.14.

TABLE 3.14 Experimental Values of the Incidence Angle Modifier

θ (degree)	0	30	45	60	65	80
K_θ	1.00	0.98	0.94	0.80	0.73	0.40

If U_L is the overall loss coefficient of the solar collector, the product FU_L may be expressed by a linear form using the performance parameters a_1 and a_2 as

$$FU_L = a_1 + a_2(t_{cm} - t_a). \tag{3.75}$$

The collector efficiency factor F of the tested collector, of which collector the configuration of the absorber and the sheet and tube dimensions are fully defined, is given as [15]

$$F = \frac{1}{sU_L/\pi D_i U_i + sU_L/\lambda_b + s/(D_e + (s - D_e)\varepsilon_f)}, \tag{3.76}$$

where s is the distance between the tubes; D_i and D_e are the internal and external diameters of the tube, respectively; U_i is the internal heat transfer coefficient between the fluid and the tube wall; λ_b is the thermal conductivity of the bond; and ε_f is the fin efficiency. The fin efficiency ε_f is given by the following equation:

$$\varepsilon_f = \frac{\tanh[\omega(s - D_e)/2]}{\omega(s - D_e)/2} \tag{3.77}$$

in which ω is given by equation

$$\omega = \sqrt{\frac{U_L}{\lambda_s \delta_s}}, \tag{3.78}$$

where λ_s is the thermal conductivity of the sheet material and δ_s is the thickness of the sheet. The thermal resistance of the bond was assumed to be negligible.

All of the parameters involved in the previous equations are known except from the F and U_L. In order to calculate their values, an iterative procedure has been adopted using Eqs. (3.75)–(3.78). Knowing the values of F and U_L, the heat removal factor F_R is calculated as [15]

$$F_R = \frac{mc_p}{A_c U_L}\left[1 - \exp\left(-\frac{A_c U_L F}{mc_p}\right)\right] \tag{3.79}$$

where m is the collector mass flow rate, in kg/s; c_p is the specific heat of the fluid, in J/(kg K); and A_c is the collector aperture area, in m^2.

At each time-step, the mean absorber plate temperature t_{pm} is being calculated, using the following equation [15]:

$$t_{pm} = t_{c,i} + \frac{Q_u}{A_c F_R U_L}(1 - F_R), \tag{3.80}$$

where $t_{c,i}$ is the collector inlet temperature and Q_u is the usable energy collected by the solar thermal collector.

Since the absorber plate is in contact with the back sheet of the PV module, the calculated mean absorber plate temperature is being imported to the PV model as the temperature of the PV cells.

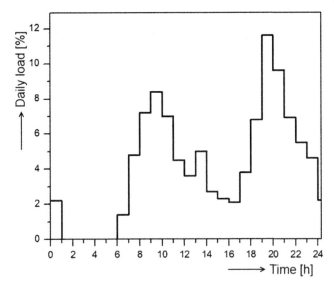

FIGURE 3.44   Daily load profile of the system.

The daily load profile of the systems examined in this study is shown in Fig. 3.44 and is simulated using the TRNSYS Type-14 water draw-forcing function. This profile is used by the well-known *f*-chart method as a typical hot-water usage model [62]. The solar system is designed to provide 200 L/day of hot water at 50°C.

The Type 77 module of the TRNSYS program is used to simulate the mains water temperature at a depth of 1 m, which generally is the depth at which the water utility pipes are buried.

The water storage tank is assumed to be stratified and is simulated by the use of a Type 60 module. The function of the pump is simulated by a Type 3 and its control by a Type 2 module, respectively. Since the amount of energy consumed by the pump annually was calculated to be virtually inconsequential, it has been neglected during the simulation. A Type 31 module has been used for the simulation of the piping in the system. A mixing valve is being simulated with a Type 11 used as a tee-piece in the storage outlet fluid. Another Type 11 is being used as a temperature-controlled liquid flow diverter in the mains cold water line. A Type 6 simulates the function of an auxiliary unit which operates as a DHW-heating device to maintain the outlet water temperature at 50°C. The efficiency of the auxiliary heaters using either heating oil or natural gas is considered to be 91%, while the efficiency of resistive heaters using electricity is considered to be 100%.

3.6.2.2 PV System

The PV system forms a simple grid-connected configuration, consisting of PV/T panels and an inverter. The model for the PV panel is based on the five parameters equivalent circuit model, as given by Type 180 module of the TRNSYS program. These five required parameters were provided by the experimental testing of the PV panel and are read from an external data file. The model accepts as input the cell temperature, which is calculated, as described previously, at each time-step. As it is a standard feature of every inverter used in this study, the model assumes that the PV array always operates at its maximum power point, meaning that the power output of the module is equal to its maximum power.

Appropriate commercially available inverters have been chosen according to the installed power of each system, in order to convert the direct current produced by the PV array to alternating current at the voltage used by the utility power grid. The simulation of the inverter is based on its efficiency versus power curve, adapted from the commercial data sheet. A type 81 TRNSYS module reads the data and performs the interpolation. Module mismatch losses of 3%, cable losses of 2%, and an annual module performance degradation of 1% were taken into account.

3.6.3 Analysis of Energy Indicators

In Fig. 3.45 is illustrated the cumulative annual solar radiation on the surface of the three PV/T systems [62]. The *useful solar heat output* per *unit collector area* q_u calculated using Eq. (3.55) is shown in Fig. 3.46, where the installation at Athens gives a better value compared to the others. This is due to the high solar radiation I_T received on the collector area (Fig. 3.45) and the low collector area for the given load.

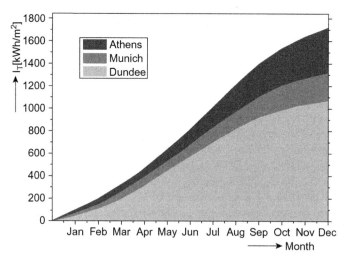

FIGURE 3.45 Cumulative annual solar radiation for the three PV/T systems.

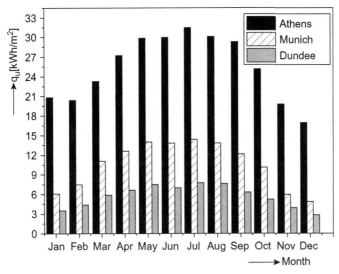

FIGURE 3.46   Useful solar heat output per unit collector area of the three PV/T systems.

The *thermal efficiency* η_c of the solar system is another important performance indicator calculated using Eq. (3.13). The comparison between the systems suggests that the annual difference in thermal efficiency (Table 3.15) is related to the variation of the climatic conditions and load, with the lower annual efficiency being that of the system at Dundee (6.5%), where the solar

TABLE 3.15 Annual Performance Indicators of the PV/T Systems

Indicator	Location		
	Athens	Munich	Dundee
I_T (kWh/m^2)	1725.2	1314.1	1065.0
Q_{load} (kWh)	2905.8	3203.8	3389.3
q_u (kWh/m^2)	304.9	127.6	69.5
η_c (%)	17.7	9.7	6.5
f (%)	59.6	50.9	49.5
E_{el} (kWh)	940.4	1618.0	2633.4
Y (kWh/kW$_p$)	1390.3	1063.1	916.1
PR (%)	79.8	80.2	85.2
η_{PV} (%)	9.6	9.6	10.2

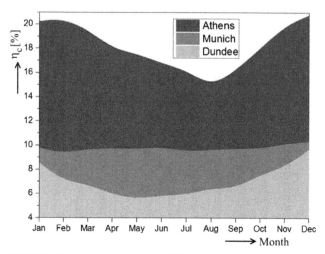

FIGURE 3.47 Solar thermal system efficiency of the three PV/T systems.

radiation and the load per square meter are the lower. The monthly variations of the solar thermal system efficiency η_c (Fig. 3.47) are due to many different interacting variables, such as the tap water temperature and the climatic conditions of the area. The high drop of the efficiency of the system installed in Athens during the summer months is due to the high temperature of the water intake and environment, as well as due to the high levels of solar radiation. These parameters combined the result to a high storage tank temperature, which diminishes the thermal efficiency of the solar system.

The *solar fraction* (*f*), defined as the percentage of the total thermal load satisfied by solar energy and calculated using Eq. (3.56), is a potentially misleading indicator of performance because the solar contribution is not taken into account. Therefore, while the annual solar fraction of the installations at Munich and Dundee is quite the same, the actual performance of the installations is totally different (Table 3.15).

The PV part of the hybrid systems installed at these three locations can be compared by evaluating their performance indices such as yield, PR, and PV efficiency.

The *PV system yield Y* (kWh/kW$_p$) indicates the energy generated with respect to the system size (Eq. [3.57]), and for this reason, it is a convenient way to compare the energy generated by PV systems of different size. Therefore, the system at Athens displayed the highest annual yield (1390.3 kWh/kW$_p$) in comparison with the other two systems due to high solar radiation and small installed power. In contrast, the system at Dundee has the lowest annual yield (916.1 kWh/kW$_p$) even though it generates more energy per annum in comparison with the other two systems (Fig. 3.48); that is because of the large surface area of PV panels. In Dundee, the climate of the

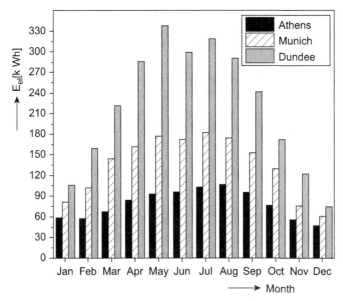

FIGURE 3.48 Electricity produced by the three PV/T systems.

area is such that from the middle of May to the middle of September, the sky typically is overcast. This also has an adverse effect on the electrical generation of the PV system (Fig. 3.48) and, consequently, to the maximum yield (Fig. 3.49) and thus electricity generation is greater in May than during the

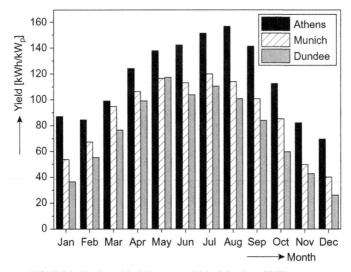

FIGURE 3.49 Monthly PV system yield of the three PV/T systems.

summer months. The maximum PV system yield of each installation occurs in a different month, mainly depending on the solar radiation received by the PV/T collectors and the PV cells' temperatures.

The PR is a dimensionless quantity that indicates the overall effect of losses on the rated output. By itself, it does not represent the amount of energy generated with respect to the system size, as it can be seen from Table 3.15, where the system installed in Athens, despite the low PR (79.8%) and high solar radiation, generates more energy with respect to the system size in comparison to the installation at Dundee, which displays a high PR (85.2%) and low solar radiation. The high PR at Dundee is caused mainly by the lower ambient temperature and solar radiation in comparison with the other two locations.

The *PV conversion efficiency* η_{PV} (Eq. [3.58]) is nearly the same for the three systems because the PV efficiency is insensitive to the solar irradiance within the practical working range (within their productive operational range). The slightly higher conversion efficiency at Dundee is also mainly due to the prevailing lower ambient temperatures in comparison with those of two other cities.

3.6.4 Conclusions

The present study determines that PV/T systems are an interesting option for residential applications in several countries and under various climatic conditions.

The competition between PV/T systems designed for DHW production and water heating systems firing conventional fuel and or electricity strongly depends on the climatic conditions and the pricing policy of the country.

The system can cover a significant portion of the DHW load, while simultaneously feeding the grid with electrical energy.

The price of the PV/T collector is expected to be reduced because of mass production and competition, which will lead to a lower initial capital cost and consequently to the economic viability of the system in areas with less favorable climatic conditions.

REFERENCES

[1] Kalogirou SA. Solar thermal collectors and applications. Progress in Energy and Combustion Science 2004;30:231—95.

[2] ASHRAE Handbook, HVAC applications. Atlanta (GA): American Society of Heating, Refrigerating and Air Conditioning Engineers; 2015.

[3] Henning HM. Solar-assisted air-conditioning in buildings: a handbook for planners. IEA solar heating & cooling program. Springer Verlag; 2007.

[4] Garg HP, Chakravertty S, Shukla AR, Agnihotri RC. Advanced tubular solar energy collector—a state of the art. Energy Conversion and Management 1983;23(3):157—69.

[5] Kim Y, Seo T. Thermal performances comparisons of the glass evacuated tube solar collectors with shapes of absorber tube. Renewable Energy 2007;32:772—95.

[6]   Kim JT, Ahn HT, Han H, Kim HT, Chun W. The performance simulation of all glass vacuum tubes with coaxial fluid conduit. International Communications in Heat and Mass Transfer 2007;34:58—97.

[7]   Han H, Kim JT, Ahn HT, Lee SJ. A three-dimensional performance analysis of all glass vacuum tubes with coaxial fluid conduit. International Communications in Heat and Mass Transfer 2008;35:589—96.

[8]   Shah LJ, Furbo S. Vertical evacuated tubular-collectors utilizing solar radiation from all direction. Applied Energy 2004;78:371—95.

[9]   Shah LJ, Furbo S. Theoretical flow investigations of an all glass evacuated tubular collector. Solar Energy 2007;81:822—8.

[10] Yan SY, Tian R, Hou S, Zhang LN. Analysis on unsteady state efficiency of glass evacuated solar collector with an inserted heat pipe. Journal of Engineering Thermophysics 2008;2:323—35.

[11] Tyagi VV, Kaushik SC, Tyagi SC. Advancement in solar photovoltaic/thermal (PV/T) hybrid collector technology. Renewable and Sustainable Energy Reviews 2012;16:1383—98.

[12] Hollands KGT. A concentrator for thin-film solar cells. Solar Energy 1971;13(2):149—63.

[13] Selcuk MK. Analysis, development and testing of a fixed tilt solar collector employing reversible vee-trough reflectors and vacuum tube receivers. Solar Energy 1979;22(5):413—26.

[14] Kalogirou S, Eleftheriou P, Lloyd S, Ward J. Design and performance characteristics of a parabolic-trough solar collector system. Applied Energy 1994;47:341—54.

[15] Duffie JA, Beckman WA. Solar engineering of thermal processes. Hoboken (NJ): Wiley & Sons, Inc.; 2013.

[16] Fernandez-Garcıa A, Zarza E, Valenzuela L, Perez M. Parabolic-trough solar collectors and their applications. Renewable and Sustainable Energy Reviews 2010;14:1695—721.

[17] Bostan I, Dulgheru V, Sobor I, Bostan V, Sochireanu A. Conversion systems of renewable energies. Chisinau: Tehnica-Info Publishing House; 2007 [in Romanian].

[18] Mills D. Advances in solar thermal electricity technology. Solar Energy 2004;76:19—31.

[19] Mills D, Morrison Graham L. Compact linear Fresnel reflector solar thermal power plants. Solar Energy 2000;68:263—83.

[20] Radosevich LG, Skinrood AC. The power production operation of Solar One, the 10 MWe solar thermal central receiver power plant. Journal of Solar Energy Engineering, Transactions ASME 1989;111(2):144—51.

[21] Jordan RC, Liu BYH. Applications of solar energy for heating and cooling of buildings. Atlanta (GA): ASHRAE Publication GRP 170; 1977.

[22] ASHRAE/ANSI Standard 93. Methods of testing to determine the thermal performance of solar collectors. Atlanta (GA): American Society of Heating, Refrigeration, and Air Conditioning Engineers; 2003.

[23] EN 12975-1. Thermal solar systems and components, solar collectors, Part 1: general requirements. Brussels: European Committee for Standardization; 2000.

[24] EN 12975-2. Thermal solar systems and components, solar collectors, Part 2: test methods. Brussels: European Committee for Standardization; 2001.

[25] Cooper PI, Dunkler RV. A now-linear flat-plate collector model. Solar Energy 1981;26 (2):133—40.

[26] Calise F, Dentice d'Accadia M, Palombo A. Transient analysis and energy optimization of solar heating and cooling systems in various configurations. Solar Energy 2010;84(3):432—49.

[27] Cabrera FJ, Fernandez-Garcıa A, Silva RMP, Perez-Garcıa M. Use of parabolic trough collectors for solar refrigeration and air-conditioning applications. Renewable and Sustainable Energy Reviews 2013;20:103—18.

[28] Kalogirou SA. Parabolic trough collectors for industrial process heat in Cyprus. Energy 2002;27:813−30.

[29] Price H, Lupfert E, Kearney D, Zarza E, Cohen G, Gee R. Advances in parabolic trough solar power technology. Journal of Solar Energy Engineering, Transactions ASME 2002;424:109−25.

[30] Kloben. 2013. http://www.kloben.it/products/view/1.

[31] Van Leeuwen J. Feasibility study solar refrigeration system for the cooling requirements of a beach bar. Report P&E 2430. Process & Energy Department, Delft University of Technology; 2010.

[32] Späte F, Ladener H. Solaraniagen. Staufen bei Freiburg. Ökobuch; 2008.

[33] Gebreslassie BH, Guillen-Gosalbez G, Jimenez L, Boer D. A systematic tool for the minimization of the life cycle impact of solar assisted absorption cooling systems. Energy 2010;35:3849−62.

[34] Twidell J, Weir T. Renewable energy resources. New York: Taylor & Francis; 2006.

[35] REN21 Renewable. Global status report; 2013. 2013. http://www.ren21.net/REN21Activities/GlobalStatusReport.aspx.

[36] Masson G, Latour M, Biancardi D. Global market outlook for photovoltaic until 2016. European Photovoltaic Industry Association 2012.

[37] Miles RW, Hynes KM, Forbes I. Photovoltaic solar cells: an overview of state-of the-art cell development and environmental issues. Progress in Crystal Growth and Characterization of Materials 2005;51:1−42.

[38] Rauschenbach HS. The principle and technology of photovoltaic energy conversion. New York: Litton Educational Publishing Inc.; 1980.

[39] IEA. Solar photovoltaic roadmap. Paris, France: International Energy Agency Publication; 2010.

[40] Beckman WA, Clark DR, Klein SA. A method for estimating the performance of photo-voltaic system. Solar Energy 1984;33(6):551−5.

[41] Huld T, Gottschalg R, Beyer H-G, Topic M. Mapping the performance of PV modules, effects of module type and data averaging. Solar Energy 2010;84:324−38.

[42] Fereira I, Kim D-S. Techno-economic review of solar cooling technologies based on location-specific data. International Journal of Refrigeration 2014;39:23−37.

[43] Klein SA, Beckman WA. Loss-of-load probabilities for stand-alone photovoltaic systems. Solar Energy 1987;39:499.

[44] Gordon J. Optimal sizing of stand-alone photovoltaic power systems. Solar Cells 1987;20(4):295−313.

[45] Catalina T. Utilisation of renewable energy sources in buildings. Bucharest: Matrix Rom Publishing House; 2015 [in Romanian].

[46] Arif Hasan M, Sumathy K. Photovoltaic thermal module concepts and their performance analysis: a review. Renewable and Sustainable Energy Reviews 2010;14:1845−59.

[47] Chow TT. A review on photovoltaic/thermal hybrid solar technology. Applied Energy 2010;87:365−79.

[48] Prakash J. Transient analysis of a photovoltaic/thermal solar collector for co-generation of electricity and hot air/water. Energy Conversion and Management 1994;35:967−72.

[49] Sarhaddi F, Farahat S, Ajam H, Behzadmehr A, Mahdavi Adeli M. An improved thermal and electrical model for a solar photovoltaic thermal (PV/T) air collector. Applied Energy 2010;87:2328−39.

[50] Coventry JS. Performance of a concentrating photovoltaic/thermal solar collector. Solar Energy 2005;78:211−22.

[51] Kostic LjT, Pavlovic TM, Pavlovic ZT. Optimal design of orientation of PV/T collector with reflectors. Applied Energy 2010;87:3023–9.

[52] Hjothman M, Yatim B, Sopian K, Abubakar M. Performance analysis of a double pass photovoltaic/thermal (PV/T) solar collector with CPC and fins. Renewable Energy 2005;30:2005–17.

[53] Mittelman G, Kribus A, Mouchtar O, Dayan A. Water desalination with concentrating photovoltaic/thermal (CPVT) systems. Solar Energy 2009;83:1322–34.

[54] Fang G, Hu H, Liu X. Experimental investigation on the photovoltaic-thermal solar heat pump air-conditioning system on water-heating mode. Experimental Thermal and Fluid Science 2010;34:736–43.

[55] Anderson TN, Duke M, Morrison GL, Carson JK. Performance of a building integrated photovoltaic/thermal (BIPVT) solar collector. Solar Energy 2009;83:445–55.

[56] Ji J, Lu JP, Chow TT, He W, Pei G. A sensitivity study of a hybrid photovoltaic/thermal water-heating system with natural circulation. Applied Energy 2007;84:222–37.

[57] Shan F, Cao L, Fang G. Dynamic performances modeling of a photovoltaic-thermal collector with water heating in buildings. Energy and Buildings 2013;66:485–94.

[58] Skoplaki E, Palyvos JA. On the temperature dependence of photovoltaic module electrical performance: a review of efficiency/power correlations. Solar Energy 2009;83:614–24.

[59] Winterton RHS. Where did the Dittus and Boelter equation come from. International Journal of Heat and Mass Transfer 1998;41:4–5.

[60] Tiwari A, Sodha MS, Chandra A, Joshi JC. Performance evaluation of photovoltaic/thermal solar air collector for composite climate of India. Solar Energy Materials and Solar Cells 2006;90:175–89.

[61] Sarbu I, Sebarchievici C. Ground-source heat pumps: fundamentals, experiments and applications. Oxford: Elsevier; 2015.

[62] Axaopoulos PJ, Fylladitakis ED. Performance and economic evaluation of a hybrid photovoltaic/thermal solar system for residential applications. Energy and Buildings 2013;65:488–96.

[63] Meteonorm version 5.1. Bern, Switzerland: Meteonorm Software; 2004.

[64] EUROFINS–Modulo Uno SpA, Volther Powertherm conformity test statement, http://solimpeks.com.au/downloads/Eurofins%20%20conformity%20test%20Powertherm.pdf.

Chapter 4

Thermal Energy Storage

4.1 GENERALITIES

One of the main aspects of solar systems is storage. Thermal energy storage (TES) is a technology that stocks thermal energy by heating or cooling a storage medium so that the stored energy can be used at a later time for heating and cooling applications and power generation. TES systems are used particularly in buildings and industrial processes. In these applications, approximately half of the energy consumed is in the form of thermal energy, the demand for which may vary during any given day and from one day to next. Therefore, TES systems can help balance energy demand and supply on a daily, weekly, and even seasonal basis. Advantages of using TES in an energy system are the increase of the overall efficiency and better reliability, but it can also lead to better economics, reducing investment and running costs, and less pollution of the environment and less carbon dioxide (CO_2) emissions [1]. TES is becoming particularly important for electricity storage in combination with concentrated solar power (CSP) plants where solar heat can be stored for electricity production when sunlight is not available.

In Europe, it has been estimated that around 1.4 million GWh/year could be saved and 400 million tons of CO_2 emissions avoided, in the building and industrial sectors by more extensive use of heat and cold storage [2].

Storage density, in terms of the amount of energy per unit of volume or mass, is an important issue for applications in order to optimize a solar ratio (how much of the solar radiation is useful for the heating/cooling purposes), efficiency of the appliances (solar thermal collectors and absorption chillers), and room consumption. For these reasons, it is worth to investigate the possibility of using phase-change materials (PCMs) in solar system applications. The potential of PCMs is to increase the energy density of small-sized water storage tanks, reducing solar storage volume for a given solar fraction or increasing the solar fraction for a given available volume [3].

It is possible to think of thermal storage in the hot and/or in the cold side of the plant. The former allows the storage of hot water from the collectors (and from the auxiliary heater) to be supplied to the generator of the absorption chiller (in cooling mode) or directly to the users (in heating mode). The latter allows the storage of cold water produced by the absorption chiller to be

Solar Heating and Cooling Systems. http://dx.doi.org/10.1016/B978-0-12-811662-3.00004-9

99

supplied to the cooling terminals inside the building. It is usual to identify the three situations just described as, respectively, "hot", "warm", and "cold" storage because of the different temperature ranges. Typically, a hot tank may work at 80–90°C, a warm tank at 40–50°C, and a cold tank at 7–15°C.

While heat storages in the hot side of solar plants are always present because of the heating and/or domestic hot water (DHW) production, cold storages are justified in bigger size plants. Cold storages are used not only to get economic advantages from the electricity tariffs (in case of electric compression chiller) depending on the time-of-the-day but also to lower the cooling power installed and to allow more continuous operation of the chiller [4].

Support for research and development (R&D) of new storage materials, as well as policy measures and investment incentives for TES integration in buildings, industrial applications, and variable renewable power generation, is essential to foster its deployment.

This chapter is focused on the analysis of TES technologies that provide a way of valorizing solar heat and reducing the energy demand of buildings. The principles of several energy storage methods and calculation of storage capacities are described. Sensible heat storage technologies including the use of water, underground, and packed-bed are briefly reviewed. Latent-heat storage (LHS) systems associated with PCMs for use in solar heating and cooling of buildings, solar water-heating and heat-pump systems, and thermochemical heat storage are also presented. Additionally, a three-dimensional heat-transfer simulation model of latent-heat TES is developed to investigate the quasi-steady state and transient heat transfer of PCMs. The numerical simulation results using paraffin RT20 are compared with available experimental data for cooling and heating of buildings. Finally, outstanding information on the performance and costs of TES systems are included.

4.2 CLASSIFICATION AND CHARACTERISTICS OF STORAGE SYSTEMS

The main types of TES of solar energy are presented in Fig. 4.1. An energy storage system can be described in terms of the following characteristics:

- *Capacity* defines the energy stored in the system and depends on the storage process, the medium and the size of the system;
- *Power* defines how fast the energy stored in the system can be discharged (and charged);
- *Efficiency* is the ratio of the energy provided to the user to the energy needed to charge the storage system. It accounts for the energy loss during the storage period and the charging/discharging cycle;
- *Storage period* defines how long the energy is stored and lasts hours to months (i.e., hours, days, weeks, and months for seasonal storage);

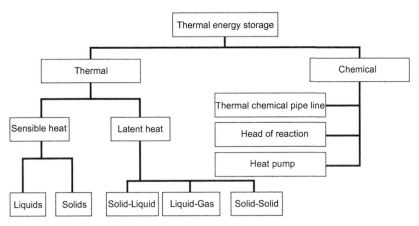

FIGURE 4.1   Types of thermal energy storage of solar energy.

- *Charge and discharge times* defines how much time is needed to charge/ discharge the system; and
- *Cost* refers to either capacity (€/kWh) or power (€/kW) of the storage system and depends on the capital and operation costs of the storage equipment and its lifetime (i.e., the number of cycles).

Capacity, power, and discharge time are interdependent variables, and in some storage systems, capacity and power can also depend on each other. Typical parameters for TES systems are shown in Table 4.1 [5], including capacity, power, efficiency, storage period, and cost. High-energy storage density and high power capacity for charging and discharging are desirable properties of any storage system. It is well known that there are three methods for TES at temperatures from −40°C to more than 400°C: sensible heat, latent heat associated with PCMs, and thermochemical storage (TCS) associated with chemical reactions (Fig. 4.2) [6].

The choice of storage medium depends on the nature of the process. For water heating, energy storage as sensible heat of stored water is logical. If air-heating collectors are used, storage in sensible or latent-heat effects in particulate storage units is indicated, such as sensible heat in a pebble-bed heat exchanger. In passive heating, storage is provided as a sensible heat in building the elements. If photovoltaic or photochemical processes are used, storage is logically in the form of chemical energy.

4.3 SENSIBLE HEAT STORAGE

Sensible heat storage (SHS) (Fig. 4.2A) is the simplest method based on storing thermal energy by heating or cooling a liquid or solid storage medium (e.g., water, sand, molten salts, or rocks), with water being the cheapest option.

TABLE 4.1 Typical Parameters of TES Systems

TES System	Capacity (kWh/t)	Power (MW)	Efficiency (%)	Storage Period	Cost (€/kWh)
Sensible (hot water)	10–50	0.001–10	50–90	days/ months	0.1–10
PCM	50–150	0.001–1	75–90	hours/ months	10–50
Chemical reactions	120–250	0.01–1	75–100	hours/ days	8–100

The most popular and commercial heat storage medium is water, which has a number of residential and industrial applications. Underground storage of sensible heat in both liquid and solid media is also used for typically large-scale applications. SHS has two main advantages: it is cheap and without the risks derived from the use of toxic materials.

SHS system utilizes the heat capacity and the change in temperature of the storage medium during the process of charging and discharging. The amount of heat stored depends on the specific heat of the medium, the temperature change, and the amount of storage material [7].

$$Q_s = \int_{t_i}^{t_f} mc_p dt = mc_p(t_f - t_i) \tag{4.1}$$

where Q_s is the quantity of heat stored, in J; m is the mass of heat storage medium, in kg; c_p is the specific heat, in J/(kg·K); t_i is the initial temperature, in °C; and t_f is the final temperature, in °C.

The SHS capacities of some selected solid–liquid materials are shown in Table 4.2. Water appears to be the best SHS liquid available because it is inexpensive and has a high specific heat. However, above 100°C, oils, molten

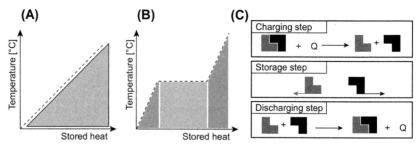

FIGURE 4.2 Methods of thermal energy storage (A) sensible heat; (B) latent heat; (C) thermochemical reactions.

TABLE 4.2 List of Selected Solid–Liquid Materials for Sensible Heat Storage

Medium	Fluid Type	Temperature Range (°C)	Density (kg/m³)	Specific Heat (J/(kg·K))
Rock	–	20	2560	879
Brick	–	20	1600	840
Concrete	–	20	1900–2300	880
Water	–	0–100	1000	4190
Calorie HT43	Oil	12–260	867	2200
Engine oil	Oil	≤160	888	1880
Ethanol	Organic liquid	≤78	790	2400
Propane	Organic liquid	≤97	800	2500
Butane	Organic liquid	≤118	809	2400
Isotunaol	Organic liquid	≤100	808	3000
Isopentanol	Organic liquid	≤148	831	2200
Octane	Organic liquid	≤126	704	2400

salts, liquid metals, etc. are used. For air-heating applications, rock bed-type storage materials are used.

4.3.1 Water Tank Storage

The use of hot-water tanks is a well-known technology for TES. Hot-water tanks serve the purpose of energy saving in water heating systems based on solar energy and in cogeneration (i.e., heat and power) energy supply systems. State-of the-art projects [8] have shown that water tank storage is a cost-effective storage option and that its efficiency can be further improved by ensuring optimal water stratification in the tank and highly effective thermal insulation. Today's R&D activities focus, for example, on evacuated super-insulation with a thermal conductivity of 0.01 W/(mK) at 90°C and 0.1 mbar and on optimized system integration. A typical system in which a water tank is used is shown in Fig. 4.3.

The energy storage capacity of a water (or other liquid) storage unit at uniform temperature (i.e., fully mixed, or no stratified) operating over a finite temperature difference is given by Eq. (4.1) redefined as

$$Q_s = mc_p \Delta t_s \qquad (4.2)$$

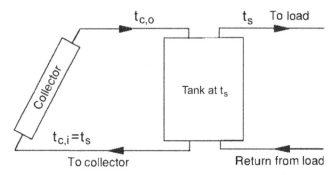

FIGURE 4.3   A typical system using water tank storage.

where Q_s is the total heat capacity for a cycle operating through the temperature range Δt_s and m, c_p are the mass and the specific heat, respectively of water in the unit. The temperature range over which such a unit can operate is limited at the lower extreme for most applications by the requirements of the process. The upper limit may be determined by the process, the vapor pressure of the liquid, or the collector heat loss.

An energy balance on the no stratified tank is

$$mc_p \frac{dt_s}{d\tau} = Q_u - Q_L - U_s A_s(t_i - t_a) \tag{4.3}$$

where Q_u and Q_L are rates of addition or removal of energy from the collector and to the load; U_s is the heat loss coefficient of storage tank; A_s is the storage tank surface area; t_a is the ambient temperature for the tank; τ is the time.

Eq. (4.3) is to be integrated over time to determine the long-term performance of the storage unit and the solar process. Useful long-term analytical solutions are not possible due to the complex time dependence of some of the terms. There are many possible numerical integration methods. Using simple Euler integration is usually satisfactory (i.e., rewriting the temperature derivative as $(t_s - t_i)/\Delta\tau$ and solving for the tank temperature at the end of a time increment):

$$t_s = t_i + \frac{\Delta\tau}{mc_p}[Q_u - Q_L - U_s A_s(t_i - t_a)] \tag{4.4}$$

Eq. (4.4) can be used to predict the water storage temperature as a function of time. Once the tank temperature is known, other temperature-dependent quantities can be estimated.

Hot-water storage systems used as buffer storage for DHW supply are usually in the range of 500 L to several cubic meters (m^3). This technology is also used in solar thermal installations for DHW combined with building heating systems (combisystems). Large hot-water tanks are used for seasonal storage of solar thermal heat in combination with small district heating systems. These systems can have a volume up to several thousand cubic meters. Charging temperatures are in the range of 80—90°C. The usable temperature difference can be enhanced by the use of heat pumps for discharging (down to temperatures around 10°C).

4.3.2 Underground Storage

Underground thermal energy storage (UTES) is also a widely used storage technology, which makes use of the ground (e.g., the soil, sand, rocks, and clay) as a storage medium for both heat and cold storages.

Means must be provided to add energy to and remove it from the medium. This is done by pumping heat transfer fluids (HTFs) through pipe arrays in the ground. The pipes may be vertical U-tubes inserted in wells (*boreholes*) that are spaced at appropriate intervals in the storage field or they may be horizontal pipes buried in trenches (see Chapter 9). The rates of charging and discharging are limited by the area of the pipe arrays and the rates of heat transfer through the ground surrounding the pipes. If the storage medium is porous, energy transport may occur by evaporation and condensation and by movement of water through the medium, and a complete analysis of such a store must include consideration of both heat and mass transfers. These storage systems are usually not insulated, although insulation may be provided at the ground surface.

Boreholes (ground heat exchangers) are also frequently used in combination with heat pumps where the ground heat exchanger extracts low-temperature heat from the soil.

Aquifer storage is closely related to ground storage, except that the primary storage medium is water, which flows at low rates through the ground. Water is pumped out of and into the ground to heat it and extract energy from it. Water flow also provides a mechanism for heat exchange with the ground itself. As a practical matter, aquifers cannot be insulated. Only aquifers that have low natural flow rates through the storage field can be used. A further limitation may be in chemical reactions of heated water with the ground materials. Aquifers, as with ground storage, operate over smaller temperature ranges than water stores. Most applications deal with the storage of winter cold to be used for the cooling of large office buildings and industrial processes in the summer.

Cavern storage and pit storage are based on large underground water reservoirs created in the subsoil to serve as TES systems. Caverns are the same in their principles of operation as the tanks discussed in the previous section. Energy is added to or removed from the store by pumping water into or out of the storage unit. The major difference will be in the mechanisms for heat loss and possible thermal coupling with the ground. These storage options are technically feasible, but applications are limited because of the high investment costs.

For high-temperature (i.e., above 100°C) SHS, the technology of choice is based on the use of liquids (e.g., oil or molten salts, the latter for temperatures up to 550°C). For very high temperatures, solid materials (e.g., ceramics, concrete) are also taken into consideration. However, most of such high-temperature-sensible TES options are still under development or demonstration.

4.3.3 Pebble-Bed Storage

A pebble-bed (packed-bed) storage unit uses the heat capacity of a bed of loosely packed particulate material to store energy. A fluid, usually air, is circulated through the bed to add or remove energy. A variety of solids may be used, rock and pebble being the most widely used materials.

A pebble-bed storage unit is shown in Fig. 4.4. In operation, flow is maintained through the bed in one direction during addition of heat (usually downward) and in the opposite direction during removal of heat. Note that heat cannot be added and removed at the same time; this is in contrast to water storage systems, where simultaneous addition to and removal from storage is possible.

A major advantage of a packed-bed storage unit is its high degree of stratification. The pebbles near the entrance are heated, but the temperature of the pebbles near the exit remains unchanged and the exit-air temperature remains very close to the initial bed temperature. As time progresses, a temperature front passes through the bed. When the bed is fully charged, its temperature is uniform.

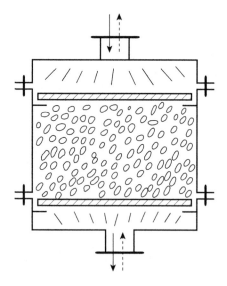

FIGURE 4.4 Pebble-bed storage system.

A packed bed in a solar heating system does not normally operate with constant inlet temperature. During the day, the variable solar radiation, ambient temperature, collector inlet temperature, load requirements, and other time-dependent conditions result in a variable collector outlet temperature.

Many studies are available on the heating and cooling of packed beds. The first analytical study was by Schumann [9] and the basic assumptions leading to this model are one-dimensional plug flow, no axial conduction or dispersion, constant properties, no mass transfer, no heat loss to the environment, and no temperature gradients within the solid particles. The differential equations for the fluid and bed temperatures (t_f, t_b) are:

$$\rho_f c_{p,f} \varepsilon \frac{\partial t_f}{\partial \tau} = -\frac{m_f c_{p,f}}{A} \frac{\partial t_f}{\partial x} + k_v (t_b - t_f) \tag{4.5}$$

$$\rho_b c_{p,b} (1 - \varepsilon) \frac{\partial t_b}{\partial \tau} = k_v (t_f - t_b) \tag{4.6}$$

where ρ_f is the fluid density; $c_{p,f}$ is the specific heat of fluid; ε is the bed void fraction; m_f is the fluid mass; A is the bed cross-sectional area; k_v is the volumetric (per unit bed volume) heat transfer coefficient between the bed and the fluid; and τ is the time.

For an air-based system, the first term on the left-hand side of Eq. (4.5) can be neglected and the equations can be written as [10]:

$$\frac{\partial t_f}{\partial (x/L)} = NTU(t_b - t_f) \tag{4.7}$$

$$\frac{\partial t_b}{\partial \Theta} = NTU(t_f - t_b) \tag{4.8}$$

$$NTU = \frac{k_v AL}{m_f c_{p,f}} \tag{4.9}$$

and the dimensionless time is

$$\Theta = \frac{\tau m_f c_{p,f}}{\rho_b c_{p,b} (1 - \varepsilon) AL} \tag{4.10}$$

where A is the bed cross-sectional area; L is the bed length; and NTU is the effectiveness.

Analytical solutions to these equations exist for a step change in inlet conditions and for a cyclic operation. For the long-term study of solar energy systems, these analytical solutions are not useful and numerical techniques as finite-difference method must be employed.

4.4 LATENT-HEAT STORAGE

The energy storage density increases and hence the volume is reduced, in the case of LHS (Fig. 4.2B). The heat is mainly stored in the phase-change process (at a quite constant temperature) and it is directly connected to the latent heat of the substance. The use of a LHS system using PCMs is an effective way of storing thermal energy and has the advantages of high-energy storage density and the isothermal nature of the storage process.

LHS is based on the heat absorption or release when a storage material undergoes a phase change from solid to liquid or liquid to gas or vice versa. The storage capacity Q_s, in J, of the LHS system with a PCM medium [7] is given by

$$Q_s = \int_{t_i}^{t_m} mc_p dt + mf\Delta q + \int_{t_m}^{t_f} mc_p dt \qquad (4.11)$$

$$Q_s = m[c_{ps}(t_m - t_i) + f\Delta q + c_{pl}(t_f - t_m)] \qquad (4.12)$$

where t_m is the melting temperature, in °C; m is the mass of PCM medium, in kg; c_{ps} is the average specific heat of the solid phase between t_i and t_m, in kJ/(kg·K); c_{pl} is the average specific heat of the liquid phase between t_m and t_f, in J/(kg K); f is the melt fraction; and Δq is the latent heat of fusion, in J/kg. For example, Glauber's salt (Na$_2$SO$_4$ · 10H$_2$O) has $c_{ps} \approx 1950$ J/(kg·°C), $c_{pl} \approx 3550$ J/(kg·°C), and $\Delta q = 2.43 \times 10^5$ J/kg at 34°C.

The measurement techniques presently used for latent heat of fusion and melting temperature of PCMs are: (1) differential thermal analysis (DTA), and (2) differential scanning calorimeter (DSC) [11]. In DSC and DTA techniques, sample and reference materials are heated at constant rates. The temperature difference between them is proportional to the difference in heat flow between the two materials and the record is the DSC curve. The recommended reference material is alumina (Al$_2$O$_3$). Latent heat of fusion is calculated using the area under the peak and melting temperature is estimated by the tangent at the point of greatest slope on the face portion of the peak.

Morrison and Abdel-Khalik [12] developed a model applicable to PCMs in small containers, where the length in flow direction is L, the cross-sectional area of the material is A, and the wetted perimeter is P. The HTF passes through the storage unit in the x direction at the mass flow rate m and with inlet temperature $t_{f,i}$.

The model can be based on three assumptions: (1) during flow, axial conduction in the fluid is negligible; (2) the Biot number is low enough that temperature gradients normal to the flow can be neglected; and (3) heat losses from the bed are negligible.

An energy balance on the material gives:

$$\frac{\partial u}{\partial \tau} = \frac{\lambda_s}{\rho_s}\frac{\partial^2 t_s}{\partial x^2} + \frac{UP}{\rho_s A}(t_f - t_s) \qquad (4.13)$$

where u, t_s, λ_s, and ρ_s are the specific internal energy, temperature, thermal conductivity, and density of the PCM; t_f, U are the circulating fluid temperature and overall heat-transfer coefficient between the fluid and PCM; and τ is the time.

An energy balance on the fluid is:

$$\frac{\partial t_f}{\partial \tau} + \frac{m}{\rho_f A_f} \frac{\partial t_f}{\partial x} = \frac{UP}{\rho_f A_f c_{p,f}}(t_s - t_f) \tag{4.14}$$

where ρ_f, A_f, and $c_{p,f}$ are the density, flow area, and specific heat of the fluid.

The equation and boundary conditions for PCM storage can be simplified for particular cases. It has been shown that axial conduction during flow is negligible, and if the fluid capacitance is small, Eqs. (4.13) and (4.14) become [10]:

$$\frac{\partial u}{\partial \Theta} = NTU(t_s - t_f) \tag{4.15}$$

$$\frac{\partial t_f}{\partial (x/L)} = NTU(t_f - t_s) \tag{4.16}$$

where ratio $\Theta = \tau m c_{p,f}/\rho_s AL$ and effectiveness $NTU = UPL/(mc_{p,f})$.

As depicted in Fig. 4.1, the phase-change process takes place in different modes: solid−solid, liquid−gas, and solid−liquid. In the first case, heat is stored by transition between different kinds of crystallization forms. For liquid−gas systems, latent heat is very high, but there are some problems in the storage control due to the high volume variations during a phase change. The most widespread are the solid−liquid PCMs that have a limited volume variation during exchange (generally less than 10%) and a fairly high melting latent heat. Melting processes involve energy densities of 100 kWh/m³ (e.g., ice) compared to a typical 25 kWh/m³ for SHS options. PCMs can be used for both short-term (daily) and long-term (seasonal) energy storages, using a variety of techniques and materials. Possible applications of PCMs are:

- implementation in gypsum board, plaster, concrete, or other wall-covering material being part of the building structure to enhance the TES capacity, with main utilization in peak-load shifting (and saving) and solar energy [13]. In this application typical operating temperature is 22−25°C but it can vary as a function of climate and heating/cooling loads;
- cold storage for cooling plants (operating temperature 5−18°C) [14];
- warm storage for heating plants (45−60°C) [15];
- hot storage for solar cooling and heating (>80°C) [16].

Any latent-heat energy storage system, therefore, possesses the following three components at least:

1. a suitable PCM with its melting point in the desired temperature range,
2. a suitable heat exchange surface, and
3. a suitable container compatible with the PCM.

4.4.1 Characteristics of PCMs

The main properties of PCMs are:

- thermophysical (latent heat of transition and thermal conductivity should be high, density and volume variation during a phase transition should be respectively high and low in order to minimize the storage volume);
- kinetic and chemical (supercooling should be limited to a few degrees, materials should have long-term chemical stability, compatibility with materials of construction, no toxicity, no fire hazard); and
- economics (low-cost and large-scale availability of the PCMs are also very important).

A large number of PCMs (organic, inorganic, and eutectic) are available in any required temperature range. PCMs are classified as different groups depending on the material nature (paraffin, fatty acids, salt hydrates, etc.) (Fig. 4.5).

Considering real applications in thermal energy store, the most widespread materials are paraffin's (organics), hydrated salts (inorganic), and fatty acids (organics). In cold storage, ice water is quite used as well. Table 4.3 shows some of the most relevant PCMs in different temperature ranges with their melting temperature, enthalpy, and density.

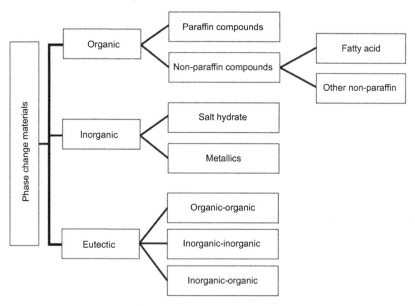

FIGURE 4.5   Classification of PCMs.

TABLE 4.3 Phase-Change Material Properties

PCM	Melting Temperature (°C)	Melting Enthalpy (kJ/kg)	Density (g/cm³)
Ice	0	333	0.92
Na-acetate trihidrate	58	250	1.30
Paraffin	−5−120	150−240	0.77
Erytritol	118	340	1.30

4.4.1.1 Organic PCMs

They can melt and solidify a lot of times without phase segregation and consequent degradation of their latent melting heat they crystallize with little or no supercooling and usually noncorrosiveness. The two main groups are:

- paraffin waxes: they consists of a mixture of mostly straight chain n-alkenes $CH_3-(CH_2)-CH_3$. The crystallization of the (CH_3)-chain releases a large amount of latent heat. Both the melting point and latent heat of fusion increase with the chain length. Due to cost consideration, however, only technical grade paraffins may be used as PCMs in LHS systems. Paraffin is safe, reliable, predictable, less expensive, noncorrosive, and available in a large temperature range (5−80°C) [17,18];
- non-paraffin organic: they are the most numerous of the PCMs with highly varied properties. A number of esters, fatty acids, alcohols, and glycols suitable for energy storage have been identified [19]. Some of the main features of these organic materials are high heat of fusion, inflammability, low thermal conductivity, low flash points, and instability at high temperatures.

4.4.1.2 Inorganic PCMs

These PCMs do not supercool appreciably and their melting enthalpies do not degrade with cycling. The two main types are:

- salt hydrate. They are alloys of inorganic salts (AB) and n kmol of water forming a typical crystalline solid of general formula $AB \cdot nH_2O$ whose solid−liquid transition is actually a dehydration and hydration of the salt. A salt hydrates usually melts to either to a salt hydrate with fewer moles of water, that is,

$$AB \cdot nH_2O \rightarrow AB \cdot mH_2O + (n - m)H_2O \qquad (4.17)$$

or to its anhydrous form:

$$AB \cdot nH_2O \rightarrow AB + nH_2O \qquad (4.18)$$

At the melting point the hydrate crystals breakup into anhydrous salt and water, or into a lower hydrate and water. One problem with most salt hydrates is that of incongruent melting caused by the fact that the released water of crystallization is not sufficient to dissolve all the solid-phase present. Due to density difference, the lower hydrate (or anhydrous salt) settles down at the bottom of the container.

Salt hydrates have been extensively studied in heat-storage applications because of their positive characteristics: high latent heat of fusion per unit volume, relatively high thermal conductivity (almost double of that of the paraffin's), and they are not very corrosive and compatible with plastics. As an example, main characteristics of some salt hydrates of the Phase Change Material Product Limited (UK) are depicted in Table 4.4 [20]. Some relevant disadvantages are incongruent melting and supercooling, which can be tackled in different ways (adding thickening agents, by mechanical stirring, by encapsulating the PCM to reduce separation, etc.). Other problem faced with salt hydrates is the spontaneous of salt hydrates with lower number of water moles during the discharge process. Adding chemicals can prevent the nucleation of lower salt hydrates, which preferentially increases the solubility of lower salt hydrates over the original salt hydrates with higher number of water moles.

- Metallic's: This category includes the low melting metals and metal eutectics. They are scarcely used in heat storage applications because of their low melting enthalpy per unit weight, even if they have high melting enthalpy per unit volume and high thermal conductivity. Some of the features of these materials are as follows: (1) low heat of fusion per unit weight (2) high heat of fusion per unit volume, (3) high thermal conductivity, (4) low specific heat, and (5) relatively low vapor pressure. A list of some selected metallic's is given in Table 4.5.

4.4.1.3 Eutectics

Eutectics are a minimum-melting composition of two or more components, each of which melts and freeze congruently forming a mixture of the component crystals during crystallization; eutectics nearly always melt and freeze without segregation. For this reason, they are a promising type of PCM for the future, even if actually they are less diffused than the other groups.

4.4.1.4 PCM's Containment

Containment of PCM is worth in order to contain the material in liquid and solid phases, to prevent its possible variation in chemical composition by interaction with environment, to increase its compatibility with other materials of the storage, to increase its handiness, and to provide suitable surface for heat transfer. Types of containment studied are bulk storage in tank heat exchangers, macroencapsulation, and microencapsulation. The main characteristic of PCM bulk systems is the need for a more extensive heat transfer than that found in

TABLE 4.4 Main Thermophysical Characteristics of Some Salt Hydrates

PCM Type	Phase Change Temperature (°C)	Density (kg/m³)	Latent Heat Capacity (kJ/kg)	Volumetric Heat Capacity (MJ/m³)	Specific Heat Capacity (kJ/(kg·K))	Thermal Conductivity (W/(m·K))
S89	89	1550	151	234	2.480	0.670
S44	44	1584	100	158	1.610	0.430
S7	7	1700	150	255	1.850	0.400

TABLE 4.5 Melting Point and Latent Heat of Fusion for Metallic's

No.	Material	Melting Point (°C)	Latent Heat (kJ/kg)
1	Gallium—gallium antimony eutectic	29.8	–
2	Gallium	30.0	80.3
3	Cerro-low eutectic	58.0	90.9
4	Bi—Cd—In eutectic	61.0	25.0
5	Cerro-bend eutectic	70.0	32.6
6	Bi—Pb—In eutectic	70.0	29.0
7	Bi—In eutectic	72.0	25.0
8	Bi—Pb eutectic	125.0	–

non-PCM tanks because the heat storage density of the PCM is higher compared to other storage media. The different approaches extensively used are inserting fins or using high conductivity particles, metal structures, fibbers in the PCM side, direct contact heat exchangers, or rolling cylinder method [21].

The two other possibilities are macro- and microencapsulating [22]. Macroencapsulating consists of including the PCM in some tube, sphere, panel, cylinder, or other, and is the most widespread. The choice of the material (plastic or metallic-aluminum or steel) and the geometry affect the thermal performance of the heat storage. Microencapsulating consists of microsphere (diameter less than 1 mm) of PCM encapsulated in a very thin and high molecular weight polymer. The spheres are then incorporated in some compatible material.

4.4.2 PCMs Used for Thermal Storage in Buildings

Storage concepts applied to the building sector have been classified as active or passive systems [23]. Passive thermal energy systems can enhance effectively the naturally available heat energy sources in order to maintain the comfort conditions in buildings and minimize the use of mechanically assisted heating or cooling systems [24].

The use of active thermal energy systems provides a high degree of control of the indoor conditions and improves the way of storing the heat energy. These systems are usually integrated in buildings to provide free cooling or to shift the thermal load from on-peak to off-peak conditions in several applications, such as DHW applications [25] or heating, ventilation, and air-conditioning (HVAC) systems [26].

4.4.2.1 Passive Technologies

The use of TES as a passive technology has the objective to provide thermal comfort with the minimum use of HVAC energy. When high thermal-mass materials are used in buildings, passive sensible storage is the technology that allows the storage of high quantity of energy, giving thermal stability inside the building. Materials typically used are rammed earth, alveolar bricks, concrete, or stone.

Standard solar walls, also known as Trombe walls, and solar water walls also use sensible storage to achieve energy savings in buildings [27]. Trombe wall (Fig. 4.6) (from the name of the French researcher that first proposed it in 1979) is a wall with high thermal capacity, shielded by a glass pane. A greenhouse effect is created, reducing thermal losses from the wall, heating the air between wall and glass that can be introduced into the room with a natural draught due to the chimney effect of the heated air.

The temperature of the wall increases as energy is absorbed, and time-dependent temperature gradients are established in the wall. Energy is lost through the glazing and is transferred from the room side of the wall to the room by radiation and convection. This storage wall can be considered as a set of N nodes connected together by a thermal network, each with a temperature and capacitance [28]. Heat is transferred by radiation across the gap and by

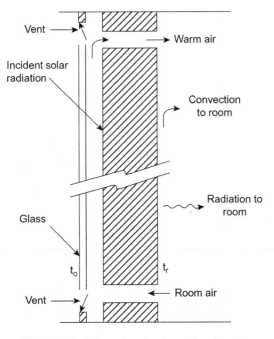

FIGURE 4.6 Schematics of a storage Trombe wall.

convection between air flowing in the gap and the absorbing surface and the inner glazing.

Energy balances are written for each node of thickness Δx, resulting in a set of ordinary differential equations with terms that represent its time-dependent temperature and energy flows to all adjacent nodes. The general energy balance for any node i in the wall is

$$\frac{\mathrm{d}t_i}{\mathrm{d}\tau} = \frac{\lambda}{\rho c_p \Delta x^2}(t_{i-1} + t_{i+1} - 2t_i), \quad i = 2, \ldots, N-1 \tag{4.19}$$

where λ is the thermal conductivity of wall; ρ is the wall density; c_p is the specific heat of wall; and τ is the time.

Equations for nodes one and N must take into account the node half-thickness and the convection and radiation heat transfer. The set of N equations are simultaneously solved for the time-dependent temperatures at each of the nodes, and from this the energy stored in the wall (relative to a base temperature t_{room}) can be calculated.

If there is airflow through vents and to the room, the energy added to the room by this mechanism will be $m_a c_{p,a}(t_o - t_r)$, where t_o is the outer glazing temperature and t_r is the room temperature.

PCM can be incorporated in construction materials using different methods, such as direct incorporation, immersion, encapsulation, microencapsulation, and shape-stabilization. In direct incorporation and immersion, the potential leakage has to be assessed. When the PCM is encapsulated or added in a shape-stabilized new material, a new layer appears in the construction system of the wall.

Traditionally, wallboards have been studied as one of the best options to incorporate PCM to building walls. A new approach in PCM-wallboards is the addition of an aluminum honeycomb in containing a microencapsulated PCM wallboard (Fig. 4.7) [29]. Similarly, PCM can also be impregnated or mixed with concrete or mortar [30]. One of the objectives pursued here is to maintain the concrete mechanical properties, while increasing its specific heat capacity. Another approach to incorporate PCM in building walls is to mix it with insulation materials. In masonry wall, the PCM incorporation can be, for example, within clay bricks (Fig. 4.8) [31]. In the PCM shutter concept, shutter-containing PCM is placed outside of the window areas. During daytime, when they are opened to the outside, the exterior side is exposed to solar radiation, heat is absorbed, and PCM melts. At night, we close the shutter, slide the windows and heat from the PCM radiates into the rooms.

4.4.2.2 Active Technologies

The use of TES in building active systems is an attractive and versatile solution for several applications for new or retrofitted buildings, such as the implementation of RES in the HVAC for space heating/cooling, the

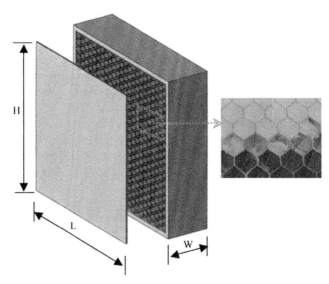

FIGURE 4.7   Microencapsulated PCM honeycomb wallboard.

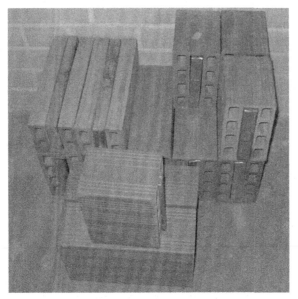

FIGURE 4.8   Clay bricks including PCM macrocapsules.

improvement in the performance of the current installations or the possible application of peak load-shifting strategies [32].

The integration of the TES in the building can be done using the core of the building (core, floor, and walls), in external solar facades, in suspended ceilings, ventilation system, PV systems, and water tanks. One of the main applications of TES in active building systems is the use of free cooling, when the storage is charged with low night-outdoor temperatures and this stored cold is discharged when required by the cooling demand [33].

Furthermore, TES have been used in building the solar systems in order to convert an intermittent energy source and meet the heating and DHW demand. The most popular solar TES building systems is extended to integrate solar air collectors in building walls [34] or use PCM in ventilated facades (Fig. 4.9) [35].

Within this context, the utilization of heat pumps with TES systems are presented as a promising technology to shift electrical loads from high-peak to off-peak periods, thus serving as a powerful tool in demand-side management.

TES has a very big potential as a key technology to reduce the energy demand of buildings and/or to improve the energy efficiency of their energy systems.

4.4.3 Advantages and Disadvantages of PCMs

The main advantages and drawbacks of PCM versus water SHSs are [36]:

- the possibility to reduce the tank volume for a given amount of energy stored, that is, true only if the storage is operated in a very narrow temperature range of around the phase-transition temperature;

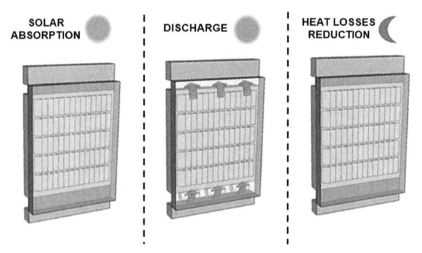

FIGURE 4.9   Operational mode of the ventilated facade with PCM.

- less on—off cycles of auxiliary heaters (for plants with storage in the hot side) and chillers (for plants with storage in the cold side);
- higher investment costs; and
- higher risks due to leak of stability and erosion of material encapsulating PCMs.

These materials make use of the latent heat between the solid- and liquid-phase changes, and must be encapsulated or stabilized for a technical use in any building system, either active or passive. This can be achieved using a direct inclusion in the wall [37], by impregnation in a porous material as gypsum [38], by using microencapsulation techniques [39], by using a shape-stabilization or slurries of PCM suspended on a thermal fluid [40]. The encapsulation is a key issue for the implementation of these technologies in the buildings and must be designed to avoid leakage and corrosion.

4.4.4 Heat Transfer in LHS Systems and Materials

The study of heat transfer characteristics of melting and solidification process is also one of the most attractive areas in the contemporary heat transfer research.

4.4.4.1 Enthalpy Formulation

By introducing an enthalpy method, the phase change problem becomes much simpler and has the following advantages: (1) the governing equation is similar to the single-phase equation; (2) there is no condition to be satisfied at the solid—liquid interface as it automatically obeys the interface condition, and (3) the enthalpy formulation allows a mushy zone between the two phases. Phase-change problems are usually solved with finite difference or finite element methods in accordance with the numerical approach. The enthalpy formulation is one of the most popular fixed-domain methods for solving the Stefan problem [17]. The major advantage is that the method does not require explicit treatment of the moving boundary.

For a phase-change process involving either melting or freezing, energy conservation can be expressed in terms of total volumetric enthalpy and temperature for constant thermophysical properties, as follows:

$$\frac{\partial H}{\partial \tau} = \nabla(\lambda_k(\nabla t)) \tag{4.20}$$

where H is the total volumetric enthalpy, in J/m^3; τ is the time, in s; λ_K is the thermal conductivity of phase k in PCM, in $W/(m \cdot {}^\circ C)$; and t is the temperature. Total volumetric enthalpy is the sum of sensible and latent heat of the PCM, that is,

$$H(t) = h(t) + \rho_l f(t) L \tag{4.21}$$

where h is the sensible volumetric enthalpy, in J/m^3; ρ_l is the density of liquid PCM, in kg/m^3; f is the melt fraction; and L is the latent heat of fusion, in J/kg.

The sensible volumetric enthalpy has the expression

$$h = \int_{t_m}^{t} \rho_k c_k dt \tag{4.22}$$

where ρ_k is the density of phase k in PCM, in kg/m^3; c_k is the specific heat of phase k in PCM, in $J/(kg \cdot {}^\circ C)$; t is the temperature, in $^\circ C$; and t_m is the melting temperature, in $^\circ C$.

In the case of isothermal phase change, the liquid fraction of melt is given by

$$f = \begin{cases} 0, & \text{if } t < t_m \text{ (solid)} \\ 0 - 1, & \text{if } t = t_m \text{ (mushy)} \\ 1, & \text{if } t > t_m \text{ (liquid)} \end{cases} \tag{4.23}$$

Following Eqs. (4.21) and (4.22), the enthalpy of PCM is

$$H = \int_{t_m}^{t} \rho_s c_s dt, \ t < t_m \text{ (solid)} \tag{4.24}$$

$$H = \rho_l f L, \ t = t_m \text{ (melting)} \tag{4.25}$$

$$H = \int_{t_m}^{t} \rho_l c_l dt + \rho_l L, \ t > t_m \text{ (liquid)}. \tag{4.26}$$

Solving Eqs. (4.24)−(4.26) for the PCM temperature, one gets

$$t = \frac{t_m + H}{\rho_s c_s}, \ H < 0 \text{ (solid)} \tag{4.27}$$

$$t = t_m, \quad 0 \leq H \leq \rho_l L \text{ (interface)} \tag{4.28}$$

$$t = \frac{t_m + (H - \rho_l L)}{\rho_l c_l}, \quad H > \rho_l L \text{ (liquid)} \tag{4.29}$$

Using Eqs. (4.21) and (4.22), an alternative form of Eq. (4.20) for a two-dimensional heat transfer in the PCM can be obtained as

$$\frac{\partial h}{\partial \tau} = \frac{\partial}{\partial x}\left(\alpha \frac{\partial h}{\partial x}\right) + \frac{\partial}{\partial y}\left(\alpha \frac{\partial h}{\partial y}\right) - \rho_l L \frac{\partial f}{\partial \tau} \tag{4.30}$$

and for the heat-exchanger container material it is

$$\frac{\partial h_f}{\partial \tau} = \frac{\partial}{\partial x}\left(\alpha_f \frac{\partial h_f}{\partial x}\right) + \frac{\partial}{\partial y}\left(\alpha_f \frac{\partial h_f}{\partial y}\right) \tag{4.31}$$

where τ is the time, in s; x and y are the space coordinates, in m; α is the diffusivity of PCM, in m^2/s; α_f is the thermal diffusivity of container material, in m^2/s; and h_f is the sensible volumetric enthalpy of container material, in J/m^3.

4.4.4.2 Numerical Solution

In order to obtain the algebraic equations using the control volume technique developed by Voller [41], it was necessary to divide the domain into elementary control volumes and then integrate the equation in these control volumes. Eq. (4.30) is solved using a fully implicit finite difference method. The finite difference equation for the PCM is obtained on integrating Eq. (4.30) over each control volume. The discretization of Eq. (4.30) for $\Delta x = \Delta y$ leads to the following scheme (Fig. 4.10):

$$h_P = h_P^o + \alpha R(h_E - 4h_P + h_W + h_N + h_S) + \rho_l L\left(f_P^o - f_P^k\right) \tag{4.32}$$

$$a_E h_E + a_W h_W + a_P h_P + a_N h_N + a_S h_S = Q \tag{4.33}$$

in which

$$a_E = a_W = a_N = a_S = -\alpha R$$
$$a_P = 1 - a_E - a_W - a_P - a_N - a_S \tag{4.34}$$

and

$$Q = h_P^o + \rho_l L\left(f_P^o - f_P^k\right), \quad R = \frac{dt}{(dx)^2} \tag{4.35}$$

where W, E, P, N, S are the west, east, center, north and south nodes.

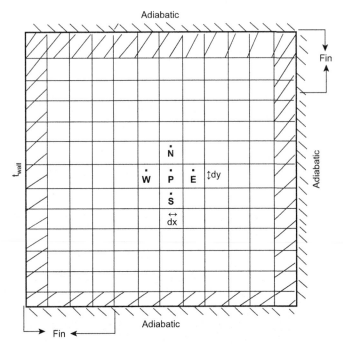

FIGURE 4.10 Two-dimensional domain.

Eq. (4.33) can be solved using a three-diagonal matrix algorithm. The central feature of the present fixed grid enthalpy method is the source term Q. Here, h_P^o and f_P^o refer to the enthalpy and the melt fraction, respectively, from the previous time step. The source term Q keeps track of the latent-heat evolution, and its driving element is the melt fraction. Its value is determined iteratively from the solution of the enthalpy equation. Hence, after the $(k+1)$th numerical solution of the enthalpy equation of the Pth node, Eq. (4.33) may be rearranged as

$$a_P h_P = -a_E h_E - a_W h_W - a_N h_N - a_S h_S + h_P^o + \rho_l L \left(f_P^o - f_P^k \right). \qquad (4.36)$$

If a phase change is occurring about the Pth node (i.e., $0 \le f \le 1$), the $(k+1)$th estimate of the melt fraction needs to be updated such that the left side of Eq. (4.36) is zero, that is,

$$0 = -a_E h_E - a_W h_W - a_N h_N - a_S h_S + h_P^o + \rho_l L \left(f_P^o - f_P^k \right). \qquad (4.37)$$

Eq. (4.37) may be rearranged as

$$f_P^{k+1} = \frac{-a_E h_E - a_W h_W - a_N h_N - a_S h_S + h_P^o}{\rho_l L + f_P^o}. \qquad (4.38)$$

The melt fraction update Eq. (4.38) is applied at every node after the $(k+1)$th solution of Eq. (4.33) for sensible volumetric enthalpy h, along with an under/over correction, that is,

$$f = \begin{cases} 0, & \text{if} (f)^{k+1} \le 0 \\ 1, & \text{if} (f)^{k+1} \ge 1 \end{cases}. \qquad (4.39)$$

Convergence at a given time step is declared when the difference in the total enthalpy fields falls below a given tolerance, that is,

$$\frac{\left| H^{k+1} - H^k \right|}{\rho_k c_k} \le \varepsilon. \qquad (4.40)$$

The value of ε can be set to 10^{-4}. The thermophysical properties of various heat exchanger materials [42] are given in Table 4.6.

4.4.4.3 A Three-Dimensional Heat Transfer Simulation Model of LHS

Centralized LHS system offers potential benefits in energy efficiency, in load shifting, and in emergency heating/cooling load systems. This section reports the development of a mathematical model of a centralized finned LHS for analyzing the thermal performance of melting process for both quasi-steady state and transient conjugate heat-transfer problems [43]. The developed model utilizes PCM technology that stores and retrieves energy at almost a constant temperature and is subjected to constant convective boundary

TABLE 4.6 Thermophysical Properties of Various Container Materials

Name of Material	Thermal Conductivity (W/(m·°C))	Density (kg/m³)	Specific Heat (kJ/(kg·°C))
Glass	0.78	2700	0.840
Stainless steel	7.70	8010	0.500
Tin	64	7304	0.226
Aluminum mixed	137	2659	0.867
Aluminum	204	2707	0.896
Copper	386	8954	0.383

conditions of free air stream based on a specific ventilation air flow rate. The numerical solution is conducted using the commercial software package FLUENT-12 [44]. The numerical simulation results using paraffin RT20 are compared with available experimental data for cooling and heating of buildings.

Numerical technique is employed to simulate PCM heat transfer within a certain range of phase change temperature, which typically uses the enthalpy porosity theory to deal with solid–liquid interface. The porosity effect is found to be similar to the liquid volume fraction of the porous media at mushy region [45]. Based on multiphase flow models such as volume of fluid (VOF) method, mixture and Euler model, only VOF and solidification/melting model can be applied simultaneously.

For simplicity, the following assumptions are considered:

- The axial conduction and viscous dissipation in the fluid are negligible and are ignored.
- PCM and porous matrix material are considered homogenous and isotropic.
- The thermophysical properties of the PCM and transfer fluid are independent of temperature; however, the properties of the PCM could be different in the solid and liquid phases.
- The PCM is considered at a single mean melting temperature t_m.
- The effect of radiation heat transfer is negligible.

Governing equations. Density and dynamic viscosity of the liquid PCM depend on its temperature. The density ρ of PCM, in kg/m³, is approximated by [46]

$$\rho = \frac{\rho_1}{\phi(t - t_1) + 1} \tag{4.41}$$

where ρ_1 is the reference density of PCM at the melting temperature t_1, and ϕ is the expansion factor. The value of $\phi = 0.001 \text{K}^{-1}$ can be selected [46].

The dynamic viscosity μ of the liquid PCM, in kg/(m·s), can be calculated using equation

$$\mu = \exp\left(-4.25 + \frac{1790}{t}\right) \tag{4.42}$$

where t is the temperature of PCM, in K.

The energy equation is written in terms of the sensible enthalpy as following:

$$h = \int_{t_{ref}}^{t} c_p dt \tag{4.43}$$

Or

$$\frac{\partial \rho h}{\partial x} + \text{div}(\rho \bar{u} h) = \text{div}\left(\frac{\lambda}{c_p} \text{grad } h\right) + Q_h \tag{4.44}$$

where h is the sensible enthalpy, in J/kg; $\bar{u} = (u, v, w)$ is the velocity, in m/s; λ is the thermal conductivity, in W/(m·K); c_p is the specific heat at constant pressure, in J/(kg·K); and Q_h is the latent-heat source term. In order to describe the fluid flow in the full liquid and mushy regions, the conservation equations of momentum and mass are required. In the enthalpyporosity approach, the energy equation source term Q_h, which is accounting for the latent-heat term, could be written in the following form:

$$Q_h = \frac{\partial(\rho \, \Delta H)}{\partial t} + \text{div}(\rho \, \bar{u} \, \Delta H) \tag{4.45}$$

where $\Delta H = F(t)$, the latent-heat content, in J/kg, is recognized as a function F of temperature t. The value of $F(t)$ can be generalized as follows:

$$F(t) = \begin{cases} L, & t \geq t_{liquid} \\ L(1 - f_s), & t_{liquid} \geq t \geq t_{solid} \\ 0, & t < t_{solid} \end{cases} \tag{4.46}$$

where L is the latent heat of phase change, in J/kg, and f_s is the local solid fraction. Assuming a Newtonian laminar flow, the continuity and momentum equations are

$$\frac{\partial \rho}{\partial \tau} + \text{div}(\rho \bar{u}) = 0 \tag{4.47}$$

$$\frac{\partial(\rho u)}{\partial \tau} + \text{div}(\rho \bar{u} u) = \text{div}(\mu \text{ grad } u) - \frac{\partial p}{\partial x} + Au \tag{4.48}$$

$$\frac{\partial(\rho v)}{\partial \tau} + \text{div}(\rho \bar{u} v) = \text{div}(\mu \text{ grad } v) - \frac{\partial p}{\partial y} + Av + Q_b \tag{4.49}$$

$$\frac{\partial(\rho w)}{\partial \tau} + \mathrm{div}(\rho \bar{u} w) = \mathrm{div}(\mu \ \mathrm{grad} \ w) - \frac{\partial p}{\partial z} + Aw \qquad (4.50)$$

where $\bar{u} = (u, v, w)$ is the velocity, p is the effective pressure, and μ is the viscosity.

The parameter A in Eqs. (4.48)–(4.50) represents a sources term for PCM. Based on Darcy's law, Carman-Koseny [47] suggested the following equation for calculating A:

$$A = \frac{C(1 - \varepsilon)^2}{\varepsilon^3 + \omega} \qquad (4.51)$$

where ε is the porosity. Value of C is related to the morphological properties of porous medium, and it is an assumed constant of 1.6×10^5. The constant ω is used to avoid dividing over zero and is set to be 10^{-3}.

The buoyancy source term, Q_b is introduced by [45]

$$Q_b = \frac{\rho_{\mathrm{ref}} \beta_t (h - h_{\mathrm{ref}})}{c_p} \qquad (4.52)$$

where β_t is the thermal expansion coefficient, in K^{-1}; c_p is the specific heat, in J/(kg K); h_{ref} and ρ_{ref} are reference values of enthalpy, in J/kg, and density, in kg/m^3, respectively. Darcy's law is used to describe the flow of the fluid through the porous medium as

$$\bar{u} = -\frac{k}{\mu} \mathrm{grad} \ p \qquad (4.53)$$

where k is the permeability, which is considered as a function of the porosity.

The total enthalpy H of the PCM is the sum of sensible heat, $h = c_p t$, and latent heat ΔH as

$$H = h + \Delta H \qquad (4.54)$$

The liquid fraction f can be expressed as

$$f = \begin{cases} 0 & t < t_{\mathrm{liquid}} \\ 1 & t > t_{\mathrm{liquid}} \\ \dfrac{t - t_{\mathrm{solid}}}{t_{\mathrm{liquid}} - t_{\mathrm{solid}}} & t_{\mathrm{solid}} < t < t_{\mathrm{liquid}} \end{cases} \qquad (4.55)$$

The latent-heat content of the PCM can be written in the following form

$$\Delta H = fL. \qquad (4.56)$$

In the enthalpyporosity technique, the mushy region is treated as a porous medium. For the purpose of the methodology development, it is worthwhile to consider the whole cavity as a porous medium. In fully solidified regions, the porosity (ε) is set to be equal zero and takes the value $\varepsilon = 1$ in fully liquid regions, while in mushy regions ε lies between 0 and 1. Accordingly, the flow velocity is linked to the porosity state and is defined as $u = \varepsilon u_i$ where u_i is the flow velocity.

Physical model: The schematic diagram of the three-dimensional physical model of the centralized storage unit filled with PCMs is shown in Fig. 4.11. Two different fins, made of aluminum, are used to increase the thermal performance of the storage unit and are connected to the metal box from both sides of lower and upper faces. The fins on the external side of the box are to increase the exposed area for convective heat flux whilst fins inside the box are aimed at boosting the thermal conduction heat flux. The box is filled with paraffin RT20 with a melting point of 22°C, heat storage capacity of 172 kJ/kg within an operating temperature range of 11−26°C, and specific heat capacity of 1.8 and 2.4 kJ/(kg·K) for solid and liquid, respectively [48].

Paraffin RT20 is chemically stable and commercially available compared with the other materials. In addition, the phase change temperature range of paraffin RT20 is suitable to regulate the indoor air temperature within the range of the comfort condition inside the building [49]. The centralized LHTES system is integrated into a mechanical ventilation system with an advanced control unit through suitable air supply ducts for free cooling of a

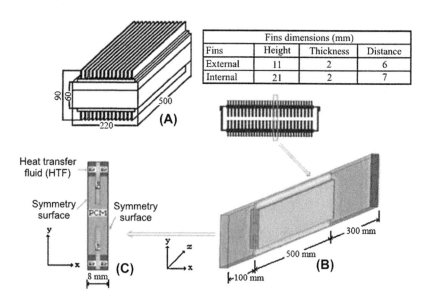

FIGURE 4.11 Geometrical configuration of LHTES. (A) schematic of LHTES system; (B) three-dimensional computational domain; (C) cross-section of computational domain.

low-energy building. Thus, the swing of indoor air temperature is stabilized during the day-time.

Model description: The three-dimensional computational domain storage unit filled with PCMs where the aluminum fins are arranged orthogonal to the axis of the unit. The HTF flows in the vicinity of such unit. The model has three zones: (1) air box with airflow around the fins and system, (2) PCM box (fluid/solid), and (3) fins box (solid).

All boxes are coupled to each other as one geometrical body. Air and PCM are coupled so that the energy transfers from air to fin and then from fin to PCM. Due to a symmetrical structure of the considered unit, the computational domain is simplified to only one symmetry unit cell in which the planes of symmetry are in the middle of the fin and are in middle between the two subsequent fins, as illustrated in Fig. 4.11.

Model validation: To verify the model, we compared the numerical simulation of evolution of fraction of melted phase with Gua and Viskanta experiment [50]. Gua and Viskanta measured the performance of a physical model in a rectangular cavity of a length of 88.9 mm and height of 63.5 mm filled with another PCM (gallium). The melting temperature of gallium is assumed to be 29.8°C. At the initial time, the left wall temperature suddenly rises to 38°C and remains constant while the temperature at the right wall was maintained constant at 28.3°C. Table 4.7 gives properties of paraffin, gallium, air, and aluminum. The model employed a single precision, unsteady solver to solve the implicit scheme of second order, time step was set to 0.2 s. The numerical solution was performed using FLUENT-12. Computational domain of two-dimensional physical model was meshed to 44 × 32 using GAMBIT software. Fig. 4.12 shows the solid—liquid interface positions for 2, 6, 10, and 17 min of melting process.

The numerical simulation results are in good agreement with the experimental results as presented in Fig. 4.12. The numerical model shows a great potential to predict the fluid-flow and heat transfer performances of a centralized LHS system.

4.5 CHEMICAL ENERGY STORAGE

The TCS uses thermochemical materials (TCMs) that store and release heat by a reversible endothermic/exothermic reaction process (Fig. 4.2C). During the charging process, heat is applied to the material A, resulting in a separation of two parts B + C. The resulting reaction products can be easily separated and stored until the discharge process is required. Then, the two parts B + C are mixed at a suitable pressure and temperature conditions and energy is released.

The products B and C can be stored separately, and thermal losses from the storage units are restricted to sensible heat effects, which are usually small compared to heats of reaction.

Thermal decomposition of metal oxides for energy storage has been considered by Simmons [51]. These reactions may have the advantage that the

TABLE 4.7 Properties of Paraffin, Gallium, Air, and Aluminum Used for Calculation

Materials	ρ (kg/m³)	λ (W/(m·K))	c_p (J/(kg·K))	t_{PCM} (°C)	L (J/kg)	μ (kg/(m·s))
Paraffin	$740/(0.001 \times (t - 293.15) + 1)$	0.15	RT20(DSC)	20–22	172.000	$0.001 \times \exp(-4.25 + 1970/t)$
Gallium	6093	32	381.5	29.78	80.160	1.81×10^{-3}
Air	$1.2 \times 10^{-5}\, t^2 - 0.01134\, t + 3.498$	0.0242	1006.43	–	–	1.7894×10^{-5}
Aluminum	2719	2024	871	–	–	–

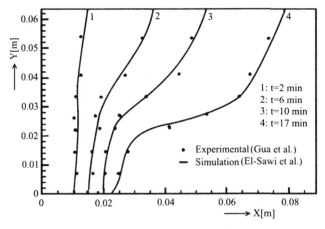

FIGURE 4.12 Comparison between experimental and numerical results.

oxygen evolved can be used for other purposes or discarded and oxygen from the atmosphere used in the reverse reactions. Two examples are the decomposition of potassium oxide

$$4KO_2 \leftrightarrow 2K_2O + 3O_2,$$

which occurs over a temperature range of 300–800°C with a heat of decomposition of 2.1 MJ/kg, and lead oxide

$$2PbO_2 \leftrightarrow 2PbO + O_2,$$

which occurs over a temperature range of 300–350°C with a heat of decomposition of 0.26 MJ/kg. There are many practical problems yet to be faced in the use of these reactions.

Energy storage by thermal decomposition of $Ca(OH)_2$ has been extensively studied by Fujii et al. [52]. The reaction is $Ca(OH)_2 \leftrightarrow CaO + H_2O$. The forward reaction will proceed at temperatures above about 450°C; the rates of reaction can be enhanced by the addition of zinc or aluminum. The product CaO is stored in the absence of water. The reverse exothermic reaction proceeds easily.

An example of a photochemical decomposition reaction is the decomposition of nitrosyl chloride, which can be written as

$$NOCl + photons \rightarrow NO + Cl.$$

The atomic chlorine produced forms chlorine gas, Cl^2, with the release of a substantial part of the energy added to the NOCl in decomposition. Thus, the overall reaction is

$$2NOCl + photons \rightarrow 2NO + Cl_2.$$

The reverse reaction can be carried out to recover part of the energy of the photons entering the reaction.

TABLE 4.8 Some Chemical Reactions for Thermal Energy Storage

	Reaction	Temperature (°C)	Energy Density (kJ/kg)
Methane steam reforming	$CH_4 + H_2O = CO + 3H_2$	480–1195	6053
Ammonia dissociation	$2NH_3 = N_2 + 3H_2$	400–500	3940
Thermal dehydrogenation of metal hydrides	$MgH_2 = Mg + H_2$	200–500	3079 (heat) 9000 (H_2)
Dehydration of metal hydroxides	$CA(OH)_2 = CAO + H_2O$	402–572	1415
Catalytic dissociation	$SO_3 = SO_2 + \frac{1}{2} O_2$	520–960	1235

Processes that produce electrical energy may have storage provided as chemical energy in electrical storage batteries or their equivalent.

Thermochemical reactions, such as adsorption (i.e., adhesion of a substance to the surface of another solid or liquid), can be used to store heat and cold, as well as to control humidity. The high storage capacity of sorption processes also allows thermal energy transportation.

Table 4.8 lists some of the most interesting chemical reactions for TES [53]. While sorption storages can only work up to temperatures of about 350°C, chemical reactions can go much higher.

4.6 COOL THERMAL ENERGY STORAGE

Cool thermal energy storage (CTES) has recently attracted increasing interest in industrial refrigeration applications, such as process cooling, food preservation, and building air-conditioning systems. CTES appears to be one of the most appropriate methods for correcting the mismatch that occurs between the supply and demand of energy. Cool energy storage requires a better insulation tank as the energy available in the cool state is expensive, compared to the heat available in a hot storage tank. Cheralathan et al. [54] investigated the performance of an industrial refrigeration system integrated with CTES. The authors have indicated significant savings in capital and operating cost, in thermal storage-integrated systems. The size of the PCM-based CTES system was also considerably reduced when compared with that of a chilled water system.

The sorption phenomenon can also be applied for TES. In that case, a heat source promotes the dissociation (endothermic process) of a working pair, whose substances can be stored separately. When they come into contact again, heat is released (exothermic process). Therefore, the energy can then be stored with virtually no loss because the heat is not stored in a sensible or latent form but rather as potential energy, as long as the substances are kept separate.

Typical applications involve adsorption of water vapor to silica-gel or zeolites (i.e., microporous crystalline alumina-silicates). Of special importance for use in hot/humid climates or confined spaces with high humidity are open sorption systems based on lithium-chloride to cool water and on zeolites to control humidity.

Adsorption TES is a promising technology that can provide an excellent solution for long-term TES in a more compact and efficient way. Solar thermal energy or waste heat from several processes can be used to regenerate the adsorbent and promote the energy storage [55].

The adsorption cycle has already been used in several research projects to promote TES. In 1990, Kaubek and Maier-Laxhuber [56] patented an adsorption apparatus to be used as an electric-heating storage, working with the zeolite/water pair and reporting a 30% savings in the energy consumption. The system can be used as an air-heating device or combined with a hot-water tank. In the first case, the adsorbent bed is heated by electric heating rods during desorption phase, regenerating the adsorbent and releasing the condensation heat into the space to be heated. In the latter case, the condensation heat is released into a water tank during the desorption phase, while the adsorption heat is transferred to the water tank through a specific closed circuit in the adsorption stage. Hauer [57] presented a seasonal adsorption TES system, working with the silica-gel/water pair (Fig. 4.13). During the summer, while the system is charging, the heat from the solar collectors is conducted to three adsorbent beds, promoting the desorption stage. In the winter, the low temperatures in the solar collector promote the evaporation of the water in the evaporators/condensers, and the heat of adsorption is released to the building heating system.

Schwamberger et al. [58] performed a simulation study of a new cycle concept for the adsorption heat pumps operating with the zeolite/water pair (Fig. 4.14). This cycle makes use of stratified TES to improve the internal heat recovery between the adsorption and desorption phases of the cycle. During the adsorption phase, the heat of adsorption is carried away by cooler HTF from the adsorber to the storage tank, entering at a height corresponding to its temperature. During the desorption phase, the adsorber is heated by supplying warm fluid from the storage tank, while the cooler fluid returning from the adsorber is stored in the storage tank. An external heat source helps to preserve the stratification effect in the tank. The heat rejected in the condenser and in an external heat sink can be used for space heating.

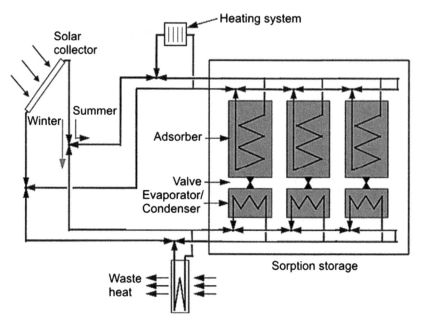

FIGURE 4.13 Seasonal adsorption thermal storage system.

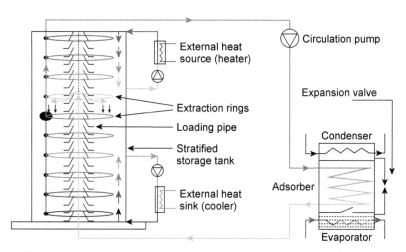

FIGURE 4.14 Schematic of an adsorption heat pump working with the zeolite/water pair.

A cascade storage system offers vast potential for the improvement of a solar cooling-system performance. In a cascaded storage system, PCMs with different melting temperatures are arranged in a series to store heat in different temperatures. In comparison with a conventional single PCM-based storage system, a cascaded multiple PCM-based storage system would improve solar

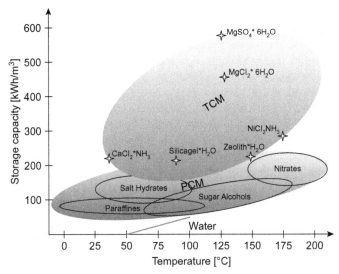

FIGURE 4.15   Storage capacity depending on temperature for TES.

collecting efficiency as the lower temperature at the bottom of the tank is connected to the inlet of the solar collector. The numerical results from the parametric study investigated by Shaikh and Lafdi [59] indicated that the total energy charged rate can be significantly enhanced by using composite PCMs as compared to the single PCM.

Fig. 4.15 shows the different TES technologies: sensible heat (i.e., water as an example); latent heat (i.e., different materials); and thermochemical (i.e., sorption and chemical reactions).

4.7 PERFORMANCE AND COST OF THERMAL ENERGY STORAGE SYSTEMS

TES includes a number of different technologies, each one with its own specific performance, application and cost.

Important fields of application for TES systems are in the building sector (e.g., DHW, space heating, and air-conditioning) and in the industrial sector (e.g., process heat and cold). TES systems can be installed as either centralized plants or distributed devices. Centralized plants are designed to store waste heat from large industrial processes, conventional power plants, combined heat and power plants, and from renewable power plants, such as CSP. Their power capacity ranges typically from hundreds of kW to several MW. Distributed devices are usually buffer storage systems to accumulate solar heat to be used for domestic and commercial buildings (e.g., hot water, heating, and appliances). Distributed systems are mostly in the range of a few to tens of kW.

TES systems based on SHS offer a storage capacity ranging from 10 to 50 kWh/t and storage efficiencies between 50 and 90%, depending on the specific heat of the storage medium and thermal insulation technologies. PCMs can offer higher storage capacity and storage efficiencies from 75 to 90%. In most cases, storage is based on a solid−liquid phase change with energy densities of 100 kWh/m^3 (e.g., ice). TCS systems can reach storage capacities of up to 250 kWh/t with operation temperatures of more than 300°C and efficiencies from 75% to nearly 100%. The cost of a complete system for SHS ranges between 0.1 and 10 €/kWh, depending on the size, application, and thermal insulation technology. The costs for PCM and TCS systems are in general higher. In these systems, major costs are associated with the heat (and mass) transfer technology, which has to be installed to achieve a sufficient charging/discharging power. Costs of LHS systems based on PCMs range between 10 and 50 €/kWh while TCS costs are estimated to range between 8 and 100 €/kWh. The economic viability of a TES depends heavily on application and operation needs, including the number and frequency of the storage cycles.

Costs estimated of TES systems include storage materials, technical equipment for charging and discharging, and operation costs.

TES systems for sensible heat are rather inexpensive as they consist basically of a simple tank for the storage medium and the equipment to charge/discharge. Storage medium (e.g., water, soil, rocks, concrete, or molten salts) are usually relatively cheap. However, the container of the storage material requires effective thermal insulation, which may be an important element of the TES cost.

In the case of UTES systems, boreholes and heat exchangers to activate the underground storage are the most important cost elements. Specific costs range from 0.1 to 10 €/kWh [2] and depend heavily on local conditions.

PCM storage and TCS systems are significantly more complex and expensive than the storage systems for sensible heat. In most cases (e.g., thermochemical reactors), they use enhanced heat and mass transfer technologies to achieve the required performance in terms of storage capacity and power, and the cost of the equipment is much higher than the cost for the storage material. The cost of systems using expensive microencapsulated PCMs, which avoid the use of heat exchange surfaces, can be even higher.

The difference between the pure PCM and the complete TES system is even higher for active PCM installations. As an example, the costs of a calcium-chloride storage for the heat rejected from a thermally driven absorption chiller includes the cost of calcium chloride, which is rather inexpensive (0.3 €/kg) and the cost of a container, heat exchanger, and other components that is around 65 €/kWh [2].

Materials for TCS are also expensive as they have to be prepared (e.g., pelletized or layered over supporting structures). Also expensive are the containers and the auxiliary TCS equipment for both heat and mass transfer during energy charging and discharging. TCS systems can be operated as either open systems (i.e., basically packed beds of pellets at ambient pressure) or closed systems. Open systems are often the cheapest option while closed systems need sophisticated heat exchangers.

The overall economic evaluation of a TES system depends significantly on the specific application and operation needs, including the number and frequency of storage cycles.

TES technologies face some barriers to market entry and cost is a key issue. Other barriers relate to material properties and stability, in particular for TCS. Each storage application needs a specific TES design to fit specific boundary conditions and requirements. R&D activities focus on all TES technologies.

TES market development and penetration varies considerably, depending on the application fields and regions. Thus, TES potential for cogeneration and district heating in Europe is associated with the building stock. The implementation rate of cogeneration is 10.2%, while the implementation of TES in these systems is assumed to be 15%. As far as TES for power applications is concerned, a driving sector is the CSP where almost all new power plants in operation or under construction are equipped with TES systems, mostly based on molten salt. This is perhaps the most important development filed for large, centralized TES installations.

REFERENCES

[1] Dincer I, Rosen MA. Thermal energy storage, systems and application. Chichester, UK: John Wiley & Sons; 2002.

[2] IEA IRENA. The energy technology systems analysis programme (ETSAP): technology brief E17. 2013. www.irena.org/Publications.

[3] Medrano M, Yilmaz MO, Nogués M, Martorell I, Roca J, Cabeza LF. Experimental evaluation of commercial heat exchangers for use as PCM thermal storage systems. Applied Energy 2009;86:2047−55.

[4] Lazzarin R. Solar cooling plants: how to arrange solar collectors, absorption chillers and the load. International Journal of Low Carbon Technology 2007;2(4):376−90.

[5] Hauer A. In: Storage Technology Issues and Opportunities, International Low-Carbon Energy Technology Platform, Strategic and Cross-Cutting Workshop "Energy Storage − Issues and Opportunities", Paris, France; 2011.

[6] de Garcia A, Cabeza CF. Phase change materials and thermal energy storage for buildings. Energy and Buildings 2015;103:414−9.

[7] Lane GA. Solar heat storage-latent heat materials, vol. I. Boca Raton, FL: CRC Press; 1983.

[8] IEA. Energy conservation through energy storage (ECES) programme, International Energy Agency, Brochure, http://www.iea-eces.org/files/090525_broschuere_eces.pdf.

[9] Schumann TEW. Heat transfer: a liquid flowing through a porous prism. Journal of Franklin Institute 1929;208:405−16.

[10] Duffie JA, Beckman WA. Solar engineering of thermal processes. Hoboken, NJ: Wiley & Sons, Inc.; 2013.

[11] Buddhi D, Sawhney RL, Seghal PN, Bansal NK. A simplification of the differential thermal analysis method to determine the latent heat of fusion of phase change materials. Journal of Physics D: Applied Physics 1987;20:1601−5.

[12] Morrison DJ, Abdel-Khalik SI. Effects of phase-change energy storage on the performance of air-based and liquid-based solar heating systems. Solar Energy 1978;20:57−67.

[13] Khudhair AM, Farid MM. A review on energy conservation in building applications with thermal storage by latent heat using phase change materials. Energy Conversion and Management 2004;45:263−75.

[14] He B, Setterwall F. Technical grade paraffin waxes as phase change materials for cool thermal storage and cool storage systems capital cost estimation. Energy Conversion and Management 2002;43:1709−23.

[15] Farid MM, Khudhair AM, Razak SA, Al-Hallaj S. A review on phase change energy storage: materials and applications. Energy Conversion and Management 2004; 45:1597−615.

[16] Bruno F. Using phase change materials (PCMs) for space heating and cooling in buildings, EcoLibrium. Journal of Australian Institute of Refrigeration, Air Conditioning and Heating 2005;4(2):26−31.

[17] Sharma A, Tyagi VV, Chen CR, Buddhi D. Review on thermal energy storage with phase change materials and applications. Renewable and Sustainable Energy Reviews 2009;13:318−45.

[18] Himran S, Suwono A, Mansoori GA. Characterization of alkenes and paraffin waxes for application as phase change energy storage medium. Energy Sources 1994;16(1):117−28.

[19] Abhat A. Development of a modular heat exchanger with integrated latent heat storage. In: Germany Ministry of Science and Technology Bonn, Report no. BMFT FBT 81−050; 1981.

[20] Noro M, Lazzarin RM, Busato F. Solar cooling and heating plants: an energy and economic analysis of liquid sensible vs. phase change material (PCM) heat storage. International Journal of Refrigeration 2014;39:104−16.

[21] Felix Regin A, Solanki SC, Saini JS. Heat transfer characteristics of thermal energy storage system using PCM capsules: a review. Renewable and Sustainable Energy Reviews 2008;12:2438−58.

[22] Cabeza LF, Castell A, Barreneche C, De Gracia A, Fernandez AI. Materials used as PCM in thermal energy storage in buildings: a review. Renewable and Sustainable Energy Reviews 2011;15(3):1675−95.

[23] Gil A, Medrano M, Martorell I, Lazzaro A, Dolado P, Zalba B, Cabeza LF. State of the art on high temperature thermal energy storage for power generation. Part 1-Concepts, materials and modellization. Renewable and Sustainable Energy Reviews 2010;14:31−55.

[24] Parameshwaran R, Kalaiselvam S, Harikrishnan S, Elayaperumal A. Sustainable thermal energy storage technologies for buildings: a review. Renewable and Sustainable Energy Reviews 2012;16:2394−433.

[25] de Gracia A, Oro E, Farid MM, Cabeza LF. Thermal analysis of including phase change material in a domestic hot water cylinder. Applied Thermal Engineering 2011;31:3938−45.

[26] Agyenim F, Hewitt N. The development of a finned phase change material (PCM) storage system to take advantage of off-peak electricity tariff for improvement in cost of heat pump operation. Energy and Buildings 2010;42:1552−60.

[27] Saadatian O, Sopian K, Lim CH, Asim N, Sulaiman MY. Trombe walls: a review of opportunities and challenges in research and development. Renewable and Sustainable Energy Reviews 2012;16:6340−51.

[28]   Ohanessian P, Charters WWS. Thermal simulation of a passive solar house using a Trombe-Michel wall structure. Solar Energy 1978;20:275—81.

[29]   Lai C, Hokoi S. Thermal performance of an aluminum honeycomb wallboard incorporating microencapsulated PCM. Energy and Buildings 2014;73:37—47.

[30]   Desai D, Miller M, Lynch JP, Li VC. Development of thermally adaptive engineered cementitious composite for passive heat storage. Construction and Building Materials 2014;67:366—72.

[31]   Silva T, Vicente R, Soares N, Ferreira V. Experimental testing and numerical modelling of masonry wall solution with PCM incorporation: a passive construction solution. Energy and Buildings 2012;49:235—45.

[32]   Sun Y, Wang S, Xiao F, Gao D. Peak load shifting control using different cold thermal energy storage facilities in commercial buildings: a review. Energy Conversion and Management 2013;71:101—14.

[33]   Waqas A, Ud Din Z. Phase change material (PCM) storage for free cooling of buildings-a review. Renewable and Sustainable Energy Reviews 2013;18:607—25.

[34]   Fraisse G, Johannes K, Trillat-Berdal V, Achard G. The use of a heavy internal wall with a ventilated air gap to store solar energy and improve summer comfort in timber frame houses. Energy and Buildings 2006;38:293—302.

[35]   de Gracia A, Navarro L, Castell A, Ruiz-Pardo A, Alvarez S, Cabeza LF. Experimental study of a ventilated facade with PCM during winter period. Energy and Buildings 2012;58:324—32.

[36]   Chidambaram LA, Ramana AS, Kamaraj G, Velraj R. Review of solar cooling methods and thermal storage options. Renewable and Sustainable Energy Reviews 2011;15:3220—8.

[37]   Zhang P, Ma ZW, Wang RZ. An overview of phase change material slurries: MPCS and CHS. Renewable and Sustainable Energy Reviews 2010;14:598—614.

[38]   Nomura T, Okinaka N, Akiyama T. Impregnation of porous material with phase change material for thermal energy storage. Materials Chemistry and Physics 2009;115:846—50.

[39]   Tyagi VV, Kaushik SC, Tyagi SK, Akiyama T. Development of phase change materials based microencapsulated technology for buildings: a review. Renewable and Sustainable Energy Reviews 2011;15:1373—91.

[40]   Delgado M, Lazzaro A, Mazo J, Zalba B. Review on phase change material emulsions and microencapsulated phase change material slurries: materials, heat transfer studies and applications. Renewable and Sustainable Energy Reviews 2012;16:253—73.

[41]   Voller VR. Fast implicit finite-difference method for the analysis of phase change problems. Numerical Heat Transfer — Part B 1990;17:155—69.

[42]   Garg HP. Treatise on solar energy: fundamentals of solar energy, vol. 1. New York: John Wiley & Sons; 1982.

[43]   El-Sawi A, Haghighat F, Akbari H. Centralized latent heat thermal energy storage system: model development and validation. Energy and Buildings 2013;65:260—71.

[44]   ANSYS FLUENT 12.0 User's Guide. USA: ANSYS, Inc.; 2009.

[45]   Brent AD, Voller VR, Reid KJ. Enthalpy-porosity technique for modeling convection-diffusion phase change: application to the melting of a pure metal. Numerical Heat Transfer 1988;13:297—318.

[46]   Humphries WR, Griggs EI. A design handbook for phase change thermal control and energy storage devices. NASA STI/Recon Technical Report N 78. 1977.

[47]   Carman PC. Fluid flow through granular beds. Transactions of the Institute of Chemical Engineers 1937;15:150—6.

[48] Stritih U, Butala V. Experimental investigation of energy saving in buildings with PCM cold storage. International Journal of Refrigeration 2010;33:1676–83.

[49] Tatsidjodoung P, Le Pierres N, Luo L. A review of potential materials for thermal energy storage in building applications. Renewable and Sustainable Energy Reviews 2013;18:327–49.

[50] Gua C, Viskanta R. Melting and solidification of a pure metal on a vertical wall. Journal of Heat Transfer 1986;108:174–81.

[51] Simmons JA. Reversible oxidation of metal oxides for thermal energy storage. Proceedings of the International Solar Energy (ISES) Meeting 1976;8:219.

[52] Fujii I, Tsuchiya K, Higano M, Yamada J. Studies of an energy storage system by use of the reversible chemical reaction: $CaO + H_2O \leftrightarrow Ca(OH)_2$. Solar Energy 1985;34:367–77.

[53] Garg HP. Solar thermal energy storage. Boston, Lancaster: D. Reidel Publishing Company; 1985.

[54] Cheralathan M, Verlaj R, Renganarayanan S. Performance analysis on industrial refrigeration system integrated with encapsulated PCM-based cool thermal energy storage system. International Journal of Energy Research 2007;31:1398–413.

[55] Fernandes MS, Brites GJVN, Costa JJ, Gaspar AR, Costa VAF. Review and future trends of solar adsorption refrigeration systems. Renewable and Sustainable Energy Reviews 2014;39:102–23.

[56] Kaubek F, Maier-Laxhuber P. Adsorption apparatus used as an electro-heating storage. 1990. United States Patent US 4956977 A. 18 Set.

[57] Hauer A. Adsorption systems for TES – design and demonstration projects. In: Paksoy HO, editor. Thermal energy storage for sustainable energy consumption. Netherlands: Springer; 2007.

[58] Schwamberger V, Joshi C, Schmidt FP. Second law analysis of a novel cycle concept for adsorption heat pumps. In: Proceedings of the International Sorption Heat Pump Conference (ISHPC11), Padua, Italy; 2011. p. 991–8.

[59] Shaikh S, Lafdi K. Effect of multiple phase change materials (PCMs) slab configurations on thermal energy storage. Energy Conversion and Management 2006;47:2103–17.

Chapter 5

Solar Water and Space-Heating Systems

5.1 GENERALITIES

The concept of low-energy building is based on the reduction of the primary energy demand through a high insulation level, the use of high efficiency heating/cooling systems, and the integration of renewable energy sources (RES) into the building plant. Each system design aims to increase the solar fraction (SF) value and to reduce the consumed auxiliary energy, which is usually selected as fossil-fueled sources.

Solar heating systems are a type of renewable energy technology that has been increasingly used in the past decade across Europe to provide heating, air-conditioning (A/C), and domestic hot water (DHW) for buildings. These systems have enabled the use of low-temperature terminal units, such as radiators and radiant systems.

One of the great things about solar energy is the use of both simple and complex strategies to capture it and the utilization for space and water heating. There are two strategies for capturing the power of the sun: active and passive solar heating.

Passive solar systems require little, if any, nonrenewable energy to make them function [1]. Every building is passive in the sense that the sun tends to warm it by day, and it loses heat at night. Passive systems incorporate solar collection, storage, and distribution into the architectural design of the building. Passive solar heating and lighting design must consider the building envelope and its orientation, the thermal storage mass, and window configuration and design.

Active solar systems use either liquid or air as the collector fluid. Active systems must have a continuous availability of electricity to operate pumps and fans. A complete system includes solar collectors, energy storage devices, and pumps or fans for transferring energy to storage or to the load. Active solar energy systems have been combined with heat pumps for water and/or space heating. Freeman et al. [2] present information on performance and estimated energy saving for solar-heat pumps.

Solar Heating and Cooling Systems. http://dx.doi.org/10.1016/B978-0-12-811662-3.00005-0

A solar thermal system consists of a solar collector, a heat exchanger, storage, a backup system, and a load. The load can be for space heating or hot water. This system may serve for both space heating and DHW production.

This chapter provides a description of main types of solar space and water heating systems, concentrating on classifications, system components, and operation principles. It is also focused on active and combisystems. Important information on the space-heating/cooling load calculation and the selection of the solar thermal systems are discussed. The *f*-chart method applicable to evaluate space and water heating in many climates and conditions and *Transient System Simulation* (TRNSYS) program is briefly described. Additionally, some installation, operation, and maintenance instructions for solar heating systems are analyzed and examples of DHW systems and combisystems application are presented. Finally, valuable information on the solar district heating and solar energy use for industrial applications is provided.

5.2 SOLAR WATER HEATING SYSTEMS

The share of energy used to heat the DHW has been increasing significantly as the thermal performance of building an envelope has been improving. The heating energy for DHW depends on the use of water, which has decreased significantly during the last few decades. Twenty five years ago the average use was around 200 l/day/pers in residential buildings; now it is 120−140 l/day/pers. The reduction is partly due to water-saving faucets, new washing methods and, even more importantly, the common use of a shower instead of a bath tub for a better personal hygiene. About 40% of the domestic water is used as hot water. It can correspond to 30−40 kWh/m^2 of energy use in residential buildings, which is significant (25%) in relation to energy use in the EU buildings that is typically in the range of 100−150 kWh/m^2. The shares are even higher in low-energy and nearly zero-energy buildings: approximately 50% in single-family buildings and more than 50% in multifamily buildings.

Solar heating of DHW is, in most cases, the most cost-effective use of renewable energy. Water use profiles affect the sizing of the collector, storage tanks, and backup heating. An important issue to keep in mind with solar water heating systems is the possibility of low-temperature level (below 55°C) in the system, which may allow the growth of *Legionella* bacteria in the plumbing system.

A solar water heating system (Fig. 5.1) includes a solar collector that absorbs solar radiation and converts it to heat, which is then absorbed by a heat transfer fluid (HTF) (water, antifreeze, or air) that passes through the collector. The HTF is stored or used directly.

For the storage water heaters, the required heating rate Q_{hw}, in W, can be computed using equation:

$$Q_{hw} = \frac{G_{hw}\rho c_p \Delta t}{\eta},$$

(5.1)

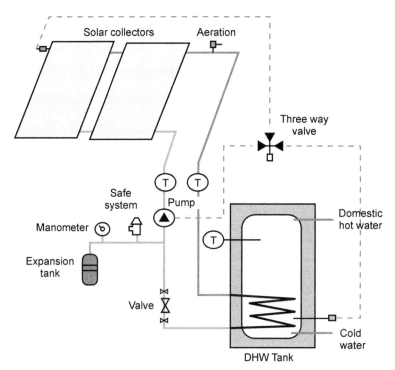

FIGURE 5.1   Components of water heating system.

where G_{hw} is the maximum flow rate of hot water, in m^3/s; ρ is the water density, in kg/m^3; c_p is the specific heat of water, in J/(kg K); Δt is the temperature rise, in K; and η is the heater efficiency. An exact calculation of the heat demand for DHW can be made using international or national norms.

The volume of hot-water storage tank V_{ST}, in m^3, can be computed using equation

$$V_{ST} = \Delta\tau \frac{Q_{hw}}{\rho c_p \Delta t}, \tag{5.2}$$

where $\Delta\tau$ is the heating time of water, in s.

The volume of DHW storage tank V_{ST}, in liters, can be expressed as

$$V_{ST} = \frac{2V_{dhw}N_p(t_{dhw} - t_{cw})}{t_{ST} - t_{cw}}, \tag{5.3}$$

where V_{dhw} is the specific DHW volume, in liters/pers; N_p is the person number; t_{dhw} is the DHW temperature, in °C; t_{ST} is the water temperature in storage tank, in °C; and t_{cw} is the cold water temperature, in °C.

Portions of the solar energy system are exposed to the weather, so they must be protected from freezing. The system must also be protected from overheating caused by high insolation levels during periods of low energy demand.

5.2.1 Types of Solar Water Heating Systems

In a solar water heating system, water is heated directly in the collector or indirectly by an HTF that is heated in the collector, passes through a heat exchanger, and transfers its heat to the domestic or service water. The HTF is transported by either natural or forced circulation. Natural circulation occurs by natural convection (thermosiphon), whereas forced circulation uses pumps or fans. Except for thermosiphon system which need no control, solar domestic and service water heaters are controlled by differential thermostats.

Five types of solar systems are used to heat domestic and service hot water: thermosiphon, direct circulation, indirect circulation, integral collector storage, and site built. Recirculation and drain down are two methods used to protect direct solar water heaters from freezing. Drains down systems are direct-circulation water heating systems in which potable water is pumped from storage to the collector array where it is heated. Circulation continues until usable solar heat is no longer available. When a freezing condition is anticipated or a power outage occurs, the system drains automatically by isolating the collector array and exterior piping from the city water pressure and using one or more valves for draining.

5.2.1.1 Direct and Indirect Systems

Direct or *open loop* systems (Fig. 5.2) circulate potable water through the collectors. They are relatively cheap but can have the following drawbacks:

- They offer little or no overheat protection unless they have a heat export pump;
- They offer little or no freeze protection, unless the collectors are freeze-tolerant; and

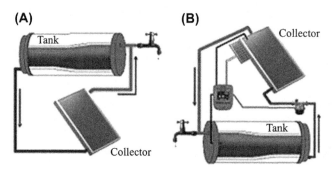

FIGURE 5.2   Direct systems: (A) passive system with tank above collector; (B) active system with pump and controller driven by a PV panel.

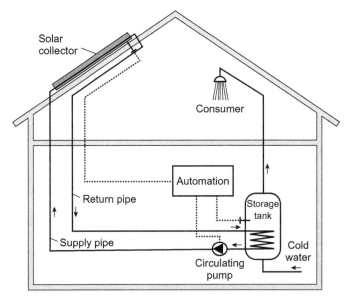

FIGURE 5.3   Indirect DHW heating system.

- Collectors accumulate scale in hard-water areas, unless an ion-exchange softener is used.

Until the advent of freeze-tolerant solar collectors, they were not considered suitable for cold climates since, in the event of the collector being damaged by a freeze, pressurized water lines will force water to gush from the freeze-damaged collector until the problem is noticed and rectified.

Indirect or *closed loop* systems (Fig. 5.3) use a heat exchanger that separates the potable water from the HTF that circulates through the collector. The two most common HTFs are water and an antifreeze-water mix that typically uses nontoxic propylene glycol. After being heated in the panels, the HTF travels to the heat exchanger, where its heat is transferred to the potable water. Though slightly more expensive, indirect systems offer freeze protection and typically offer overheat protection as well.

To produce DHW with temperature of 45°C from cold water at temperature of 10°C, the absorber plate must be reach the temperature of 50−70°C to transfer efficiently the heat to HTF and DHW.

5.2.1.2 Passive and Active Systems

Passive systems (Fig. 5.2A) rely on heat-driven convection or heat pipes to circulate water or heating fluid (heat carrier) in the system. Passive solar water-heating systems cost less and have extremely low or no maintenance, but the efficiency of a passive system is significantly lower than that of an active system. Overheating and freezing are major concerns.

Active systems (Fig. 5.2B) use one or more pumps to circulate water and/or heating fluid in the system. Though slightly more expensive, active systems offer several advantages:

- The storage tank can be situated lower than the collectors, allowing increased freedom in system design and allowing preexisting storage tanks to be used;
- The storage tank can be hidden from view;
- The storage tank can be placed in conditioned or semi-conditioned space, reducing heat loss;
- Drain-back tanks can be used;
- Superior efficiency; and
- Increased control over the system.

Modern active solar water systems have electronic controllers that offer a wide range of functionality, such as the modification of settings that control the system, interaction with a backup electric or gas-driven water heater, calculation and logging of the energy saved by solar water heating system, safety functions, remote access, and informative displays, such as temperature readings.

The most popular pump controller is a *differential controller* (DC) that senses temperature differences between the water leaving the solar collector and the water in the storage tank near the heat exchanger. In a typical active system, the controller turns the pump on when the water in the collector is approximately 8−10°C warmer than the water in the tank and it turns the pump off when the temperature difference approaches 3−5°C. This ensures the water always gains heat from the collector when the pump operates and prevents the pump from cycling on and off too often.

Some active solar water heating systems use energy obtained by a small PV panel to power one or more variable-speed DC-pump(s). To ensure proper performance and longevity of the pump(s), the DC-pump and PV panel must be suitably matched. Some PV-pumped solar thermal systems are of the antifreeze variety and some use freeze-tolerant solar collectors. The solar collectors will almost always be hot when the pump(s) are operating (i.e., when the sun is bright), and some do not use solar controllers. Sometimes, however, a DC is used to prevent the operation of the pumps when there is sunlight to power the pump but the collectors are still cooler than the water in storage. One advantage of a PV-driven system is that solar hot-water can still be collected during a power outage if the sun is shining.

An active solar water heating system can be equipped with a bubble pump instead of an electric pump. A bubble pump circulates the HTF between collector and storage tank using solar power, without any external energy source, and is suitable for FPC as well as TTC. In a bubble pump system, the closed HTF circuit is under reduced pressure, which causes the liquid to boil at low temperature as it is heated by the sun. The HTF typically arrives at the

FIGURE 5.4   Integrated collector storage system.

heat exchanger at 70°C and returns to the circulating pump at 50°C. In frost-prone climates the HTF is waterpropylene glycol solution, usually in the ratio of 60—40%.

5.2.1.3 Passive Direct Systems

Integral collector storage (ICS) system (Fig. 5.4) uses a tank that acts as both storage and solar collector. Batch heaters are basically thin rectilinear tanks with a glass-side facing the position of the sun at noon. They are simple and less costly than plate and tube collectors, but they sometimes require extra bracing if installed on a roof, suffer from significant heat loss at night since the side facing the sun is largely uninsulated, and are only suitable in moderate climates [3].

Convection heat storage (CHS) system is similar to an ICS system, except that the storage tank and collector are physically separated and transfer between the two is driven by convection. CHS systems typically use standard FPCs or ETCs, and the storage tank must be located above the collectors for convection to work properly. The main benefit of a CHS system over an ICS system is that heat loss is largely avoided since (1) the storage tank can be better insulated, and (2) since the panels are located below the storage tank, heat loss in the panels will not cause convection, as the cold water will prefer to stay at the lowest part of the system.

5.2.1.4 Active Indirect Systems

Pressurized antifreeze systems use a water-antifreeze (glycol) solution for HTF in order to prevent freeze damage. Though effective at preventing freeze damage, antifreeze systems have many drawbacks:

- If the HTF gets too hot (e.g., when the homeowner is on vacation) the glycol degrades into acid. After degradation, the glycol not only fails to provide freeze protection, but also begins to eat away at the solar loop's components—the collectors, the pipes, the pump, etc. Due to the acid and excessive heat, the longevity of parts within the solar loop is greatly reduced.

- The waterglycol HTF must be replaced every 3—8 years, depending on the temperatures it has experienced.
- Some jurisdictions require double-walled heat exchangers even though propylene glycol is nontoxic.
- Even though the HTF contains glycol to prevent freezing, it will still circulate hot water from the storage tank into the collectors at low temperatures (e.g., below 4°C), causing substantial heat loss.

Drain-back systems are generally indirect active systems that circulate the HTF (water or waterglycol solution) through the closed collector loop to a heat exchanger, where its heat is transferred to the potable water. Circulation continues until usable energy is no longer available. When the pump stops, HTF drains by gravity to a storage or tank (Fig. 5.5). In a pressurized system, the tank also serves as an expansion tank, so it must have a temperature- and pressure relief valve to protect against excessive pressure. In an unpressurized system, the tank is open and vented to the atmosphere [4].

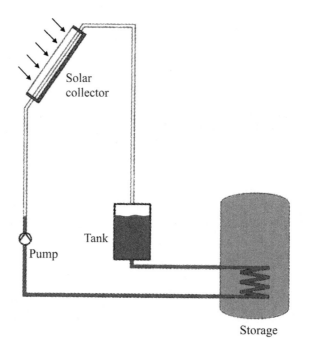

FIGURE 5.5   Operation principle of drain-back system.

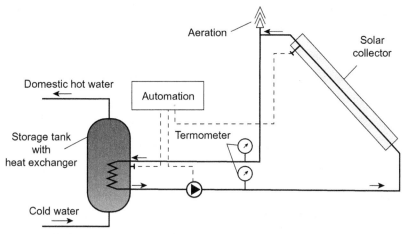

FIGURE 5.6   Closed circuit with storage tank and heat exchanger system.

5.2.2 Examples of Solar DHW Systems

Technique development in the field of solar energy in the last 20–25 years has generated the emergence of a diversified range of DHW solar systems. As an example, three constructive variants used in practice for closed circuit and with heat exchanger solar systems are presented [5]:

- The standard variant for a DHW solar system is presented in Fig. 5.6. The solution is the simplest and cheapest system with forced circulation, thus being the most common installation. The circulating pump transports the HTF between solar collector and heat exchanger in the storage tank (coil), when the fluid temperature in solar collector is higher than the DHW temperature in storage tank;
- For medium and large plants, two lower volume storage tanks instead of a large-volume one are used, and a three-way valve, driven depending on HTF and tank water temperature (Fig. 5.7), is used to control the water heating in the two storage tanks. This constitutes an advantageous operational solution (variable consumptions). The storage tanks can be both for DHW or one for DHW and another one for heating (preheating) of distribution fluid (heat carrier) in a heating system;
- Another constructive variant is represented by solar collector use for domestic water heating as well as for heating of swimming pool water by means of a heat exchanger (Fig. 5.8). For each square meter of swimming pool with normal depth are necessary 0.5–0.7 m^2 of solar collector.

5.3 SOLAR SPACE-HEATING SYSTEMS

Because of the reduction of the world fossil-fuel reserves and strict environmental protection standards, one main research direction in the construction

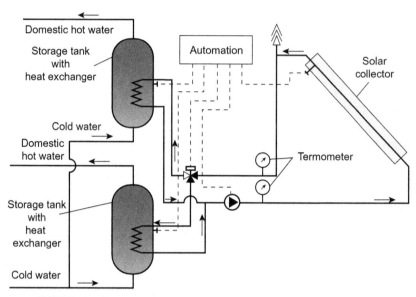

FIGURE 5.7   Closed circuit with two storage tanks and heat exchanger system.

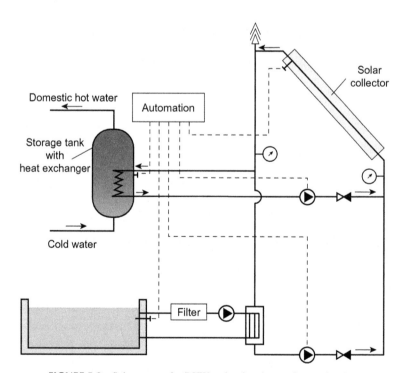

FIGURE 5.8   Solar system for DHW and swimming pool water heating.

field has become the reduction of energy consumption, including materials, technology, and building plans with lower specific energy need, on one hand, and equipment with high performance on the other hand.

Thermal energy obtained from the sun with a solar thermal system can be used for space heating. The solar heating systems fall into two principal categories: passive and active.

Passive systems may be divided into several categories. In a *direct-gain* passive system, the solar collector (windows) and storage (e.g., floors, walls) are part of the occupied space and typically have the highest percent of the heating load met by the solar. To reduce heat losses, thermal mass should be well insulated from the outdoor environment or ground [4]. Thermal mass can be defined as a material's ability to absorb, store, and release heat. When the sun's rays enter the building in the winter months, the heat energy is absorbed by the materials inside the building that have a high thermal mass. This would include materials that are dense, such as stone, brick and concrete, or ceramic tile. These materials absorb and hold onto heat during the period of time that the sun shine on them and then slowly release that heat throughout the nighttimes, keeping the home at a more stable and comfortable temperature. *Indirect-gain* passive systems use the south-facing wall surface or the roof to absorb solar radiation, which causes a rise in temperature that, in turn, conveys heat into the building in several ways. The glazing reduces heat loss from the wall back to the atmosphere and increases the system's collection efficiency. In another indirect-gain passive system, a metal roof/ceiling supports transparent plastic bags filled with water. Movable insulation above these water-filled bags is rolled away during the winter day to allow the sun to warm the stored water. The water then transmits heat indoors by convection and radiation.

Active solar space-heating systems (Fig. 5.9) use solar energy to heat an HTF (liquid or air) in collector circuit and then transfer the solar heat directly to the interior space or to a storage tank for later use. Liquid systems are more often used when storage is included, and are well suited for radiant heating systems, boilers with hot-water radiators, and even absorption heat pumps. Thus, the distribution fluid (hot water) that is used by existing heating system (e.g., radiator, radiant floor) is circulated through a heat exchanger in the storage tank. As the hot water passes through the heat exchanger, it is warmed, and then returned to a heating system. If the solar system cannot provide adequate space heating, an auxiliary or backup system provides the additional heat. Both liquid and air systems can supplement forced air systems.

5.3.1 Components and Control of System

Solar space-heating systems are designed to provide large quantities of hot water for residential and commercial buildings. A typical system includes several components:

- *Solar collectors* absorb sunlight to collect heat. There are different types of solar collectors described in Chapter 3.

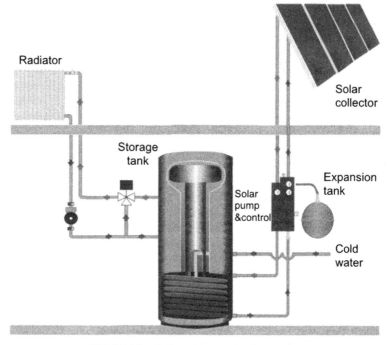

FIGURE 5.9   Active solar space-heating system.

- *Pumps* are used in active systems to circulate the HTF through the solar collectors. A controller is used to turn the pump on when the sun is out and off for the rest of the time. Pump size depends on the size of the system and the height the water has to rise. A good pump will last at least 20 years.
- *Heat exchanger* is used in closed-loop systems to transfer the heat from HTF to distribution fluid (hot water) without mixing the liquids.
- *Storage tank* holds the hot water. The tank can be a modified water heater, but it is usually larger and very well insulated. Systems that use fluids other than water usually heat the water by passing it through a coil of tubing in the tank, which is full of hot fluid. Specialty or custom tanks may be necessary in systems with very large storage requirements. They are usually stainless steel, fiberglass, or high-temperature plastic. Concrete and wood (hot tub) tanks are also options.
- *Monitoring system* used by system owners to measure and track system performance.

Two types of control schemes are commonly used on solar collectors on building-scale applications: on–off and proportional. With an on–off controller, a decision is made to turn the circulating pumps on or off depending on whether or not useful output is available from the collectors. With a proportional controller, the pump speed is varied in an attempt to maintain a specified temperature level at the collector outlet. Both strategies have advantages and disadvantages, largely depending on the ultimate use of the collected energy.

The most common control scheme requires two temperature sensors, one in the bottom of the storage unit and one on the absorber plate at the exit of a collector (or on the pipe near the plate). Assume the collector has low heat capacity. When HTF is flowing, the collector transducer senses the exit fluid temperature. When the fluid is not flowing, the mean plate temperature t_p is measured. A controller receives this temperature and the temperature at the bottom of the storage unit. This storage temperature will be called $t_{s,i}$; when the pump turns on, the temperature at the bottom of storage will equal the inlet fluid temperature if the connecting pipes are lossless. Whenever the plate temperature at no-flow conditions exceeds $t_{s,i}$ by a specific amount Δt_{on}, the pump is turned on.

When the pump is on and the measured temperature difference falls below a specified amount Δt_{off}, the controller turns the pump off. Care must be exercised when choosing both Δt_{on} and Δt_{off} to avoid having the pump cycle on and off.

The turn-off criterion must satisfy the following inequality or the system will be unstable [6]:

$$\Delta t_{off} \leq \frac{A_c F_R U_L}{mc_p} \Delta t_{on}, \qquad (5.4)$$

where A_c is the solar collector area, in m^2; F_R is the collector heat removal factor; U_L is the overall heat loss coefficient, in $kW/(m^2 K)$; m is the mass flow rate of fluid, in kg/s; and c_p is the specific heat of fluid, in $kJ/(kg\ K)$.

5.3.2 Types of Solar Space-Heating Systems

There are two basic types of active solar heating systems: those that use hydronic collectors, and those that use air collectors. The collector is the device that is exposed to the sun, and warms air or a fluid for use by the system.

Hydronic (Liquid-Based) Systems. These systems are an extension of solar water heating systems. Hydronic collectors warm a liquid (water or an antifreeze solution) by circulating them through a manifold that is exposed to the sun. Solar liquid collectors are most appropriate for central heating.

Air Systems. These systems are designed to collect and transfer heat to warm the air circulating in the building. As the name suggests, air collectors similarly pass air through a sun-exposed collector, rather than a liquid.

Both hydronic and air systems absorb and collect solar radiation to heat a liquid or air and then use the heated liquid or air to transfer heat directly to an interior building space or to a storage system from which the heat can be distributed.

5.3.3 Selection and Thermal Load of a Solar Heating System

Selecting the appropriate solar heating system depends on factors such as the site, design, and heat demand of building.

The local climate, the type and efficiency of the collector(s), and the collector area determine how much heat a solar heating system can provide. It is usually most economical to design an active system to provide 40–80% of the building's heat demand.

To determine solar process dynamics, load dynamics must be known. There are no general methods for predicting the time dependence of loads; detailed information on heat requirements (energy and temperature), hot-water requirements, cooling needs, and so on, must be available. Lacking data, it may be possible to assume typical standard load distributions. A commonly used DHW load pattern is shown in Fig. 5.10 [7].

Hot-water use is concentrated in the morning and evening hours; to supply these loads, stored solar energy must be used. Minor changes in time dependence of loads do not have major effect on annual performance of solar DHW systems.

Very detailed models of daily DHW use have been proposed that include such variables as the number of occupants and their ages, the presence or absence of dishwashers and clothes washers, the size of storage tank, the season of the year, the ambient and delivery temperatures, and whether or not the occupants pay for the hot water [8].

The loads to be met by water heating systems can be considered to include three parts, the sensible heat requirements of the water, loss from storage tanks, and loss from the distribution system:

$$Q_{wh} = mc_p(t_{s,o} - t_{s,i}) + (UA)_{ST}(t_{ST} - t_a) + (UA)_{flow}L_{pipe}\Delta t_m \qquad (5.5)$$

in which:

$$\Delta t_m = \frac{(t_{in} - t_a) - (t_{out} - t_a)}{\ln[(t_{in} - t_a)/(t_{out} - t_a)]}, \qquad (5.6)$$

where Q_{wh} is the water-heating load, in kW; m is the mass flow rate of fluid, in kg/s; c_p is the specific heat of fluid, in kJ/(kg K); $t_{s,o}$ is the delivery temperature of hot water, in °C; $t_{s,i}$ is the supply temperature of storage tank, in °C; $(UA)_{ST}$ is the tank-loss coefficient-area product, in kW/K; t_{ST} is the water temperature in

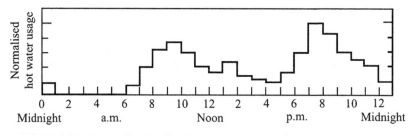

FIGURE 5.10   A normalized profile of hourly hot-water use for a domestic application.

the tank, in °C; t_a is the ambient air temperature, in °C; $(UA)_{flow}$ is the flowing heat loss coefficient per meter of pipe, in kW/(m·K); L_{pipe} is the length of hot-water pipe, in m; Δt_m is the log-mean temperature difference, in K; t_{in} is the water temperature entering pipe, in °C; and t_{out} is the water temperature leaving pipe, in °C.

The determination of the heat demand for space heating can be performed according to European standard EN 12,831 or national standards—for example, DIN 4701 and EnEV 2002 (Germany); ÖNORM 7500 (Austria); SIA 380-1 and SIA 384.2 (Switzerland); and SR 1907 (Romania). The determination of the cooling demand can be performed according to European standard EN 15,243 or national standards, for example VDI 2078 (Germany) and SR 6648-1,2 (Romania).

To estimate the space-heating load, *the degree-day method* can be used [9], which is based on the principle that the energy loss from a building is proportional to the difference in temperature between indoors and outdoors.

The degree-day method and its generalizations can provide a simple estimate of annual loads, which can be accurate if the indoor temperature and internal gains are relatively constant and if the heating or cooling systems operate for a complete season. The balance point temperature t_{ech} of a building is defined as that value of the outdoor temperature t_a at which, for the specified value of the indoor temperature t_i, the total heat loss is equal to the heat gain Q_{hg} from the sun, occupants, lights, and so forth, as

$$Q_{hg} = U_b(t_i - t_{ech}), \tag{5.7}$$

where U_b is the heat transfer coefficient of the building, in W/K.

Heating is needed only when t_a drops below t_{ech}. The rate of energy consumption of the space-heating system is

$$Q_{sh} = \frac{U_b}{\eta}[t_{ech} - t_a(\tau)]_{t_a < t_{ech}}, \tag{5.8}$$

where η is the efficiency of the heating system; τ is the time.

If η, t_{ech}, and U_b are constants, the annual heating energy E_h, in W, can be written as an integral

$$E_h = \frac{U_b}{\eta} \int [t_{ech} - t_a(\tau)]_+ dt, \tag{5.9}$$

where the plus sign (+) below the bracket indicates that only positive values are counted.

This integral of the temperature difference conveniently summarizes the effect of outdoor temperature on a building. In practice, it is approximated by summing averages over short-time intervals (daily) and the result N_h, in (K·days) is called degree-days as

$$N_h = (1 \text{ day}) \sum_{days} (t_{ech} - t_a). \tag{5.10}$$

Here, the summation is to extend over the entire year or over the heating season. The balance point temperature t_{ech} is also known as the base of the degree-days. In terms of degree-days, the annual heating energy is

$$E_h = \frac{U}{\eta} N_h. \tag{5.11}$$

Cooling degree-days can be calculated using an equation analogous to Eq. (5.10) for heating degree-days as

$$N_{cool} = (1 \text{ day}) \sum_{days} (t_a - t_{ech}). \tag{5.12}$$

The variable-base model is used since the balance point temperature varies widely from one building to another due to widely differing personal preferences for thermostat settings and setbacks and because of different building characteristics. The basic idea is to assume a typical probability distribution of temperature data, characterized by its average $\bar{t}_{aj}$ and by its standard deviation σ. Erbs et al. [10] developed a model that needs as input only the average $\bar{t}_{aj}$ for each month of the year. The standard deviation σ_j, in °C, for each month is then estimated from the correlation

$$\sigma_j = 1.45 - 0.029 \bar{t}_{aj} + 0.0664 \sigma_{an} \tag{5.13}$$

in which

$$\sigma_{an} = \sqrt{\frac{1}{12} \sum_{j=1}^{12} \left(\bar{t}_{aj} - \bar{t}_{a,an}\right)^2} \tag{5.14}$$

where σ_{an} is the standard deviation of the monthly temperature about the annual average $\bar{t}_{a,an}$.

The monthly heating degree-days $N_{h,j}$ for any location are well approximated by [9] as

$$N_{h,j} = \sigma_j n^{1.5} \left[\frac{\theta_j}{2} + \frac{\ln\left(e^{-a\theta_j} + e^{a\theta_j}\right)}{2a} \right] \tag{5.15}$$

in which

$$\theta_j = \frac{t_{ech} - \bar{t}_{aj}}{\sigma_j \sqrt{n}}, \tag{5.16}$$

where θ_j is a normalized temperature variable, n is the number of days in the month, and $a = 1.698$.

The annual heating degree-days can be estimated with equation

$$N_h = \sum_{j=1}^{12} N_{h,j}. \tag{5.17}$$

The computer program GRAZIL has been elaborated based on a variable-base model for personal computer (PC) microsystems, using the Engineering Equation Solver (EES) software package.

Systems providing less than 40% of heat demand are rarely cost-effective except when using solar air heater collectors that heat one or two rooms and require no heat storage. A well-designed and insulated building that incorporates passive solar heating techniques will require a smaller and less costly heating system of any type, and may need very little auxiliary heat other than solar.

Besides the fact that designing an active system to supply enough heat 100% of the time is generally not practical or cost-effective, most building codes and mortgage lenders require a backup heating system. Auxiliary or backup systems supply heat when the solar system cannot meet heating demands. Backups can range from a wood stove to a conventional central heating system.

5.3.4 Designing and Simulating Solar Heating Systems

5.3.4.1 Design Methods

Design methods for solar thermal systems can be put in three general categories, according to the assumptions on which they are based and the ways in which the calculations are done.

The first category applies to systems in which the collector operating temperature is known or can be estimated and for which critical radiation levels can be established. An example in this category is the utilizability method, based on analysis of hourly weather data to obtain the fraction of the total month's radiation that is above a critical level.

The second category of design methods includes those that are correlations of the results of a large number of detailed simulations such as f-chart method [11].

The third category of design methods is based on short-cut simulations. In these methods, simulations are done using representative days of meteorological data and the results are related to longer term performance.

The *f-chart method* provides a means for estimating the fraction of a total heating load that will be supplied by solar energy for a given solar heating system. The f-chart method requires the following data:

- monthly average daily radiation on a horizontal surface;
- monthly average ambient air temperatures;
- collector thermal performance curve slope and intercept from standard collector tests ($F_R U_L$ and $F_R(\alpha\tau)_n$); and
- monthly space- and water-heating loads.

The method is a correlation of the results of many hundreds of thermal performance simulations of solar heating systems. The resulting correlations give f, the fraction of the monthly heating load (for space heating and hot water) supplied by solar energy as a function of two dimensionless parameters: one, X is related to the ratio of collector losses to heating loads, and the other, Y is related to the ratio of absorbed solar radiation to heating loads.

The f-charts assume three standard system configurations [4,6], liquid and air systems for space (and hot water) heating and systems for DHW only. The system configurations that can be evaluated by the f-chart method are expected to be common in residential applications.

Computer simulations correlate dimensionless variables and the long-term performances of the systems. The fraction f of the monthly space- and water-heating loads supplied by solar energy is empirically related to two dimensionless groups. The two dimensionless groups are

$$X = \frac{F_R U_L A_c \Delta \tau}{Q_L} \frac{F_r}{F_R} (t_{ref} - \bar{t}_a) \tag{5.18}$$

$$Y = \frac{F_R (\alpha\tau)_n \bar{I}_T N A_c}{Q_L} \frac{F_r}{F_R} \frac{(\overline{\alpha\tau})}{(\alpha\tau)_n}, \tag{5.19}$$

where F_R is the collector heat removal factor; F_r is the collector heat-exchanger efficiency factor; U_L is the collector overall heat-loss coefficient, in $W/(m^2 \cdot K)$; A_c is the collector area, in m^2; $\Delta\tau$ is the total number of seconds in month; Q_L is the monthly total heating load for space heating and hot water, in J; t_{ref} is the empirically derived reference temperature (100°C); $\bar{t}_a$ is the monthly average ambient air temperature, in °C; $\bar{I}_T$ is the monthly averaged daily radiation incident on collector surface per unit area J/m^2; N is the number of days in month; $(\overline{\alpha\tau})$ is the monthly average absorptancetransmittance product; and $(\alpha\tau)_n$ is the normal absorptancetransmittance product.

$F_R U_L$ and $F_R(\alpha\tau)$ are obtained from collector test results. The ratios F_r/F_R and $(\overline{\alpha\tau})/(\alpha\tau)_n$ are calculated using methods given by Beckman et al. [11]. The value of $\bar{t}_a$ is obtained from meteorological records for the month and location desired. $\bar{I}_T$ can be calculated from the monthly averaged daily radiation on a horizontal surface. The monthly load Q_L can be determined by the methods previously discussed. Value of the collector area A_c is selected for the calculations. Thus, all the terms in these equations can be determined from the available information.

The f-chart method for liquid systems is similar to that for air systems. The fraction of the monthly total heating load supplied by the solar air-heating system is correlated with dimensionless groups X and Y as:

- for air system:

$$f = 1.04Y - 0.065X - 0.159Y^2 + 0.00187X^2 - 0.0095Y^3 \tag{5.20}$$

- for liquid system:

$$f = 1.029Y - 0.065X - 0.245Y^2 + 0.0018X^2 - 0.025Y^3. \tag{5.21}$$

This is done for each month i of the year. The solar energy contribution for the month i is the product of f_i and the total heating load Q_{Li} for this month. Finally, the fraction of the annual heating load supplied by solar energy F is the sum of the monthly solar energy contributions divided by the annual load:

$$F = \frac{\sum f_i Q_{Li}}{\sum Q_{Li}} \tag{5.22}$$

The correlation YX both for the air and liquid systems was plotted in f-chart [4,11]. The value of f is determined at the intersection of X and Y on the f-chart.

The collector heat-removal factor F_R that appears in X and Y is a function of the collector fluid flow rate. Because of the higher cost of power for moving fluid through air collectors than through liquid collectors, the capacitance rate used in air heaters is ordinarily much lower than in liquid heaters. As a result, air heaters generally have a lower F_R. Values of F_R corresponding to the expected airflow in the collector must be used to calculate X and Y.

Increased airflow rate tends to improve collector performance by increasing F_R, but it tends to decrease performance by reducing the degree of thermal stratification in the pebble bed (or water storage tank). Eq. (5.20) for air systems is based on a collector airflow rate of 10 l/(s·m²). The performance with different collector airflow rates can be estimated by using the appropriate values of F_R in both X and Y. A further modification to the value of X is required to account for the change in degree of stratification in the pebble bed.

Air system performance is less sensitive to storage capacity than that of liquid systems for two reasons: (1) air systems can operate with air delivered directly to the building in which storage is not used, and (2) pebble beds are highly stratified and additional capacity is effectively added to the cold end of the bed, which is seldom heated and cooled to the same extent as the hot end. Eq. (5.20) for air systems is for a nominal storage capacity of 0.25 m³ of pebbles per m² of collector area, which corresponds to 350 kJ/(m²·°C). Performance of systems with other storage capacities can be determined by modifying the dimensionless group X as described in [11].

5.3.4.2 TRNSYS Simulation Program

The use of simulation methods in the study of solar processes is a relatively recent development. If simulations are properly used, they can provide a wealth of information about solar systems and thermal design.

Simulations are numerical experiments and can give the same kinds of thermal performance information as can physical experiments. They are, however, relatively quick and inexpensive and can produce information on the effect of design variable changes on the system performance by a series of numerical experiments all using exactly the same loads and weather. Once simulations have been verified with experiments, new systems can be designed with confidence using the simulation methods.

In principle, all of the physical parameters of collectors, storage, and other components are the variables that need to be taken into account in the design of solar systems. For the solar heating of a building, for example, the primary system design variable is collector area, with storage capacity and other design variables being of secondary importance provided they are within reasonable bounds of good design practice. Simulations can provide information on effects of collector area (or other variables), which is essentially impossible to get by other means.

The parallels between numerical experiments (simulations) and physical experiments are strong. In principle, it is possible to compute what it is possible to measure. In practice, it may be easier to compute than to measure some variables (e.g., temperatures in parts of a system that are inaccessible for placement of temperature sensors). Simulations can be arranged to subject systems to extremes of weather, loads, or other external forces.

Some of the programs that have been applied to solar processes have been written specifically for study of solar energy systems. Others were intended for no-solar applications but have had models of solar components added to them to make them useful for solar problems. The common thread in them is the ability to solve the combinations of algebraic and differential equations that represent the physical behavior of the equipment.

Over the past two decades, hundreds of programs have been written to study energy efficiency, renewable energy, and sustainability in buildings.

This subsection includes a brief description of TRNSYS program [12], originally developed for study of solar processes and its applications but it is used for simulation of a wider variety of thermal processes.

Subroutines are available that represent the components in typical solar energy systems. A list of the components and combinations of components in the TRNSYS library is shown in Table 5.1.

Users can readily write their own component subroutines if they are not satisfied with those provided. By a simple language, the components are *connected* together in a manner analogous to piping, ducting, and wiring in a physical system. The programmer also supplies values for all of the parameters describing the components to be used. The program does the necessary simultaneous solutions of the algebraic and differential equations, which represent the components and organizes the input and output. Varying levels of complexity can be used in the calculation. For example, a flat-plate collector can be represented by constant values of $F_R U_L$ and $F_R(\alpha\tau)$ or it can be represented by values of U_L and $(\alpha\tau)$, which are calculated at each time step for the conditions that change through time as the simulation proceeds.

Current versions of TRNSYS have, in the executive program, three integration algorithms. The one that is extensively used is the Modified-Euler method. It is essentially a first-order predictor corrector algorithm using Euler's method for the predicting step and the trapezoid rule for the correcting step. The advantage of a predictorcorrector integration algorithm for solving

TABLE 5.1 Components in Standard Library of TRNSYS

- Building loads and structures
  - Energy/(degree-hour) house
  - Roof and attic
  - Detailed zone (transfer function)
  - Overhang and wing wall shading
  - Window
  - Thermal storage wall
  - Attached sunspace
  - Detailed multizone building
  - Lumped-capacitance building
- Controller components
  - Differential controller (DC) with hysteresis
  - Three-stage room thermostat
  - Iterative feedback controller
  - Proportional-integral-differential (PID) controller (in heating)
  - Microprocessor controller
  - Five-stage room thermostat
- Electrical components
  - Shepherd and Hyman battery models
  - Regulators and inverters
  - Photovoltaic (PV) thermal collector
  - Wind energy conversion system (wind turbines)
  - Photovoltaic array
  - Diesel engine generator set (DEGS)
  - DEGS dispatch controller
  - Power conditioning
  - Lead-acid battery (with gassing effects)
  - Alternating-current bus bar
- Heat exchangers
  - Constant-effectiveness heat exchanger
  - Counter flow heat exchanger
  - Cross-flow
  - Parallel flow
  - Shell and tube
  - Waste heat recovery
- Heating, ventilation, and air-conditioning (HVAC) equipment
  - Auxiliary heaters
  - Dual-source heat pumps
  - Cooling coils (simplified and detailed models)
  - Conditioning equipment
  - Part-load performance
  - Cooling towers
  - Parallel chillers
  - Auxiliary cooling unit
  - Single-effect hot-water-fired absorption chiller
  - Furnace
- Hydrogen systems
  - Electrolyzer controller

Continued

TABLE 5.1 Components in Standard Library of TRNSYS—cont'd

- Master level controller for stand alone power systems
- Advanced alkaline electrolyzer
- Compressed gas storage
- Multistage compressor
- Fuel cells (PEM and alkaline)
- Hydronics
  - Pumps
  - Tee-piece, flow mixer, flow diverter, tempering valve
  - Pressure relief valve
  - Pipe
  - Duct
  - Fans
- Output devices
  - Printer
  - Histogram plotter
  - Simulation summarizer
  - Economics
  - Online plotter
- Physical phenomena
  - Solar radiation processor
  - Collector array shading
  - Psychrometrics
  - Hourly weather data generator
  - Refrigerant properties
  - Shading by external objects
  - Effective sky temperature calculation
  - Undisturbed ground temperature profile
  - Convective heat transfer coefficient calculation
- Solar collector components
  - Flat-plate solar collector
  - Thermosiphon collector with integral storage
  - Evacuated-tube solar collector
  - Performance map solar collector
  - Theoretical flat-plate solar collector
  - Compound parabolic trough (CPC) solar collector
- Thermal storage components
  - Stratified fluid storage tank
  - Rock bed thermal storage
  - Algebraic tank (plug flow)
- Utility components
  - Data file reader
  - Time-dependent forcing function
  - Quantity integrator
  - Load profile sequencer
  - Periodic integrator
  - Unit conversion routine
  - Calling Excel worksheets
  - Calling Engineering Equation Solver (EES) routines

TABLE 5.1 Components in Standard Library of TRNSYS—cont'd

- Parameter replacement
- Formatted file data reader (TRNSYS TMY, TMY2, Energy Plus)
- Variable-volume tank
- Detailed fluid storage tank with optional heaters and variable internal time step
- Input value recall
- Holiday calculator
- Utility rate scheduler
- Calling CONTAM
- Calling MATLAB
- Calling COMIS
- Weather data reading and processing
  - Standard format files
  - User format files

simultaneous algebraic and differential equations is that the iterative calculations occurring during a single time step are performed at a constant value of time (this is not the case for the Runge—Kutta algorithms). As a result, the solutions to the algebraic equations of the system converge, by successive substitution, as the iteration required solving the differential equation progresses.

Meteorological data, including solar radiation, ambient air temperature, and wind speed, influence collector performance, and are needed to calculate system performance over time. All simulations are done with past meteorological data, and it is necessary to select a data set to use in simulations. For studies of process dynamics, data for a few days or weeks may be adequate if they represent the range of conditions of interest. For design purposes, it is best to use a full year's data or a full season's data if the process is a seasonal one. Data are available for many years for some stations, and it is necessary to select a satisfactory set. The METEONORM program [13] has a database of over 7000 worldwide stations that can generate data on monthly, daily, or hourly time scales on surfaces of any orientation. An *Hourly Weather Data Generator* based on different algorithms is available in TRNSYS.

5.3.5 Installation, Operation and Maintenance Instructions for Solar Systems

Most solar space-heating systems are installed on the roof or on the ground adjacent to the building. For best results, solar collectors should be installed such that they

- receive direct sunlight between the hours of 9 a.m. and 3 p.m. year-round; and
- face south (although they can be oriented up to 30% toward the south-east or south-west).

Solar thermal systems are more forgiving than solar PV systems. Solar PV panels need to be installed facing south and are very sensitive to shading.

A solar thermal system requires moderate annual/periodic maintenance to ensure efficient operation. Maintenance is generally provided by the installer and/or the manufacturer. Periodic visual inspection may be necessary to properly maintain your solar system.

How well an active solar heating system performs depends on effective sitting, system design, and installation, and the quality and durability of the components. The collectors and controls now manufactured are of high quality. The biggest factor now is finding an experienced contractor who can properly design and install the system. A qualified contractor can conduct annual maintenance inspections.

5.4 SOLAR COMBISYSTEMS

5.4.1 System Description

High-energy consumption in buildings is an important problem for sustainable future and many researchers tried to solve this problem via different heating/cooling systems. A *solar combisystem* (SCS) is one of these systems, which provides both solar spaces heating/cooling as well as hot water from a common array of solar thermal collectors, usually backed up by an auxiliary nonsolar heat source [14]. When a geothermal heat pump is used, the combisystem is called geosolar. The experimental studies showed that SCSs are capable of providing energy demand from 10% to 100% depending on the climatic conditions, system components, system efficiencies, and energy demand [15]. Europe has most well-developed market for different solar thermal applications [16]. Thus, depending on the size of the SCS installed, the annual space-heating contribution can range from 10% to 60% or more in ultra-low energy buildings; even up to 100% where a large interseasonal thermal store or concentrating solar thermal heat is used. Table 5.2 summarize the major studies about SCS in the literature.

Solar combisystems may range in size from those installed in individual properties to those serving several in a block heating scheme. Those serving larger groups of properties district heating tend to be called central solar heating systems.

SCSs can be classified according to two main aspects: (1) by the *heat storage category* (the way in which water is added to and drawn from the storage tank and its effect on stratification); (2) by the *auxiliary heat management category* (the way in which auxiliary heaters can be integrated into the system). These categories are described in Tables 5.3 and 5.4.

The simplest combisystems, the type A, have no "controlled storage device." Instead they pump warm water from the solar collectors through radiant floor heating pipes embedded in the concrete floor slab. The floor slab is

TABLE 5.2 Some Studies About SCS in the Literature

Researchers	Year	Investigation Topic	Results
Weiss [17]	2003	Scientific book	Basic principles of the system
Andersen et al. [18]	2004	Thermal performance of SCS in different climatic conditions	Thermal performance of an SCS mostly affected by the balance between energy input and consumption (in Denmark)
Kacan and Ulgen [19]	2012		Energy saving ratio is observed between 59% and 89% monthly and annual fractional solar consumption (FSC) value is found as approx. 83%
Asaee et al. [20]	2014		FSC values are determined between 32% and 93% for different locations in Canada
Ellehauge and Shah [21]	2000	Different system design in the market	33–50% are used for DHW and the most common system design is two closed flow-cycles for both DHW and space heating
Drück and Hahne [22]	1998	Thermal performance of combistores	Creating a temperature distribution in storage tank facilitates the output of hot water for different purposes
Kacan [23]	2011	Ph.D. thesis	Energetic and exergetic efficiency results of the system are found by 1-min time intervals and FSC value is ranged between 10% and 100% annually
Hin and Zmeureanu [24]	2014	System optimization	The payback times of (5.8–6.6 years) different System configuration is found to be not acceptable

thickened to provide thermal mass so that the heat from the pipes (at the bottom of the slab) is released during the evening.

Tools for designing SCSs are available, varying from manufacturer's guidelines to nomograms to different computer simulation software (e.g., CombiSun [25], SHWwin [26]) of varying complexity and accuracy.

Fig. 5.11 shows one of the many systems for DHW and space heating [4]. In this case, a large, atmospheric pressure storage tank is used, from which water is pumped to the collectors by pump P_1 in response to the differential thermostat T_1. Drain back is used to prevent freezing, because the amount of

TABLE 5.3 Heat Storage Categories

Category	Description
A	No controlled storage device for space heating and cooling
B	Heat management and stratification enhancement by means of multiple tanks and/or by multiple inlet/outlet pipes and/or by three- or four-way valves to control flow through the inlet/outlet pipes
C	Heat management using natural convection in storage tanks and/or between them to maintain stratification to a certain extent
D	Heat management using natural convection in storage tanks and built-in stratification devices
B/D	Heat management by natural convection in storage tanks and built-in stratifies as well as multiple tanks and/or multiple inlet/outlet pipes and/or three- or four-way valves to control flow through the inlet/outlet pipes

TABLE 5.4 Auxiliary Heat Management Categories

Category	Description
M (mixed mode)	The space-heating loop is fed from a single store heated by both solar collectors and the auxiliary heater
P (parallel mode)	The space-heating loop is fed alternatively by the solar collectors (or a solar water storage tank), or by the auxiliary heater; or there is no hydraulic connection between the solar heat distribution and the auxiliary heat emissions
S (serial mode)	The space-heating loop may be fed by the auxiliary heater, or by both the solar collectors (or a solar water storage tank) and the auxiliary heater connected in series on the return line of the space-heating loop

antifreeze required would be prohibitively expensive. DHW is obtained by placing a heat exchanger coil in the tank near the top, where, even if stratification occurs, the hottest water will be found.

An auxiliary water heater boosts the temperature of the sun-heated water when required. Thermostat T_2 senses the indoor temperature and starts pump P_2 when heat is needed. If the water in the storage tank becomes too cool to provide enough heat, the second contact on the thermostat calls for heat from the auxiliary heater.

Water-to-air heat pumps, which use sun-heated water from the storage tank as the evaporator energy source, are an alternative auxiliary heat source.

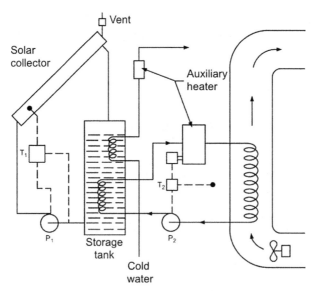

FIGURE 5.11 Schematic of solar system for DHW and space heating.

Storage-tank sizing is one of the essential problems of system optimization and determines annual SF (the annual solar contribution to the water-heating load divided by the total water-heating load). In Fig. 5.12 is presented, for two buildings, specific energy requirement q_{req} of auxiliary heat source, which has the value of 80 kWh/(m²·year) for the building with sheat demand of 100 kWh/(m²·year) and the value of 25 kWh/(m²·year) for the building with heat demand of 50 kWh/(m²·year).

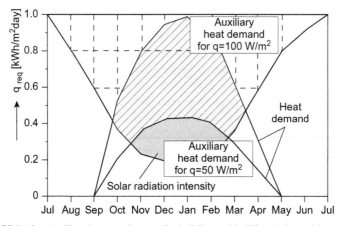

FIGURE 5.12 Auxiliary heat requirement for buildings with different thermal insulations.

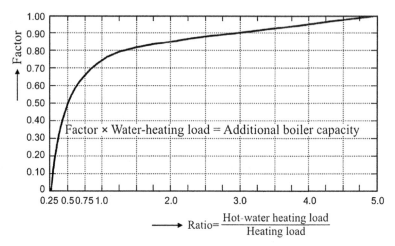

FIGURE 5.13   Sizing factor for combination heating and water-heating boilers.

When DHW is heated indirectly by a space-heating boiler, Fig. 5.13 [4] may be used to determine the additional boiler capacity required to meet the recovery demands of the domestic water-heating load. Indirect heaters include immersion coils in boilers as well as heat exchangers with space-heating media.

Because the boiler capacity must meet not only the water supply requirement but also space-heating loads, Fig. 5.13 indicates the reduction of additional heat supply for water heating if the ratio of water-heating load to space-heating load is low. The factor obtained from Fig. 5.13 is multiplied by the peak water-heating load to obtain the additional boiler output capacity required.

5.4.2 Examples of Solar Combisystems Application

Solar combisystems can be well used at isolated consumers for DHW production and building-heating. An efficient solar thermal system is one that combines solar passive heating with solar active heating and utilize long-time storage of collected heat by a buried thermal-insulated storage tank (Fig. 5.14). Collected heat is used in winter for space heating and DHW production.

Currently, a large number of installations are based on the use of solar energy in combination with the heat pump. For the most efficient use of solar energy throughout the year, the energy is stored in summer and is consumed during the winter. The heat from the sun is stored in water contained in thermal-insulated storage tanks.

The heating and hot-water system in a solar house (Fig. 5.15) from Essen, Germany, uses solar energy as a heat source through the flat-plate collector (5) [27].

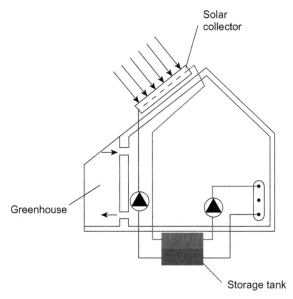

FIGURE 5.14   Solar DHW and heating system for isolated consumers.

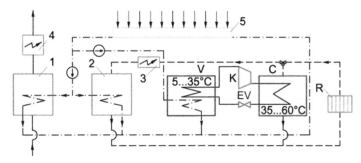

FIGURE 5.15   Solar-assisted water-to-water heat pump for heating and DHW: (1) DHW storage; (2) heating water storage; (3) additional electric heater; (4) DHW electric heater; and (5) flat solar collector.

The hot water from the solar collector circuit heats boiler (1) used for DHW production and boiler (2) used to heat the water for the radiators. When this system is not enough, a heat pump starts to operate using heated water from the solar collector, producing the hot water in condenser C for the heaters. The installation has the possibility to heat water with electric energy when this is not possible with solar energy.

Fig. 5.16 presents a radiant floor heating system with a heat pump and a vacuum tube solar collector, operating with water temperatures of 20−30°C [27]. A ground-water heat pump with horizontal collectors or vertical loops and

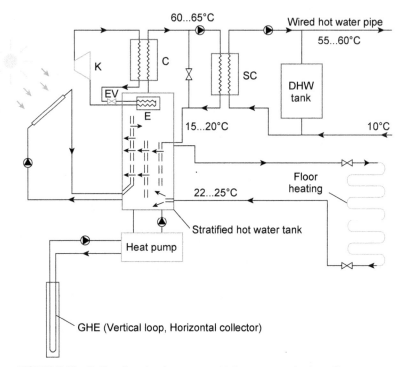

FIGURE 5.16   Radiant floor heating system with heat pump and solar collector.

solar collectors transfers its heat to a stratified hot-water tank. Reheating hot water is performed both by an additional heat pump and directly by an electric heater.

For a family house with a heat demand of approximately 8000 kWh/year, the solar collector can cover approximately 2000 kWh/year. If taking into account the circulation pumps' power, the electricity consumption of the heat pump reaches 1500 kW/year, and the coefficient of performance (COP) of the heat pump can reach a value of 4.

Initially, installation can be realized without the solar collector, leaving the possibility of installing one in the future. Instead, a heat pump for hot-water reheating can be mounted with an electric heater.

In heating-dominated climates, the single ground-coupled heat pump (GCHP) system may cause a thermal heat depletion of the ground, which progressively decreases the heat pump's entering fluid temperature. As a result, the system performance becomes less efficient. Similar to the cases of cooling-dominated buildings, the use of a supplemental heat supply device, such as a solar thermal collector, can significantly reduce the ground heat exchanger (GHE) size and the borehole installation cost. Fig. 5.17 shows the basic operating principle of the hybrid GCHP system with a solar collector.

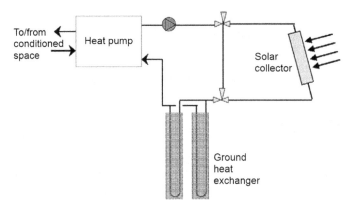

FIGURE 5.17   Schematic diagram of a GCHP system with solar collector.

The idea to couple a solar collector to the coil of pipes buried in the ground, by means of which solar energy can be stored in the ground, was first proposed by Penrod in 1956. Recently, a number of efforts have been made to investigate the performance and applications of the solar-assisted GCHP systems. Chiasson and Yavuzturk [28] presented a system simulation approach to assess the feasibility of the hybrid GCHP systems with solar thermal collectors in heating-dominated buildings. Yuehong et al. [29] conducted the experimental studies of a solar-ground heat pump system, where the heating mode is alternated between a solar energy-source heat pump and a ground-source heat pump with a vertical double-spiral coil GHE. Ozgener and Hepbasli [30] experimentally investigated the performance characteristics of a solar-assisted GCHP system for greenhouse heating with a vertical GHE.

A solar-assisted GCHP heating system with latent heat thermal energy storage (LHTES) was investigated by Zongwei et al. [31]. The hybrid heating system can implement eight different operation models according to the outdoor weather conditions by means of alternative heat source changes among the solar energy, ground heat, and the latent-heat thermal energy storage tank. Finally, it is claimed that the LHEST can improve the SF of the system, and thus the COP of the heating system can be increased.

5.5 SOLAR DISTRICT HEATING

Solar district heating is the provision of central heating and hot water from solar energy by a system in which the water is heated centrally by arrays of solar thermal collectors and distributed through district heating pipe networks (or in block or local heating systems by smaller installations). District heating is best suited to areas with a high building and population density in relatively cold climates.

For block systems, the solar collectors are typically mounted on the building roof tops. For district heating systems, the collectors may instead be installed on the ground.

Solar district heating systems can feed heat into substations in connected buildings or are directly connected to the district heating primary circuit on site.

Compared to small solar heating systems (SCSs), central solar heating systems have better priceperformance ratios due to the lower installation price, the higher thermal efficiency, and less maintenance.

District heating, considered as the most efficient method for building space heating, has been dramatically developed over the past decades. Some researchers focused on the combination of district heating and combined heat and power plants, renewable energy, industrial heat recovery, etc. [32−35].

5.5.1 Components of District Heating Systems

District heating systems consist of three primary components: the heating plant, the transmission and distribution network, and the consumer systems (Fig. 5.18).

The *central plant* (*district heating station*) may be any type of boiler, solar energy, a geothermal source, or thermal energy developed as a by-product of electrical generation. In plants serving hospitals, industrial customers, or those also generating electricity, steam is the usual choice for production in the plant and, often, for distribution to customers. For systems serving largely residential and commercial buildings, hot water is an attractive distribution medium.

The second component is the *transmission-and-distribution-pipe network* that conveys the energy. The piping is often the most expensive portion of

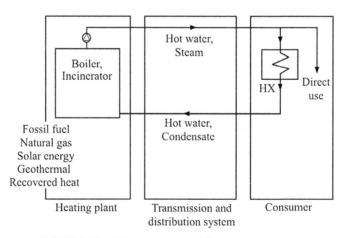

FIGURE 5.18   Major components of district heating system.

district heating and usually consists of a combination of preinsulated and field-insulated pipe in both concrete tunnel and direct burial applications. The capital investment for the transmission and distribution system, which usually constitutes most of the capital cost for the overall system, often ranges from 50% to 75% of the total cost for district heating system. Because the initial cost is high in distribution systems, it is important to optimize its use.

The third component is the *consumer system*, which includes in-building equipment. When hot water is supplied, it may be used directly by the building HVAC systems or indirectly where isolated by a heat exchanger (HX). When steam is supplied, it may be (1) used directly for heating, (2) directed through a pressure-reducing station for use in low-pressure (0−100 kPa) steam space heating and service water heating, or (3) passed through a steam-to-water heat exchanger.

5.5.2 Solar-Sourced District Heat

Heat sources in use for various district heating systems include power plants designed for combined heat and power (cogeneration or CHP), including both combustion and nuclear power plants and solar heat, geothermal heat or industrial heat pumps, which extract heat from ground, river, or lake water.

Use of solar heat for district heating has been increasing in Denmark and Germany in recent years [35]. The systems usually include interseasonal thermal energy storage for a consistent heat output day-to-day and between summer and winter. Good examples are in Vojens at 50 MW, Dronninglund at 27 MW, and Marstal at 13 MW in Denmark [36]. These systems have been incrementally expanded to supply 10−40% of their villages' annual space-heating needs. The solar-thermal panels are ground-mounted in fields. The heat storage is pit storage, borehole cluster, and the traditional water tank.

The majority of the large-scale plants supply heat to residential buildings in block and district heating systems. Typical operating temperatures range from below 30°C and above 100°C (water storage).

Most of these plants have no storage as they can utilize the district heating network as storage (as long as they provide a small amount of heat in comparison to the total load in the district heating system). Without storage, the system should cover not more than 90% of the summer load, to prevent stagnation of the plant due to lack of heat dissipation. By coverage of a high summer load and a time lag between heat production and heat dissipation, the integration of heat storage is necessary. The combination with a storage for a local hot-water supply of buildings or with a seasonal thermal storage for the store of solar energy collected in the summer time to the winter months are possible.

Water circulation in the closed circuit composed in the supply distribution network and return distribution network is assured by a circulating pump station located in the district heating plant.

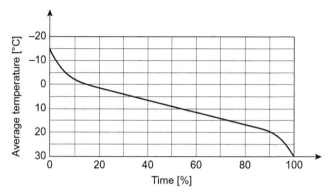

FIGURE 5.19   Daily average outdoor air temperature distribution throughout the year.

5.5.3 Energy-Saving Potential for Pumping Water in District Heating Plant

This subsection provides a comprehensive discussion about pump control in district heating plants and analyzes the energy efficiency of flow control methods.

5.5.3.1 Thermal Load of the Heating Plant

Thermal heat plants are designed to meet the consumers' energy need in the coldest period of the year. However, most of the energy demand is much lower than the designed value of the heating plant thermal load.

Fig. 5.19 shows the yearly distribution of the daily average outdoor air temperature [37]. The diagram shows that, throughout the year, the lowest temperatures represent approximately 5%. Thus, if the heating plant is designed to cover the maximum energy need, then for 95% of the year, the plant is oversized. At the same time, for approximately 25% of the year, the average temperature is higher than $+15°C$. Thus, for 25% of the year, the thermal energy provided by the heating plant is used only to produce DHW, which needs only a small portion of the installed capacity of the heating plant.

For a constant difference between the supply and return temperatures of the hot water, the delivered energy does not vary proportionally with the discharge. Generally, at constant supply temperature, the return temperature is lower when the heat demand decreases.

The main goal in operating a heating system is to provide the consumers with the heat demand according to outdoor climatic parameters. Thus, the heating system is provided with an adjustment system, which can be qualitative, quantitative, or mixed.

The quantitative adjustment requires varying the flow rate during operation, while keeping the hot-water parameters constant. This can be done by (1) fixed-speed pumps (FSPs) with different technical characteristics or (2) variable-speed pumps (VSPs).

Some heating plants comprise fixed-speed hydraulic pumps arranged in parallel and discharge controlled by the number of pumps in operation. The cost related to pump operation can be reduced by decreasing the energy consumption. An attractive alternative to reach this target is the use of VSPs instead of FSPs.

5.5.3.2 Solutions for Reducing the Pumping Energy

The power absorbed P, in kW, by a pump at a certain rotational speed is given by [38]

$$P = \frac{\gamma Q H_P}{1000\eta} = 3600 w_p Q \qquad (5.23)$$

in which

$$w_p = 0.00272 \frac{H_p}{\eta}, \qquad (5.24)$$

where γ is the specific weight of water, in N/m^3; Q is the pump flow rate, in m^3/s; H_p is the pump head for the operating point, in m; η is the general efficiency of the pumping station; and w_p is the specific pumping energy, in kWh/m^3.

Pump Operation at Fixed Speeds. Pumps used in heating plants are often centrifugal pumps, which are characterized by a head-flow (H-Q) curve that remains sufficiently flat for a wide range of flows. If the pump used is an FSP, the operating point is forced to move along the pump curve corresponding to the fixed nominal speed. There are two methods to vary the water discharge in a pipe network with an FSP:

- Bypassing part of the water discharge (the pump operates at the same pump head, water flow rate increases and the absorbed power also increases).
- Introducing a supplementary pressure loss using a control valve (the operating point is heading toward the left on the HQ curve) [39].

Although the water discharge adjustment using the control valve can lead to higher energy efficiency of the water distribution system when the nominal pump head is lower than the optimal value, this method has the following disadvantages:

- Increased wear of the control valve throttling elements;
- Noise, vibrations, and hydraulic impacts with negative effects in the system; and
- Low operation reliability of the pumps.

Pump Operation at Variable Speeds. The best method to obtain a variable water flow rate is the use of variable-speed pumps. The flow control (Fig. 5.20)

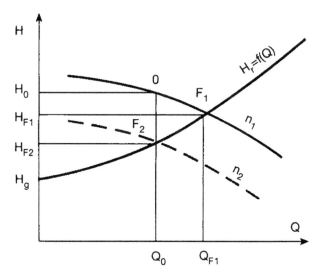

FIGURE 5.20 Flow adjustment using pump speed control.

is achieved by changing the pump curve H (at different pump speeds n_1 and n_2) on the fixed system curve H_r. Pipework curve H_r starts from point $(0, H_g)$, where H_g is the geodesic head. The operating point F_2 corresponds to the reduced pump head H_{F2}.

The relationships that link the pump characteristics (flow Q, head H, and power P) operating at different speeds (n_1, n_2) are described by the affinity laws as

$$\frac{Q_1}{Q_2} = \frac{n_1}{n_2} \tag{5.25}$$

$$\frac{H_1}{H_2} = \left(\frac{n_1}{n_2}\right)^2 \tag{5.26}$$

$$\frac{P_1}{P_2} = \left(\frac{n_1}{n_2}\right)^3. \tag{5.27}$$

The approximation introduced into the powerspeed relation implies that the efficiency will remain constant for speeds n_1 and n_2, that is, the efficiency curve will only be shifted to the left in the case of speed reduction. Eq. (5.28) provides an analytical relationship between two different speeds (n_1 and n_2) and the corresponding efficiencies (η_1 and η_2) as

$$\eta_2 = 1 - (1 - \eta_1)\left(\frac{n_1}{n_2}\right)^{0.1}. \tag{5.28}$$

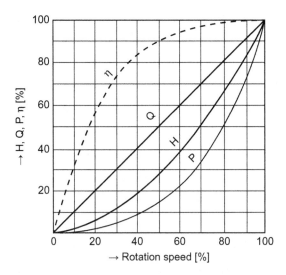

FIGURE 5.21 Variation of the centrifugal pump curves.

Therefore, as indicated by Sarbu and Borza [38], the changes in efficiency can be neglected if the changes in speed do not exceed 33% of the nominal pump speed; this approximation is particularly justified for large pumps. Fig. 5.21 shows the variation curves of H, Q, P, and η for centrifugal pumps depending on pump speed n. A 20% reduction of the pump speed will decrease the power demand by 50% at constant pump efficiency. Thus, the possibility exists to reduce the pumping energy consumption by using variable-speed drives (VSDs).

In systems with high friction loss, the most energy-efficient control option is an electronic VSD, commonly referred to as variable frequency drive (VFD). The most common form of VFD is the voltage-source, pulse-width modulated (PWM) frequency converter (often incorrectly referred to as an inverter). The principal duty of the VFD is to alter the main supply to vary the speed of the electric motor while delivering the required torque at a higher efficiency. As a result, as the pump speed changes, the pump curve is adjusted for different operating conditions. The main advantages of variable-speed pumps are seen when the operating conditions in the system are characterized by a high variability.

Pumping systems with a widely varying flow rate demand are often implemented using parallel pumps [40]. If several pumps are to operate in parallel, the rotational speed can be modified for a single pump (while the other pumps operate at nominal speed and nominal discharge), and the frequency converter automatically connects to the other pumps.

To correlate the pumped flow with the heat demand and to ensure the required pressure using minimum energy, an automatic pump speed control

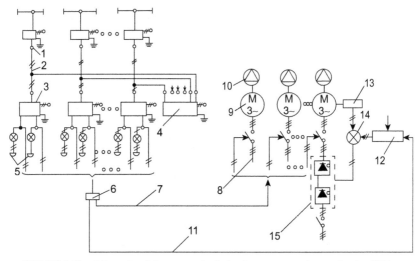

FIGURE 5.22 Schematic of the automatic device for pump speed control using VFDs.

device was designed (Fig. 5.22), which is composed of an asynchronous electric motor with a breakdown rotor associated with a static frequency converter (with thyristor). The command given by the pressure controller is taken by the frequency converter, assisted by a process computer.

Electronic pressure transducers FE1GM (1) dispatch the pressure from significant junctions of the distribution network by connecting lines (2), using electric signals of 2−10 mA c.c. or 4−20 mA c.c., to the regulator milli-ampermeter 1ARE192 (3) placed in the pumping station. Meanwhile, the electrical signals are concomitantly transmitted to an electronic recorder ELR362 A (4), in serial connection with the regulator milliampermeters, to ensure the continuous recording of a pressure diagram from 12 measuring points. Signaling boards (5) have signaling lamps and bells controlled by the regulator milliampermeters through over-and-under relays. The milli-ampermeters also use a continuousdiscontinuous programmer ELX73 (6) to control via connection lines (7) the coupling or decoupling of the electric motor (9) of the pumps (10) to the energy supply network (8). The con-tinuousdiscontinuous programmer is connected through lines (11) to a process computer (12), which assigns the minimum and maximum values of the required pressures in the system to ensure an optimal water supply at a low energy cost. Variable-speed motors are connected to the rotary transducers DT171 (13), which send their signals to the comparison-making element (14). This element takes into consideration the signal's value and controls the fre-quency converter CSFV (15) accordingly; the converter operates with one or more of the electric motors of the pumps.

The frequency converter enables a primary adjustment by coupling or decoupling pumps and a secondary adjustment in the connecting intervals by modifying the speed of the one pump.

The highest efficiency in VSP systems can be achieved if all of the system components (including the motor and VFD) have a higher efficiency at the operating points. The general expression of VSP system efficiency is given by [41]

$$\eta = \eta_m \eta_{VFD} \eta_p, \qquad (5.29)$$

where η_m is the motor efficiency; η_{VFD} is the efficiency of the variable-speed drive; and η_p is the pump efficiency.

Variable Frequency Drives. The VFDs (or variable-speed controllers) can control motor speed by varying the effective voltage and frequency, which are obtained from a fixed voltage and fixed frequency three-phase mains supply. Brushless permanent magnet (PM) motors cannot operate without a VFD due to their structure. However, from a cost and complexity point of view, there is not much of a difference between the VFDs used in induction motors (IMs) and brushless PM motors for the same power ratings.

Because a VFD can operate the motor over a wide range of speeds, knowing the efficiency values of the VFD/motor combination at each operating point is important to accurately determine the operating point of a pump. Stockman et al. [42] demonstrated that the brushless PM motor/VFD has far better efficiency values compared to the IM/VFD.

It should be noted that in variable-drive systems, additional losses are generated in the motor by the variable-frequency drive.

5.5.3.3 Throttling Control Valve Versus Variable-Speed Drive—Case Studies

Energy Efficiency of the Adjustment Methods. In Timisoara, Romania, the central heating system has equipment designed to meet the heat demand of the city. However, the equipment was produced in 1970s or 1990s, so their normal use does not meet current standards on energy efficiency or environmental protection. The European community requires that these rules be met; otherwise, the operation of the existing equipment will be stopped. However, to maintain equipment as operational requires adapting them to new requirements. Before this modernization is accomplished, the operations of more pumps used in the heat supply period were analyzed [37]. The energy efficiency of the above adjustment methods were analyzed based on pump operation for different rotational speeds (Fig. 5.23).

If the maximum discharge is 350 m^3/h and the pump head is 28 m, the absorbed power results are 42.5 kW. If the water discharge is reduced to 100 m^3/h using a throttling valve, the pump head increases to 50 m and the absorbed power will be 23 kW at a constant rotational speed of 1650 rot/min. The operating curves are marked with A-B on the *HQ* curve and with A'-B' on the power diagram.

The correspondence with the power relations is presented by the dashed curve. Thus, comparing the absorbed power *P*, using control valve and speed

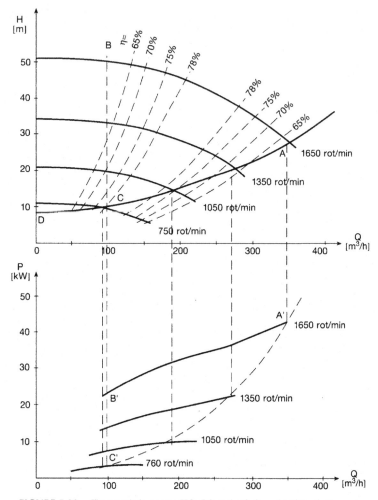

FIGURE 5.23   Characteristic curves HQ, PQ and ηQ for a heating plant pump.

control is possible. Consequently, if the yearly distribution is known, the energy consumption E can be determined. The numerical results, based on the characteristic curves from Fig. 5.23, are summarized in Table 5.5. From the results of the analysis, the yearly energy consumption decreases from 275,064 kWh to 124,173 kWh using rotational speed adjustment. The energy savings are approximately 151,000 kWh, which represents approximatelys 55%.

The use of VFDs reduces equipment and motor energy use. These reductions depend on the equipment type and the ratio to nominal rotational speed; average global losses could be considered less than 5% [43]. Thus, reducing the energy savings obtained from VFDs with these losses results in a final energy saving of approximately 50%.

TABLE 5.5 Energy Consumption Using Control Valve and Speed Control

Distribution			Control valve		Speed control	
Discharge Q (m³/h)	%	Time τ (h)	Power P (kW)	Energy E (kWh)	Power P (kW)	Energy E (kWh)
350	5	438	42.5	18,615	42.5	18,614
300	15	1314	38.5	50,589	29.0	38,106
250	20	1752	35.0	61,320	18.5	32,412
200	20	1752	31.5	55,188	10.0	17,520
150	20	1752	28.0	49,056	6.5	11,388
100	20	1752	23.0	40,296	3.5	6132
Total	100	8760	-	275,064	-	124,173

Assessment of Energy Savings with Variable-Speed Drives. The measured parameters of one pump were the hot-water discharge and the water pressure. Measurements were performed every hour over 18 days in April, a month with large variations in hot water discharge. In this period, the heat-demand adjustment was made by varying the hot-water discharge and keeping the hot-water temperature constant. Knowing the water temperature, the required powers were calculated for every hour for two adjustment methods: throttling control valve, and rotational speed variation with VFDs.

The pump characteristics were the following: type, TD 500-400-750; discharge, 3150 m³/h; and pump head, 70 m. The motor characteristics were the following: type, MIB-X 710Y; power, 800 kW; rated current, 94 A; voltage, 6000 V; rotational speed, 995 rpm; $\cos\varphi = 0.87$; mass, 6000 kg.

Heat demand depends on the outdoor air temperature and is influenced by the return network temperature. The operational flow rate of the pump was of course lower than the nominal flow rate, for which the pump was built. To achieve the desired discharge, as provided in the chart control, the valve mounted on the pump outlet needed to be closed. Thus, by reducing the water discharge, the pump head becomes higher than that from the characteristic curves at the same flow rate. The power absorbed by the electric motor is also lower.

Fig. 5.24 presents the characteristic curves $H = f(Q)$, $P = f(Q)$, and $\eta = f(Q)$. For the case of throttling valve control, the absorbed power was computed using Eq. (5.23) for each hour. The power absorbed by the electric motor with the frequency convertor is obtained from Eqs. (5.24) and (5.27):

$$P_2 = P_1 \left(\frac{Q_2}{Q_1}\right)^3, \tag{5.30}$$

where P_1 is the absorbed power of the FSP for the operating point $Q_1 = 3150$ m³/h and $H_1 = 70$ m.

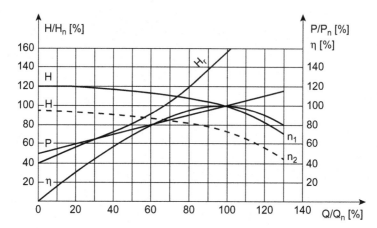

FIGURE 5.24   Power variation of a heating station pump.

Table 5.6 provides the results obtained for comparative energy analysis of the two adjustment methods. The VFD's method of speed control obtained energy savings of approximately 38% compared to the throttling control valve method.

Economical Efficiency. This study has shown that the investment cost of auxiliary equipment for safety maintenance of variable-speed pumps represents approximately 10% of the total operation costs. Thus, 90% represents the energy consumption for the total operation period, which is approximately 15–20 years. At the same time, the energy savings obtained using variable-speed pumps will lead to a shorter recovery time of the investment costs. Therefore, an electricity tariff of 0.1 €/kWh and a supplementary investment cost of 6000 € for the case analyzed above (Table 5.6) results in an investment recovery time of five years, on the basis of simple payback time economic criterion.

The electric energy used to pump hot water is approximately 35% of the total operation energy used in heating stations. The total power consumption of the heating station is approximately 10,000 kW, so the pump power consumption is 3500 kW. Energy savings from using variable-speed pumps compared with throttle control is 33%, which over the 4000 h heating season could save 4,620,000 kWh. This energy savings could reduce the financial burden of heating by approximately 500,000 € every year.

The diagram illustrated in Fig. 5.25 shows the possibility of establishing the energy savings using variable-speed pumps depending on the absorbed power, the Q/Q_{max} ratio, and the yearly operation time. Thus, the case of a 15-kW motor power, an average operating discharge of 70% of the nominal value, and an operating time of 5300 h/year (60%) would achieve an energy savings of 43,000 kWh.

TABLE 5.6 Energy Savings for Pump Speed Control

No.	Adjustment Method	Time τ [h]	Discharge Q (m³/h)	Pump Head H (m)	Power P (kW)	Energy E (kWh/day)	Specific Energy w_P (kWh/m³)
1	Control valve	6	2010	5.0	321.6	8531.2	0.167
		4	1950	5.0	312.0		
		4	2200	5.5	387.2		
		4	2150	5.0	344.0		
		6	2300	5.5	404.8		
2	Speed control	6	2010	4.6	183.6	5284.0	0.104
		4	1950	4.6	167.6		
		4	2200	5.2	240.7		
		4	2150	4.6	224.7		
		6	2300	5.3	275.1		
	Energy saving, ΔE [MWh/year]		1185.2				
	[%]		38.1				

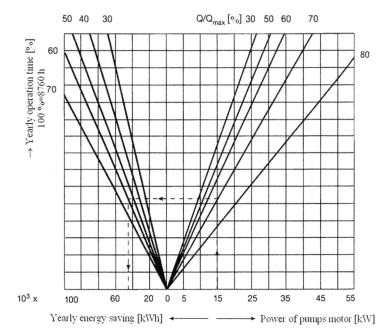

FIGURE 5.25   Yearly energy savings with variable-speed pumps.

Flow rate adjustment with variable-speed pumps is an advantageous method of water pumping in district heating stations, assuring the correlation between the heat demand and water discharge and so obtaining important energy savings that can reach up to 50%.

The water pressure is met continuously at the required values by pump speed control, obtaining an important reduction of water losses in the system. Additionally, the high-pressure values that can lead to equipment operation defects are avoided.

5.5.4 Heat Distribution

After generation, the heat is distributed to the customer via a network of insulated pipes. District heating systems consist of supply and return lines. Within the system, heat storage units may be installed to even out the peak-load demands.

5.5.4.1 Conduit Systems

Usually, the pipes are installed underground but there are also systems with aboveground pipes. The aboveground system has the lowest first cost and the lowest life-cycle cost because it can be maintained easily and constructed with materials that are readily available. Although the aboveground system is sometimes partially factory prefabricated, more typically it is entirely field-fabricated of components such as pipes, insulation, pipe supports, and insulation jackets or protective enclosures that are commercially available. Other common systems that are completely field-fabricated include direct-buried conduit, concrete surface trench, and underground systems that use poured insulation.

Direct-buried conduit, with a thicker steel casing coated in either epoxy or high-density polyethylene (HDPE) or wrapped in fiberglass-reinforced polymer (FRP), may be the only system that can be used in flooded sites where the conduit is in direct contact with groundwater. It is often used for short distances between buildings and the main distribution system.

Two designs are used to ensure that the insulation performs satisfactorily for the life of the system: air-space system and water-spread limiting system.

The *air-space conduit system* (Fig. 5.26A) should have an insulation that can survive short-term flooding without damage. The *water-spread limiting system* (Fig. 5.26B) encloses the insulation in an envelope that will not allow water to contact the insulation. The typical insulation is polyurethane foam, which will be ruined if excess water infiltration occurs. Polyurethane foam is limited to a temperature of about 120°C for a service life of 30 years or more. With this system, the carrier pipe, insulation, and casing are bonded together to form a single unit.

A life-cycle cost analysis may be run to determine the economical thickness of pipe insulation. Because the insulation thickness affects other parameters in some systems, each insulation thickness must be considered as a separate system. For example, a conduit system or one with a jacket around the insulation requires a larger conduit or jacket for greater insulation thicknesses. The cost of the extra conduit or jacket material may exceed that of the additional insulation and is therefore usually included in the analysis. It is usually not necessary to include excavation, installation, and back-fill costs in the analysis.

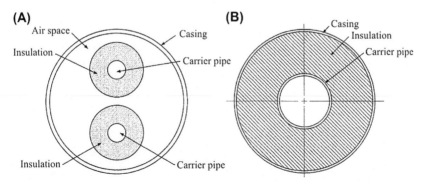

FIGURE 5.26 Conduit system (A) Two carrier pipes and annular air space; (B) Single carrier pipe and no air space.

A system's life-cycle cost LCC, in €/m, is the sum of the initial capital cost and the present value of the subsequent cost of heat lost or gained over the life of the system. The initial capital cost needs only to include those costs that are affected by insulation thickness. To calculate the life-cycle cost, the following equation can be used [3]:

$$\text{LCC} = \text{CC} + q\tau_u c_h \text{PF}, \tag{5.31}$$

where CC is the capital cost associated with pipe insulation thickness, in €/m; q is the annual average rate of heat loss, in W/m; τ_u is the utilization time for system each year, in s; c_h is the cost of heat lost from system, in €/m; and PF is the present value factor for future annual heat loss costs, dimensionless.

The present value factor is the reciprocal of the capital recovery factor CRF, which is found from the following equation:

$$\text{PF} = \frac{(1+i)^\tau - 1}{i(1+i)^\tau} = \frac{1}{\text{CRF}}, \tag{5.32}$$

where i is the interest rate and τ is the useful lifetime of system, in years.

5.5.4.2 Integration of Distributed Solar Thermal Plants

The type of feed-in scenarios can be basically divided as centralized and decentralized. The central supply is made locally at the heat source and is mostly used for local heating networks and multifamily houses.

With the integration of distributed solar thermal plants in existing district heating networks, there are several theoretical possibilities of heat supply (Fig. 5.27) [44]:

1. Extraction and feed in the district heating return (Fig. 5.27A). Due to the lowest possible temperatures, the plant has the best efficiency— often unfavorable for the district heating operators (higher heat losses, flow resistance, and low efficiency of the primary heat source). The return temperature is often fixed by a contract with the main heat producers.
2. Extraction in the district heating return and feed in the district heating supply (Fig. 5.27B). Controlled heat input on the required supply temperature (flow rate matched). This type of feed-in is preferred by the

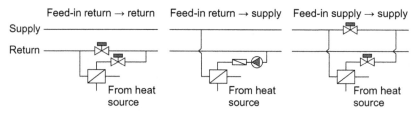

FIGURE 5.27   Hydraulic integration of solar thermal feed-in.

district heating operators (no change of return temperature, the higher part of the pumping costs contributes the solar thermal plant operator). Because of the higher feed-in temperature the collector yields are approximately 10% lower than with the feed in scenario of extraction and feed in the return.

3. Extraction and feed in the district heating supply (Fig. 5.27C). The system operates under the highest required temperature with the lowest efficiency. The pump energy is covered by district heating grid pumps. This variant is due to the high temperatures already in the collector inlet for solar thermal plants rather not in use.

5.5.4.3 Distribution Network

District heating systems can vary in size from covering entire cities with a network of large diameter primary pipes linked to secondary pipes, which in turn link to tertiary pipes that might connect to 10–50 buildings. Some district heating schemes might only be sized to meet the needs of a small village or area of a city, in which case only the secondary and tertiary pipes will be needed.

Both flowing steam and hot water incur pressure losses. Hot-water distribution systems may use intermediate booster pumps to increase the pressure at points between the heating plant and the consumer. Because of the higher density of water, pressure variations caused by elevation differences in a hot-water system are much greater than for steam systems. This can adversely affect the economics of a hot-water system by requiring the use of a higher pressure class of piping and/or booster pumps or even heat exchangers used as pressure interceptors.

Hot-water distribution systems are divided into three temperature classes:

- high-temperature hot-water (HTHW) systems supply temperatures over 175°C;
- medium-temperature hot-water (MTHW) systems supply temperatures in the range of 120–175°C; and
- low-temperature hot-water (LTHW) systems supply temperatures of 120°C or lower.

Ideally, the appropriate pipe size should be determined from an economic study of the life-cycle cost for construction and operation. In practice, however, this study is seldom performed because of the effort involved. Instead, criteria that have evolved from practice are frequently used for design. These criteria normally take the form of constraints on the maximum flow velocity or pressure drop. For steam systems, maximum flow velocities of 60–75 m/s are recommended [45]. For water systems, Europeans use the criterion that pressure losses should be limited to 100 Pa/m of pipe [46].

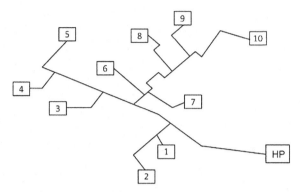

FIGURE 5.28 Topology of a district heating network.

Calculating the discharges and pressures in a pipe network with branches, loops, pumps, and heat exchangers can be difficult without the aid of a computer. Methods developed for domestic water distribution systems [47] may apply to thermal distribution systems with appropriate modifications. Most calculations are performed for a steady-state condition and have several input assumptions (e.g., temperatures, pressures, flows, and/or loads) for a specific moment in time. Some computer modeling softwares are dynamic and take-time inputs of flow and pressure to efficiently control and optimize the rotational speed of the distribution pumps.

An optimal design model of hot-water branched distribution networks (Fig. 5.28) in steady-state conditions based on the linear programming method was developed [48] and will be exposed next.

Fundamentals of the Hydraulic Calculation. The input data for hydraulic computation are the network topology, the hot-water generation scheme in heating plant, the design heat load of all consumers, and initial parameters of the heat distribution fluid.

A district heating network may be represented by a direct-connected graph composed of a finite number of arcs (pipes, pumps, heat exchangers, and fittings) connected to one another by vertices (nodes) as critical points, heat plants, consumers, and junction nodes.

The topology of a network can be completely described using the incidence matrix and the cycle matrix constructed for the associated graph. Each pipe sector has two pipes—supply pipe and return pipe—with the same dimensions.

The hydraulic calculation provides the pipe diameter and pressure loss of each pipe sector. Pressure profiles of a district heating system in dynamic and steady state conditions are drawn based on this calculus. For example, Fig. 5.29 illustrates the pressure profile for one pipe with length L, between heating plant and consumer, where $p_1 - p_2$ is the pressure loss in supply pipe, $p_2 - \bar{p}_2$ is the pressure loss in consumer's plant, and $\bar{p}_2 - \bar{p}_1$ is the pressure loss in return pipe.

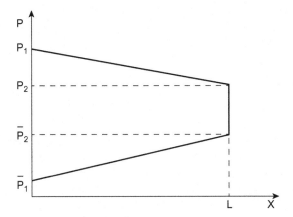

FIGURE 5.29 Pressure profile for one-pipe segment.

The pressure loss Δp_{ij} in the pipe ij of a district heating network can be calculated using the general equation:

$$\Delta p_{ij} = \frac{8\rho}{\pi^2}\left(\frac{\lambda_{ij}L_{ij}}{D_{ij}} + \zeta_{ij}\right)\frac{G_{ij}^2}{D_{ij}^4}, \tag{5.33}$$

where ρ is the water density; L_{ij}, D_{ij}, G_{ij} are the length, diameter, and discharge of pipe ij, respectively; λ_{ij} is the friction factor calculated using Colebrook–White formula; and ζ_{ij} is the minor loss coefficient of pipe ij.

Neglecting the minor pressure loss, Eq. (5.33) can be rewritten as

$$\Delta p_{ij} = \frac{8\rho\lambda_{ij}}{\pi^2}\frac{L_{ij}}{D_{ij}^5}G_{ij}^2 \tag{5.34}$$

The extreme operation conditions of the network are reached for the coldest day of the year and the consumed heat depends on the difference between the design temperature in building and the outdoor air temperature. The maximum heat load of a consumer is computed as a function of the outdoor air temperature.

The maximum heat load is expressed by water discharge consumed in the node in which the consumer is connected. When the system operates with maximum temperature difference Δt_{max} between supply and return network, the water discharge q_j concentrated in the node j is expressed as

$$q_j = \frac{Q_j}{c_p\Delta t_{max}}, \tag{5.35}$$

where Q_j is the heat load of the consumer j and c_p is the water specific heat at constant pressure.

Both the minimum and the maximum heat loads needed in two critical moments from the coldest day of the year must be considered for each consumer j. These loads determine the corresponding minimum and maximum water discharges concentrated at each node j and the average discharge q_j.

To determine annual consumed energy E for each pump, with sufficient practical precision, the following equation can be used:

$$E = \frac{1}{\eta} G \Pi \tau_u, \tag{5.36}$$

where η is the general efficiency of the pump; G is the design discharge of the considered pipe, determined using the average discharges at nodes' q_j; Π is the pumping pressure which assures network operation to average discharge q_j; and τ_u is the utilization time for heating system each year.

Optimization Model. For networks supplied by pumping, the literature suggests the use of minimum total annual costs (TACs) as a criterion [49].

For a branched heating network, the design discharges G_{ij} of the pipes are univocal determined by concentrated discharge q_j, which are known at the network nodes. These design discharges have the same value in the supply and return network but with changed sign $\left(\overline{G}_{ij} = -G_{ij}\right)$.

The discharges G_{ij} are determined for the operating condition. The series of commercial diameters that can be used $D_{k,ij} \in [D_{\max,ij}, D_{\min,ij}]$ for each pipe ij are established using the limit values of optimal diameters $D_{\max,ij}$ and $D_{\min,ij}$, computed by equation

$$D_{\max(\min),ij} = \sqrt{\frac{4G_{ij}}{\pi V_{\min(\max),ij}}}, \tag{5.37}$$

where G_{ij} is the design discharge of pipe ij and $V_{\min}$ and $V_{\max}$ are the limits of the economic velocities.

The total length of a pipe ij, with the discharge G_{ij}, may be divided into s_{ij} segments (k) of $D_{k,ij}$ diameters and $x_{k,ij}$ lengths. Taking into account Darcy-Weisbach's Eq. (5.34), the expression of the pressure difference between the two extremities of a pipe can be linearized as

$$p_i - p_j = \sum_{k=1}^{s_{ij}} \alpha_{k,ij} x_{k,ij} - \Pi_{ij} + \rho g(ZT_j - ZT_i) \tag{5.38}$$

in whichss

$$\alpha_{k,ij} = \frac{8\rho \lambda_{ij} G_{ij} |G_{ij}|}{\pi^2 D_{k,ij}^5}, \tag{5.39}$$

where Π_{ij} is the active pressure of the booster pump integrated on the pipe ij, for the discharge G_{ij}; g is the gravitational acceleration; ZT_i and ZT_j are the elevation heads at the nodes i and j, respectively.

The minimum TAC criterion can be expressed in an objective function [48], including the investment cost of network pipes and integrated pumps and the pumping energy cost, as

$$
F_o = \sum_{ij=1}^{T} \sum_{k=1}^{s_{ij}} c_{k,ij}^* x_{k,ij} + e\tau_u \sum_{j=1}^{NS} \frac{1}{\eta_j} q_j \left(p_j - \bar{p}_j \right) +
$$
$$
\sum_{ij=1}^{NP} \left[A_{ij} y_{ij} + B_{ij} \left(r_{ij} + \bar{r}_{ij} \right) + e\tau_u \frac{1}{\eta_{ij}} \left(G_{ij} \Pi_{ij} + \overline{G}_{ij} \overline{\Pi}_{ij} \right) \right] \to \min, \tag{5.40}
$$

where $c_{k,ij}^*$ is the annual specific cost of a segment k of the pipe ij, depending of the diameter $D_{k,ij}$ [49]; e is the cost of electrical energy; τ_u is the annual utilization time of heating system; NS is the number of heat sources; NP is the number of booster pumps; A_{ij} and B_{ij} are the annual fixed cost and proportional cost with power of a pump, respectively; r_{ij} and $\bar{r}_{ij}$ are the maximum powers of a booster pump integrated on the supply and return pipes ij, respectively, if $r_{ij} \neq 0$ and $\bar{r}_{ij} \neq 0$; p_j and $\bar{p}_j$ are the pressures at node j of the supply and return networks, respectively; G_{ij} and $\overline{G}_{ij}$ are the design discharges in pipes ij of the supply and return network, respectively; and $\Pi_{ij}, \overline{\Pi}_{ij}$ is the active pressure in pipe ij of the supply and return network, respectively, on which a booster pump can be installed in the nearness of the node i.

The objective function has as unknowns the decision variables $x_{k,ij}, r_{ij}, \bar{r}_{ij}, p_j, \bar{p}_j, \Pi_{ij}, \overline{\Pi}_{ij}$, and minimizes the annual total costs.

Hence, the values of the decision variables must be determined to minimize the objective function F_o, subject to:

- *constructive constraints* that are introduced to ensure that the sum of all pipe segments between any two nodes is equal to the length between those nodes:

$$
\sum_{k=1}^{s_{ij}} x_{k,ij} = L_{ij}; \quad (ij = 1, \ldots, T), \tag{5.41}
$$

where T is the number of pipes in network.

- *operational constraints* that are written in all nodes j or pipe segments ij for each of the three operation regimes corresponding to the minimum, maximum, and medium heating loads:

$$
p_i - p_j = \sum_{k=1}^{s_{ij}} \alpha_{k,ij} x_{k,ij} - \Pi_{ij} + \rho g(ZT_j - ZT_i) \tag{5.42}
$$

$$
\bar{p}_i - \bar{p}_j = -\sum_{k=1}^{s_{ij}} \alpha_{k,ij} x_{k,ij} - \overline{\Pi}_{ij} + \rho g(ZT_j - ZT_i) \tag{5.43}
$$

$$p_j - \overline{p}_j \geq \delta_j \tag{5.44}$$

$$r_{ij} \geq G_{ij}\,\Pi_{ij} \geq 0 \tag{5.45}$$

$$\overline{r}_{ij} \geq \overline{G}_{ij}\,\overline{\Pi}_{ij} \geq 0 \tag{5.46}$$

$$r_{ij} \leq My_{ij} \tag{5.47}$$

$$\overline{r}_{ij} \leq My_{ij} \tag{5.48}$$

$$h_j \leq p_j \leq H_j \tag{5.49}$$

$$\overline{h}_j \leq \overline{p}_j \leq \overline{H}_j \tag{5.50}$$

$$h_j \leq p_j + \Pi_{ij} \leq H_j \tag{5.51}$$

$$\overline{h}_j \leq \overline{p}_j + \overline{\Pi}_{ij} \leq \overline{H}_{ij} \tag{5.52}$$

$$y_{ij} = \{0,\ 1\} \tag{5.53}$$

$$X_{k,ij} \geq 0 \tag{5.54}$$

- *hydraulic constraints* that also are written for each of the three mentioned operation regimes:

$$\sum_{\substack{i \neq j \\ i=1}}^{N} G_{ij} + q_j = 0; \quad (j = 1, ..., N - NS) \tag{5.55}$$

$$\overline{G}_{ij} = -G_{ij}; \quad (ij = 1, ..., T) \tag{5.56}$$

where h_j, H_j and $\overline{h}_j$, $\overline{H}_j$ are the inferior and superior limits of the pressure at each node j of the supply and return network, respectively; δ_j is the minimum pressure difference at the node j between the supply and return pipe, which ensures the discharge q_j through the consumers' plant connected to respective node.

The constraints (5.42) and (5.43) relate the new variables with the pressure difference at the network nodes. The variables y_{ij} can have the value 0 or 1, if a booster pump must be integrated or no on the pipe ij. Through the inequality (5.47), these variables are related to the maximum pump power, where M denotes a constant with a sufficiently high value. Eqs. (5.49) and (5.51) limit the variation range of pressure in the network, either at the nodes, or at the inlet or outlet of the integrated pumps.

As the objective function (5.40) and constraints (Eqs. (5.41)–(5.56)) are linear with respect to the unknowns of system, the optimal solution can be determined according to the linear programming method [50], using the

Simplex algorithm [51]. A computer program has been elaborated based on the linear optimization model, in the FORTRAN programming language for PC-compatible microsystems.

5.5.5 Consumer Interconnections

The common fluid used for heat distribution is water or pressurized hot water. At customer level, the heat network is usually connected to the central heating system of the dwellings directly by the building HVAC system or indirectly via a heat exchanger (heat substation) that transfers energy from one fluid to another. The heat distribution fluids of both networks (hot water or steam) do not mix.

The heat substation normally has one or more of the following parts:

- Heat exchanger to split primary and secondary sides of the system;
- Control valve to regulate the flow through the heat exchanger;
- Differential pressure controller to balance the network and improve working conditions of control valve;
- Strainer to remove particles that could block the heat exchanger or control valve;
- Shut-off valve to stop the flow on primary side in case of a service or emergency;
- Heat meter to measure energy consumption and allocate costs;
- Temperature controller to control temperature on secondary side by adjusting the flow on primary side; and
- Temperature sensor to sense flow and return temperatures required for temperature control.

When energy is used directly, it may need to be reduced to pressure that is commensurate to the buildings' systems.

Direct Connection. Because a direct connection offers no barrier between the direct water and the building's own system (e.g., heating coils, fan-coils, radiator, radiant panels), water circulated at the district heating plant has the same quality as the customer's water. Direct connections, therefore, are at a greater risk of incurring damage or contamination based on the poor water quality of either party. A direct connection is often more economical than an indirect connection because the consumer is not burdened by the installation of heat exchangers, additional circulation pumps, or water treatment systems.

Fig. 5.30 shows the simplest form of hot-water direct connection, where the district heating plant pumps water through the consumer building [3]. This system includes a pressure differential regulator, a thermostatic control valve on each terminal unit, and a pressure relief valve. Most commercial systems have a flow meter installed as well as temperature sensors and transmitters to calculate the energy used. Pressure transmitters may be installed as input for the plant-circulating pump speed control. The control valve is the capacity-

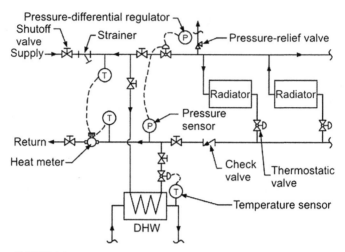

FIGURE 5.30 Direct connection of building system to district hot water.

regulating device that restricts flow to maintain either a water supply or return temperature on the consumer's side.

Particular attention must be paid to connecting high-rise buildings because they induce a static head that affects the design pressure of the entire system. Pressure control devices should be investigated carefully.

Indirect Connection. Many of the components are similar to those used in the direct connection applications, with the exception that a heat exchanger performs one or more of the following functions: heat transfer, pressure interception, and buffer between potentially different-quality water treatments.

Identical to the direct connection, the rate of energy extraction in the heat exchanger is governed by a control valve that reacts to the building load demand.

The three major advantages of using heat exchangers are (1) the static-head influences of a high-rise building are eliminated, (2) the two-water streams are separated, and (3) consumers must make up all of their own lost water. The disadvantages of using an indirect connection are the (1) additional cost of the heat exchanger and (2) temperature loss and increased pumping pressure because of the addition of another heat transfer surface.

5.5.6 Control, Operation, and Maintenance

The success of the district heating system efficiency is usually measured in terms of the temperature differential for the water system. Proper control of the district heating temperature differential is not dictated at the plant but at the consumer. If the consumer's system is not compatible with the temperature parameters of the district heating system, operating efficiency will suffer unless components in the consumer's system are modified.

Generally, maintaining a high-temperature differential (Δt) between supply and return lines is most cost-effective because it allows smaller pipes to be used in the primary distribution system. These savings must be weighed against any higher building conversion cost that may result from the need for a low primary-return temperature. Furthermore, optimization of the Δt is critical to the successful operation of the district energy system. That is the reason the customer's Δt must be monitored and controlled.

As with any major capital equipment, care must be exercised in operating and maintaining district heating systems. Both central plant equipment and terminal equipment located in the consumer's building must be operated within intended parameters and maintained on a schedule as recommended by the manufacturer.

Thermal distribution systems, especially buried systems, which are out of sight, may suffer from inadequate maintenance. To maintain the distribution system's thermal efficiency as well as its reliability, integrity, and service life, periodic preventative maintenance is strongly recommended.

5.5.7 Environmental and Economic Performance

One significant reason to pursue district energy is its environmental benefits. GHG emissions can be reduced with district energy in two ways: facilitating the use of non-carbon energy forms as solar energy for heating, and replacing less-efficient equipment in individual buildings with a more efficient central heating system.

Genchi et al. [52] use CO_2 payback time (CPT) to evaluate the environmental impact of a district heating system, which is expressed as follows:

$$\text{CPT} = \frac{CO_2 \text{ emissions during the initial construction phase}}{\text{Annual } CO_2 \text{ emissions reduction by the introduction of new systems}}.$$

$$(5.57)$$

Economics is a major factor in decision making and design. The economics of district energy depend on three main factors [53]:

- the production cost of the thermal energy;
- the cost of the thermal energy distribution network, which depends on network size and thermal loads; and
- customer connection costs.

The customer connection cost can be reduced if the district heating system is designed and developed at the same time as a community is built; this cost is higher when the project is retrofitted in a fully developed site. Persson and Werner [34] divide the total costs of a district heating system into four categories:

- Heat delivery capital cost. This cost includes the network construction cost, which is often more than half of the total delivery cost.

- Heat delivery heat-loss cost which, to some extent, is dependent on heat delivery capital cost since low heat densities have higher heat losses. This term also relates to the price of recycled heat in distribution heating.
- Heat delivery pressure-loss cost. The cost of the pressure loss during heat distribution needs to be recovered.
- Service and maintenance cost is typical of most systems.

Reihav and Werner [54] apply to evaluate the net present value (NPV) of an investment (I) in a district heating delivery system as follows:

$$\text{NPV} = \left(pQ - \frac{C_{\text{prod}}}{1-h}Q - C \right)\text{NF} - I, \tag{5.58}$$

where p is the unit price to the customer (€/GJ), Q is the annual heat delivery to the consumer (GJ/house), C_{prod} is marginal heat production cost (€/GJ), h is the annual heat loss (GJ/year), C is the annual operation and maintenance cost of the system (€/house), and NF is the net present value factor, which is dependent on the interest rate and the period (Eq. (5.32)).

5.6 SOLAR HEATING FOR INDUSTRIAL PROCESSES

The importance of energy in industrial development is very crucial since major fraction of energy is used in industrial processes. It has dominated more than 50% of total energy consumption worldwide. The delivered energy in industrial sector is utilized in four major sectors: construction, agriculture, mining, and manufacturing. Fig. 5.31 shows the industrial energy consumption trend until 2030 [55]. Over the next 25 years, worldwide industrial energy consumption is projected to grow from 51,275 ZW (1 ZettaWatt = 10^{21} Watt) in 2006 to 71,961 ZW in 2030 by an average of 1.4% per year.

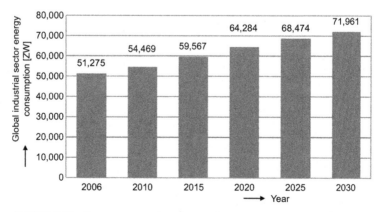

FIGURE 5.31   Energy consumption of global industrial sector during 2006–2030.

In Europe, about 27% of the total final energy demand is heat consumed by the industry. Herein, about 30% of the total industrial heat demand occurs at temperature levels below 100°C and further 27% between 100 and 400°C. A significant part of this heat, especially below 100°C, can be generated by solar thermal plants.

5.6.1 Integration of Solar Energy into Industrial System

A typical industrial energy system is composed of four main parts: power supply, production plant, energy recovery, and cooling systems. The power supply provides the energy needed for the system to operate mainly from electrical energy, heat, gas, steam, or coal. Production plant is the part of the system that executes proceedings of production.

The most important factor for solar energy integration usually is the available temperature level. This term means low temperature levels in the plant, at which a significant amount of solar energy could be transferred to the processes by the solar thermal system (e.g., via heat exchanger). The temperature at the energy integration point(s) highly influences the fluid temperatures in the solar thermal system and should be as low as possible, because collector and storage work most efficient at low temperatures. Also important is the process temperature itself. The economical best systems have process temperatures below 50°C. Process temperatures above 100°C even in open systems often increase the available temperature level, since heat recovery measures should be applied.

Integration of solar thermal systems into industrial process heat can be done in the following three ways: (1) as a heat source for direct heating of a circulating fluid (e.g., feed-up water, return of closed circuits, air preheating); (2) direct heating to process equipment with low-temperature requirement (Fig. 5.32); and (3) as an additional source for preheating of supply water for steam boilers [56]. In the first case, storage will be an important component to ensure that heat is available throughout the day. The second and third options can be used when heat supply demand is larger than can be provided by solar heating or if temperature needs are too high for a solar thermal system.

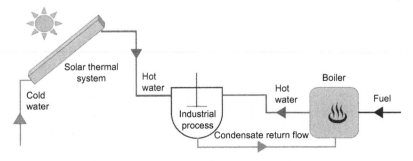

FIGURE 5.32 Direct heating to process equipment.

Integration of solar heat systems into industrial applications requires storage and control strategies to handle the noncontinuous supply of solar energy. The accurate design and sizing of a solar thermal system, taking into account the specific demand profile, can be carried out by dynamic system simulation software, such as TRNSYS program.

5.6.2 Heating Systems for Industrial Processes

Solar energy applications in industry are divided into two main categories: the solar thermal and the photovoltaic. Some of the most common applications are hot water, steam, drying and dehydration processes, washing, cleaning, pre-heating, concentration, pasteurization, sterilization, food, paper industry, textile industry, chemical industry, plastic industry, and industrial space heating [57]. Solar thermal technology can also provide an alternative to cooling processes in sectors such as the food and tobacco sector where most products' cooling is currently done by electric chillers. Table 5.7 shows few of potential industrial processes and the required temperatures for their operations [58]. Paper and food industries have the biggest heat demand. Considerable heat demand also exists in the textile and chemical industries.

Many industrial processes are involved in heat utilization with temperatures between 80 and 240°C. Industrial energy analysis shows that solar thermal energy has enormous applications in low- (i.e., 20−200°C) and, medium- (i.e., 80−240°C) temperature levels [58].

5.6.2.1 Heating of Hot Water for Washing or Cleaning

The supported system for hot-water production is an open system without heat recovery, since cleaning water is usually contaminated and cooled down by the cleaning process. Cold water (e.g., 15°C) is heated up to 60°C. In plants with stochastic cleaning water demand and very high flow rates the backup heating system is usually equipped with a hot-water storage that is heated by a boiler.

With the installation shown in Fig. 5.33, the solar thermal system can be integrated easily via an additional heat exchanger [59].

Whenever cold water has to be heated up, it is (pre-) heated by the solar thermal system before it enters the hot-water storage. A cold water by-pass at the discharging side of the solar buffer storage prevents the cold water circuit of high temperatures possibly occurring in the buffer storage (temperatures up to 90°C). At the charging side of the solar buffer storage tank, the solar heat exchanger can be by-passed in the collector circuit via a three-way valve. When the temperature at the bottom of the storage tank is lower than the outlet temperature of the collector, the circulating pump starts. First, the HTF is circulated only in the collector loop and does not enter the heat exchanger. This way, the collector circuit is heated up before the pump loading the buffer storage is activated. Otherwise, in the morning or at varying radiation conditions, heat of the storage tank could be lost to the solar circuit and there

TABLE 5.7 Industrial Processes and Ranges of Temperature

Industry	Process	Temperature (°C)
Dairy	Pressurization	60–80
	Sterilization	100–120
	Drying	120–180
	Boiler feed water	60–90
Food	Sterilization	110–120
	Pasteurization	60–80
	Cooking	60–90
	Bleaching	60–90
Textile	Bleaching, dyeing	60–90
	Drying, degreasing	100–130
	Fixing	160–180
	Pressing	80–100
Paper	Cooking, drying	60–80
	Boiler feed water	60–90
	Bleaching	130–150
Chemical	Soaps	200–260
	Synthetic rubber	150–200
	Processing heat	120–180
	Preheating water	60–90
Beverages	Washing, sterilization	60–80
	Cooking	60–70
Flours and by-products	Sterilization	60–80
Bricks and blocks	Curing	60–40
Plastics	Preparation	120–140
	Distillation	140–150
	Separation	200–220
	Extension	140–160
	Drying	180–200

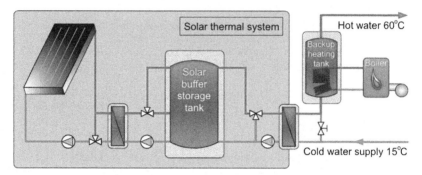

FIGURE 5.33   Water heating system with heat exchanger and serial boiler.

could be a certain risk of freezing of the heat exchanger in winter. When the storage is charged, one or more three-way valves control the inlet height of the flow in a way that the stratification of the storage is maintained as good as possible. The storage volume, of course, can also be a cascade of storages or loaded with stratification lances, if available. In industrial plants with high demand for cleaning water, the preheating of hot water with low SFs can be very economical because of the low temperature of the cold water inlet.

5.6.2.2 Heating of Make-up Water for Steam Networks

The solar support of process steam generation is usually only economical when a significant part of the steam is used in the processes directly (the steam network is an open or partly open system). Solar heating of the additional, demineralized make-up water can be attractive (Fig. 5.34) [59]; heating of the condensate return flow or the feed water directly is more expensive because of the high temperatures.

In (partially) open steam networks, the demineralized make-up water is usually mixed with the returning condensate and has to be degassed before it can enter the steam boiler. This degasification is usually done thermally using process steam from the steam boiler. With this steam, the feed water tank has

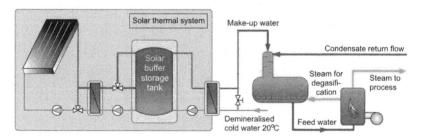

FIGURE 5.34   Heating system of make-up water for steam networks.

to be heated up to 90°C, often also up to slightly over 100°C, when the feed water tank operates at an overpressure of 0.2 or 0.3 bar. It is therefore a good solution to preheat the decalcified, additional make-up water, before it is mixed with the condensate and before the mixture has to be heated up. This way, less steam is consumed for degasification, and since the steam supports many different processes in the factory, the solar thermal system can cover a significant fraction of the overall heat demand just by adding one single heat exchanger.

This solar thermal system concept is similar to washing/cleaning. The discharging heat exchanger protects the solar buffer storage tank from corrosion and is not bypassed at the solar side, since 90°C is usually also the maximum storage temperature. No make-up water storage tank is applied, since the make-up water mass flow is usually not varying.

5.6.2.3 Heating of Industrial Baths or Vessels

In the example of Fig. 5.35, raw parts are treated in an industrial bath at a temperature of 65°C [59]. Depending on the cycle times, the raw parts cause high heat losses, since they have to be heated up by the bath and usually have a high thermal capacity. The convective heat losses only account for a small share of the heat demand. The solar thermal system heats the bath via a return-flow boost. The backup heating is done serially by a boiler. A bypass of the solar buffer storage tank allows direct solar heating of the bath.

The energy produced by the solar thermal system is usually significantly lower than the thermal demand of the bath. Since a very high temperature of 90°C is needed at the inlet of the bath heat exchanger, there is the possibility to bypass the buffer storage tank to reduce storage losses and to prevent the fluid in the heating circuit from being mixed down by lower storage temperatures. This is also important because of the minimum temperature level of 70°C in the whole system. The control of the boiler has to be able to ensure a constant inlet temperature at the necessary mass flow in this case.

When the storage is discharged, the return flow from the bath can be mixed into different heights of the storage by three-way valve(s) to ensure a good stratification when the bottom of the storage is colder than 70°C. When the

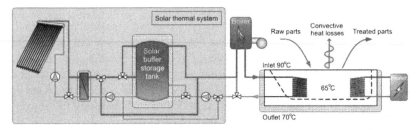

FIGURE 5.35 Solar heating system of an industrial bath.

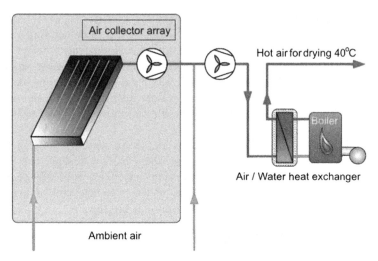

FIGURE 5.36   Heating system of air for drying process.

solar irradiation is not sufficient and the buffer storage temperature is below 70°C, a three-way valve enables the boiler to heat the bath directly without heating up the buffer storage tank.

5.6.2.4 Convective Drying with Hot Air

The process supported by the air collectors is an open drying process. Conventionally, ambient air is heated up to 40°C by an air/water heat exchanger. The solar air collector system is installed to (pre-) heat the ambient air (Fig. 5.36) [59]. The solar fan is at the hot side of the air collectors, so that leakage air flows at the collectors are not lost but used.

Compared to water collectors, the efficiency of air collectors decreases significantly when the mass flow is reduced. As a drawback, the pressure loss at the high mass flow is five to six times higher than for the low one.

Whenever heated ambient air is needed for the drying process, the conventional fan generates the necessary mass flow. This fan compensates all pressure losses in the conventional system. When the sun is not shining, the solar fan is not active and the ambient air is heated by the conventional heat exchanger. When the solar radiation is sufficient and the absorber temperature exceeds a certain value, the solar fan starts to run and generates its maximum mass flow of 100 kg/m^2. For this mass flow, the temperature lift of the solar thermal system is low but the efficiency is high. Depending on the solar irradiation, the residual temperature difference to the 40°C is added by the heat exchanger. Depending on the size and the internal field connection of the air collectors, additional cold ambient air is added automatically by the conventional fan running at constant speed to provide always the necessary mass flow rate for the process.

If the solar radiation is high, the temperature level after the admixture of the cold ambient air (after the conventional fan) can exceed 40°C. Because of that, the driving speed of the solar fan can be modulated. When the speed is reduced, the mass flow rate through the collector circuit decreases and the efficiency goes down. Because more cold ambient air is added, the temperature of 40°C can be maintained by controlling the speed of the solar fan. In this case, the conventional heater can be switched off and the electrical energy consumption of the solar fan is reduced because of the lower pressure drop in the solar circuit.

5.6.3 Solar-Powered Water Desalination Industry

Currently, one-fourth of habitants in the world are deprived of sufficient pure water. Table 5.8 shows the distribution of world population since 1950 and predictions up to 2050 [60]. As observed, the world population is more concentrated in developing countries located in Asia and Africa. Therefore, water desalination technology seems more necessary in these regions.

Solar energy can be used to desalinate sea water using small tubs embedded in the life boats called "solar stills." Solar stills are suitable for domestic applications, particularly in rural and remote areas, small islands, and big ships with no access to the grid. In this situation, solar energy is economically and technically more competitive than conventional diesel

TABLE 5.8 Distribution of World Population, in Millions, Since 1950 and Predictions up to 2050

Year	USA	EU	Africa	Asia	Total
1950	158	296	221	1377	2522
1960	186	316	277	1668	3022
1970	210	341	357	2101	3696
1980	230	356	467	2586	4440
1990	254	365	615	3114	5266
2000	278	376	784	3683	6055
2010	298	376	973	4136	6795
2020	317	371	1187	4545	7502
2030	333	362	1406	4877	8112
2040	343	349	1595	5118	8577
2050	349	332	1766	5268	8909

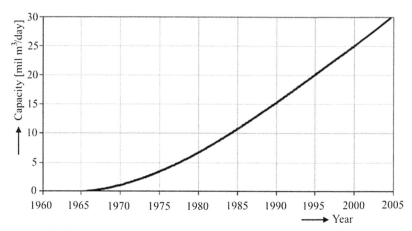

FIGURE 5.37   Global trend in installed capacity of desalination systems.

engine-powered reverse-osmosis alternatives. This method is extremely simple but it needs high initial investment, massive surface, frequent maintenance, and sensitivity to weather conditions. Hence, its application on large-scale production plants is very limited. Fig. 5.37 shows the total water desalination capacity installed worldwide between 1960 and 2005 [60].

Water desalination technologies used in industrial scale are two main categories: thermal process or phase-change technologies and membrane technologies or single-phase processes. Table 5.9 shows the different methods of the two main categories.

Solar power systems are reliable substitutes to be used as an innovative power source for water desalination plants. It is the most effective and feasible approach for such systems. In addition, they are environmentally friendly and economically competitive compared to traditional methods.

TABLE 5.9 Desalination Processes

Phase-Change Processes	Membrane Processes
1. Multistage flash	1. Reverse osmosis (without and with energy recovery)
2. Multieffect boiling	
3. Vapor compression	2. Electro dialysis
4. Freezing	
5. Humidification/dehumidification	
6. Solar stills (conventional, special, and wick-type stills)	

5.6.4 Cost of Solar Heating Systems for Industrial Processes

The cost of solar heating for industrial processes depends on process temperature level, demand continuity, project size, and the level of solar radiation of the site. For conventional flat-plate collector and evacuated tube collector, investment system costs range between 250 and 1000 €/kW in Europe, and around 200−300 €/kW in India, South Africa, and Mexico. The energy costs for feasible solar thermal systems range from 0.025 to 0.08 €/kWh, and a European roadmap targets solar heat costs of 0.03−0.06 €/kWh [61].

For concentrated systems, heating costs are in the range of 0.06−0.09 €/kWh with a target of 0.04−0.07 €/kWh for concentrating systems by 2020 [61]. Concentrated systems include parabolic dish collectors (developed and used in India) with costs ranging from 400 to 1800 $/kW, parabolic trough collectors with costs ranging from 600 to 2000 $/kW, and linear Fresnel collectors in the range of 1200−1800 $/kW. In comparison, the same technology is used in concentrated solar power plants with costs of approximately 3400−6000 $/kW [62].

Key factors for cost reductions are automation of production processes, modular designs for easier installation and integration onto industrial roofs, optimized tracking systems, standardization and certification, and material replacement of copper and steel with aluminum and polymers.

REFERENCES

[1] Yellott JI. Passive solar heating and cooling systems. ASHRAE Transactions 1977;83(2):429−36.

[2] Freeman TL, Mitchell JW, Audit TE. Performances of combined solar-heat pump systems. Solar Energy 1979;22(2):125−35.

[3] ASHRAE handbook, HVAC systems and equipment. Atlanta (GA): American Society of Heating, Refrigerating and Air Conditioning Engineers; 2016. www.ashrae.org.

[4] ASHRAE handbook, HVAC applications. Atlanta (GA): American Society of Heating, Refrigerating and Air Conditioning Engineers; 2015. www.ashrae.org.

[5] Sarbu I, Adam M. Applications of solar energy for domestic hot-water and buildings heating/cooling. International Journal of Energy 2011;5(2):34−42.

[6] Duffie JA, Beckman WA. Solar engineering of thermal processes. Hoboken (NJ): Wiley & Sons, Inc; 2013.

[7] Mutch JJ. Residential water heating, fuel consumption, economics and public policy. RAND Report R 1974;1498.

[8] Lutz DJ, Liu X, McMahon JE, Dunham C, Shown LJ, McGrue QT. Modeling patterns of hot water use in households. Report LBL-37805 UC-1600. Lawrence Berkeley National Laboratory; 1996.

[9] ASHRAE handbook. Fundamentals. Atlanta (GA): American Society of Heating, Refrigerating and Air-Conditioning; 2013. www.ashrae.org.

[10] Erbs DG, Klein SA, Beckman WA. Estimation of degree-days and ambient temperature bin data from monthly-average temperatures. ASHRAE Journal 1983;25(6):60.

[11] Beckman WA, Klein SA, Duffie JA. Solar heating design by the f-chart method. New York: Wiley-Interscience; 1977.

[12] TRNSYS 17. A transient system simulation program user manual. Madison (WI): Solar Energy Laboratory, University of Wisconsin-Madison; 2012.

[13] METEONORM> Help, version 5.1. Bern (Switzerland): Meteonorm Software; 2004. http://www.meteonorm.com.

[14] IEA Task 26. Solar combisystems. International Energy Agency; 2002. http://archive.iea-sch.org/task26/.

[15] Kacan E, Ulgen K. Energy and exergy analysis of solar combisystems. International Journal of Exergy 2014;14(3):364−87.

[16] Balaras CA, Dascalaki P, Tsekouras P, Aidonis A. High solar combisystems in Europe. ASRAE Transactions 2010;116(1):408−15.

[17] Weiss W. Solar heating systems for houses: a design handbook for solar combisystems. Germany: Cromwell Press; 2003.

[18] Andersen E, Shah IJ, Furbo S. Thermal performance of Danish solar combisystems in practice and in theory. Journal of Solar Energy Engineering 2004:744−9.

[19] Kacan E, Ulgen K. Energy analysis of solar combisystems in Turkey. Energy Conversion and Management 2012;64:378−86.

[20] Assae R, Ugursal I, Morrison I, Nen-Abdallah N. Preliminary study for solar combisystem potential in Canadian houses. Applied Energy 2014;130:510−8.

[21] Ellehauge K, Shah IJ. Solar combisystems in Denmark—the most common system designs. In: Proceedings of Euro sun conference 2000 (DK); 2000.

[22] Drück H, Hahne E. Test and comparison of hot water stores for solar combisystems. In: Proceedings of Euro sun '98 conference; 1998. p. 1−7.

[23] Kacan E. Design, application, energy and exergy analysis of solar combisystems [Ph.D. thesis]. Turkey: Ege University; 2011.

[24] Hin JNC, Zmeureanu R. Optimization of a residential solar combisystem for minimum life cycle cost, energy use and exergy destroyed. Solar Energy 2014;100:102−13.

[25] http://www.elle-kilde.dk/altener-combi/dwload.html/.

[26] http://www.iwt.tugraz.at/downloads.htm/.

[27] Sarbu I, Sebarchievici C. Ground-source heat pumps—Fundamentals, experiments and applications. Oxford (United Kingdom): Elsevier; 2015.

[28] Chiasson AD, Yavuzturk C. Assessment of the viability of hybrid geothermal heat pump systems with solar thermal collectors. ASHRAE Transactions 2003;109:487−500.

[29] Yuehong B, Guo T, Zhang L, Chen L. Solar and ground source heat pump system. Applied Energy 2004;78:231−45.

[30] Ozgener O, Hepbasli A. Performance analysis of a solar assisted ground-source heat pump system for greenhouse heating: an experimental study. Building and Environment 2005;40(8):1040−50.

[31] Zongwei H, Maoyu Z, Fanhong K, Fang W, Li Z, Tian B. Numerical simulation of solar assisted ground-source heat pump heating system with latent heat energy storage in severely cold area. Applied Thermal Engineering 2008;28(11−12):1427−36.

[32] Behnaz R, Marc R. District heating and cooling: review of technology and potential enhancements. Applied Energy 2012;93:2−10.

[33] Urban P, Sven W. District heating in sequential energy supply. Applied Energy 2012;95:123−31.

[34] Urban P, Sven W. Heat distribution and the future competitiveness of district heating. Applied Energy 2011;88:568−76.

[35]   Schmidt T, Mangold D. Large-scale thermal energy storage—Status quo and perspectives. In: First International SDH Conference, Malmö, Sweden; 2013.

[36]   Holm L. Long term experiences with solar district heating in Denmark. European Sustainable Energy Week, Brussels; 2012.

[37]   Sarbu I, Valea ES. Energy savings potential for pumping water in district heating stations. Sustainability 2015;7:5705—19.

[38]   Sarbu I, Borza I. Energetic optimization of water pumping in distribution systems. Periodica Polytechnica Mechanical Engineering 1998;2:141—52.

[39]   Rishel JB. Water pumps and pumping systems. New York (NY): McGraw-Hill; 2002.

[40]   Volk M. Pump characteristics and applications. Boca Raton (FL): Taylor&Francis Group; 2005.

[41]   Marchi A, Simpson AR, Ertugrul N. Assessing variable speed pump efficiency in water distribution systems. Drinking Water Engineering Science 2012;5:15—21.

[42]   Stockman K, Dereyne S, Vanhooydonck D, Symens W, Lemmens J, Deprez W. Iso efficiency contour measurement results for variable speed drives. Proceedings of the XIX international Conference on electrical Machines,. Rome (Italy): ICEM; 2010.

[43]   Georgescu I. Energy savings by driving with variable rotational speed of pumps and fans in automation of technological processes. Energetica 1988;36:99—109.

[44]   Streicher W, Fink C. Solarenergieeinspeisung in bestehende Fernwärmenetze. In: Internationales Symposium für Sonnenenergienutzung. Gleisdorf/Österreich, Austria; 2006. p. 153—62.

[45]   IDHA. District heating handbook. Washington D.C: International District Energy Association; 1983.

[46]   Bohm B. Energy-economy of Danish district heating systems: a technical and economic analysis. Lyngby: Laboratory of Heating and Air Conditioning. Technical University of Denmark; 1988.

[47]   Stephenson D. Pipeline design for water engineers. New York (NY): Elsevier Scientific; 1981.

[48]   Sarbu I, Brata S. Optimal design of district heating networks. In: Proceedings of the 14th Hungarian conference on district heating. Debrecen, Hungary; 1994. p. 112—7.

[49]   Sarbu I, Ostafe G. Optimal design of urban water supply pipe networks. Urban Water Journal 2016;13(5):521—35.

[50]   Bazaraa MS, Jarvis JJ, Sherali HD. Linear programming and network flows. New York (NY): Wiley; 1990.

[51]   Cunningham WH. A network simplex method. Mathematical Programming 1976;11:105—16.

[52]   Genchi Y, Kikegawa Y, Inaba A. CO_2 payback-time assessment of a regional scale heating and cooling system using a ground source heat-pump in a high energy-consumption area in Tokyo. Applied Energy 2002;71:147—60.

[53]   Marinova M, Beaudry C, Taoussi A, Trepanier M, Paris J. Economic assessment of rural district heating by bio-steam supplied by a paper mill in Canada. Bulletin of Science. Technology & Society 2008;28(2):159—73.

[54]   Reihav C, Werner S. Profitability of sparse district heating. Applied Energy 2008;85:867—77.

[55]   Abdelaziz EA, Saidur R, Mekhilef S. A review on energy saving strategies in industrial sector. Renewable and Sustainable Energy Reviews 2011;15(1):150—68.

[56]   Hafner B, Stopok O, Zahler C, Berger M, Hennecke K, Krüger D. Development of an integrated solar-fossil powered steam generation system for industrial applications. Energy Procedia 2014;48:1164—72.

[57] Muneer T, Maubleu S, Asif M. Prospects of solar water heating for textile industry in Pakistan. Renewable and Sustainable Energy Reviews 2006;10(1):1–23.

[58] Kalogirou S. The potential of solar industrial process heat applications. Applied Energy 2003;76:337–61.

[59] Heβ S, Oliva A. Solar process heat generation: Guide to solar thermal system design for selected industrial processes. Linz (Germany): Energiesparverband; 2010.

[60] Fiorenza G, Sharma VK, Braccio G. Techno-economic evaluation of a solar powered water desalination plant. Energy Conversion and Management 2003;44(14):2217–40.

[61] Ivancic A, Mugnier D, Stryi-Hipp G, Weiss W. Solar heating and cooling technology roadmap: European technology platform on renewable heating and cooling. Brussels (Belgium): European Solar Thermal Industry Federation; 2014. http://www.estif.org/fileadmin/estif/content/projects/ESTTP/Solar_H_C_Roadmap.pdf.

[62] IRENA. Renewable power generation costs in 2014. Abu Dhabi: International Renewable Energy Agency; 2015. http://costing.irena.org.

Chapter 6

Heat Distribution Systems in Buildings

6.1 GENERALITIES

To distribute the solar heat in buildings, a hydronic system (radiant panels and hot-water radiators) or a central forced-air system can be used.

In central heating systems, the hot-water supply temperature can have different values. In the recent past, the most used value in Romania, as well as in other European Union countries, was 90°C with a 20°C temperature drop, but currently, the supply temperature is typically lower than 90°C.

The assurance of the heat demand for buildings equipped with central heating installations requires systems with high efficiency not only in the heat generation process but also in the distribution of the thermal energy. One way to obtain higher efficiency of the heating systems is to use reduced temperature [1]. In addition, it is possible to use RES with higher efficiency as solar energy. Generally, flat-plate liquid collectors heat the transfer and distribution fluids to between 35 and 50°C. The system must be controlled and optimized in correspondence with the ever-changing heat demand.

The energy and exergy efficiency of central heating systems is higher at reduced hot-water temperatures [2], but based on [3], it has to be stated that this is valid only for totally balanced systems. The stability of reduced temperature central-heating system can be improved by decreasing the temperature drop level. Thus, heating systems can be obtained with a higher stability and energy efficiency by decreasing the supply temperature and the temperature drop simultaneously.

After the introduction of plastic piping, the application of water-based radiant heating with pipes embedded in room surfaces (i.e., floors, walls and ceilings) has significantly increased worldwide. Earlier applications of radiant heating systems were mainly for residential buildings because of the comfort and free use of floor space without any obstruction from installations. For similar reasons, as well as possible peak-load reduction and energy savings, radiant systems are widely applied in commercial and industrial buildings. Due to the large surfaces needed for heat transfer, the systems work with low water temperatures for heating. However, to extend the use of these types of

Solar Heating and Cooling Systems. http://dx.doi.org/10.1016/B978-0-12-811662-3.00006-2
207

generators and to benefit from their energy efficiencies to reach the targets of 20−20−20 (20% increase in energy efficiency, 20% reduction of CO_2 emissions, and 20% renewable by 2020), working with radiators, which were the most commonly used terminal units in heating systems in the past, is necessary.

There are tens of thousands of buildings to restore in Europe, the majority of which are residential. The energy challenge of the future will be in renovating existing buildings, and proposes system-engineering technologies that can be installed with minimal interventions, which will be immensely successful. Therefore, if solar technology is promoted, it must be designed to work also with radiators.

This chapter presents the heat distribution systems in buildings, including hot-water radiators, radiant panels (floor, wall, ceiling, and floor-ceiling) and room air heaters. First objective of this study is the analysis of the energy savings in central heating systems with reduced supply temperature, for different types of radiators taking into account the thermal insulation of the distribution pipes and the performance investigation of different types of low-temperature heating system with different methods. Additionally, a mathematical model for numerical modeling of the thermal emission at radiant floors is developed and experimentally validated, and a comparative analysis of the energy, environmental, and economic performances of floor, wall, ceiling, and floor-ceiling heating using numerical simulation with Transient Systems Simulation (TRNSYS) software is performed. Finally, important information for control and efficiency of SHSs is included, an analytical model for energetically analysis of the SHSs is developed, and some economic analysis indicators are presented to show the opportunity to implement these systems in buildings.

6.2 RADIATOR HEATING SYSTEM

6.2.1 Description of the System

A hot-water radiator heating system is a type of central heating. In the system, heat is generated in a boiler. For the generation of the heat, a natural gas boiler is used where the chemical energy of natural gas is transferred into the heat. Then, the heat is distributed by hot water (distribution fluid) to the radiators. The radiators heat the rooms. The radiators are installed in each heated room of the house. The hot water is circulated by a water circulation pump, which operates continuously. If the valves stop, then the hot water flows through a bypass pipe. The radiators, as a rule of thumb, are located next to the cold surfaces of the envelope. They significantly influence the thermal comfort.

The radiators can make use of the hot water generated by a solar hydronic system. One way to save energy in hot-water radiator systems is to retrofit them to provide separate zone control for different areas of large homes. Zone control is most effective when large areas of the home are not used often or are used on a different schedule than other parts of the home. A heating professional can install automatic valves on the hot-water radiators, controlled by thermostats in each zone of the house. Zone control works best in homes designed to operate in different heating zones, with each zone insulated from the others. In homes not designed for zone control, leaving one section at a lower temperature could cause comfort problems in adjacent rooms because they will lose heat to the cooler parts of the home. Zone control will also work best when the cooler sections of the home can be isolated from others by closing the doors.

The radiators release the highest amount of heat to the heated room by convection and one part by heat radiation. The convective heat transfer will lead to a lower relative humidity of the air, and, at high radiator surface temperature, dust particles can be burned, leading to lower indoor air quality. Thus, emitters should be implemented with a radiation factor as high as possible in the case of high-temperature water supplies. The values of the radiation factor are presented in Table 6.1 for typical hot-water radiators [4].

The highlights of the convective thermal field achieved with radiators, illustrated in Fig. 6.1A, are as follows:

1. a warm-air jet (1) that is raised from the area of the heater to the upper part of the room, as a result of the gravitation forces;
2. a warm-air jet (2), developed at the surface of the ceiling, also as a result of gravitation forces;

TABLE 6.1 Radiation Factor of Usual Radiators

Radiator Type	Heat Transferred by Radiation		
	Room-wards	Wall-wards	Total
Steel column radiator	0.28	0.10	0.38
Cast-iron column radiator	0.26	0.10	0.36
Panel radiator			
1/0[a]	0.38	0.18	0.56
1/1	0.25	0.11	0.36
2/0	0.23	0.10	0.33
2/1	0.20	0.08	0.28
2/2	0.17	0.07	0.24
3/3	0.14	0.04	0.18

[a]The first number represents the number of panels and the second the number of convective elements.

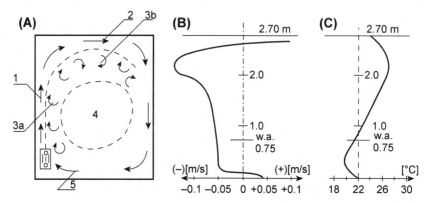

FIGURE 6.1 Convective field for the static heaters in living rooms: (A) convective currents; (B) velocity profile; (C) temperature profile. *w.a.*, work area.

3. a rotational area, for air circulation, which is more active on the vertical (3a) in the area of the warmth source and less active (3b) under the ceiling air jet;
4. a steady area (4) in the middle of the room;
5. induction currents (5) at the floor level, resulting from air cooling by vertical walls.

Computer simulation studies using TRNSYS software indicate that according to the heat flux emitted by the heater, as well as to its temperature level, it is possible to establish a velocity profile (Fig. 6.1B) related to the temperature profile (Fig. 6.1C) in rooms with a certain given geometry. The heat flux was 2163 W per one-meter radiator length, and inlet and outlet temperatures to the radiator were 90°C and 70°C, respectively. Thus, for a living room (with height $h = 2.70$ m), the air current velocity to the middle of the room is practically constant, increasing in the active circulation area and afterward it starts to decrease again at the ceiling. The air temperature increases substantially enough, from 20 to 22°C in the working area to 24–26°C under the ceiling.

With regard to the heat flux values yielded by the heaters, in situ measurements [5] have indicated that the heat flux yielded by convection has the highest value. The heat flux yielded by radiation to the convectors is lower, for radiators it is under 50%, and for radiators with metal fins it is 10–25%.

The high temperature of the hot water can lead to a lower thermal comfort level because of the asymmetric radiation [6,7].

Fig. 6.2 shows the variation of radiator surface depending on the logarithmic mean temperature difference Δt for different values of the α radiator exponent. It can be observed an increase of the radiator surface (A_R/A_{R0}) while the values of the α exponent decrease for the same temperature difference Δt. The necessary radiator surface A_R will increase for heating systems with supply/return water temperatures lower than 90/70°C.

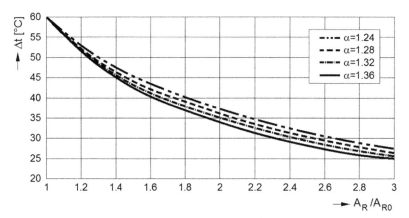

FIGURE 6.2   Variation of radiator surface.

To ensure ever-changing heat demand in a room, qualitative, quantitative, or mixed control systems are used. By qualitative control, the controlled parameter is the supply temperature and the flow rate is constant during the operation time. By quantitative control, the controlled parameter is the flow rate, the supply temperature remaining constant throughout the whole operation period.

6.2.2 Energy Saving

A radiator heating system with a thermal load of 14.0 kW that assures the heat demand for a residential building (Fig. 6.3) located in Timisoara, Romania, is considered. The classic natural gas boiler is used as a source of energy. The distribution pipe with a diameter of 40/32 mm has a length of 10 m, and the air temperature around the pipe is 10°C. The velocity of water flow in the pipe is 0.48 m/s, with nominal parameters. The effect of hot-water temperature variation in the ranges of 40–90°C on energy consumption of the heating system for qualitative and quantitative control of the heat delivered by system and different insulation levels of the distribution pipes is illustrated, applying the mathematical model developed by Sarbu [8].

The hot-water temperature drop $\Delta t_{\mathrm{w}} = t_{\mathrm{wi}} - t_{\mathrm{wo}}$ in an insulated pipe with a length of 1 m can be approximated with the following equation [8]:

$$\Delta t_{\mathrm{w}} = \frac{[8.1 + 0.045(t_{\mathrm{wi}} - t_{\mathrm{e}})(1 - \eta_{\mathrm{iz}})](t_{\mathrm{wi}} - t_{\mathrm{e}})(1 - \eta_{\mathrm{iz}})\pi D_{\mathrm{e}}}{mc_{\mathrm{p}}}, \qquad (6.1)$$

where t_{wi} and t_{wo} are the hot-water temperatures in the inlet and outlet sections of pipe, respectively; t_{e} is the air temperature around the pipe; η_{iz} is the thermal insulation efficiency; D_{e} is the external diameter of pipe; m is the mass

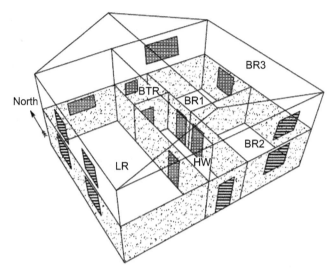

FIGURE 6.3   Residential building living room (LR); hallway (HW); bathroom (BTR); bedrooms (BR1, BR2, and BR3).

flow rate of hot water; and c_p is the specific heat of hot water. In Eq. (6.1), t_{wo} is equal to t_d (the supply hot-water temperature).

If the boiler-temperature drop and the mass flow rate are known, the energy consumption of the boiler can be determined.

In these conditions, using the geometrical interpolation method [9], the temperature drop Δt_w for insulated pipe can be written in simplified form as

- for $\eta_{iz} = 0$:

$$\Delta t_w = \left(-5 + 0.32t^{1.17}\right) \times 10^{-3} \tag{6.2}$$

- for $\eta_{iz} = 0.7$:

$$\Delta t_w = \left(-1 + 0.05t^{1.28}\right) \times 10^{-3}, \tag{6.3}$$

where t is the hot-water temperature.

Using Eqs. (6.2) and (6.3), it can be observed that the real values for the hot-water temperature drop at the boiler are higher than the theoretical values. The deviations between the real and theoretical values are lower when the hot-water temperature is much lower, and these deviations are higher when the radiator exponent is higher.

In Table 6.2, the real temperature-drop values at the boiler are summarized, for qualitative control, depending on the outdoor air temperature t_e at different values of the hot-water supply temperature t_d and for different values of η_{iz}.

The variation of real temperature-drop Δt_w, depending on the outdoor temperature t_e for different values of the hot-water supply temperature t_d and of the efficiency η_{iz}, is illustrated in Fig. 6.4 for the case of a quantitative

TABLE 6.2 Real Values of Δt_w (°C) for Qualitative Control

t_d (°C)	t_e (°C)	$\eta_{liz} = 0$		$\eta_{liz} = 0.7$	
		$\alpha = 1.24$	$\alpha = 1.36$	$\alpha = 1.24$	$\alpha = 1.36$
90	-12.8	19.67	19.67	18.98	18.98
	-8.3	17.00	17.01	16.38	16.39
	-5.0	15.04	15.05	14.48	14.48
	0	12.07	12.09	11.59	11.60
	5.0	9.09	9.12	8.70	8.71
	10.2	5.99	6.02	5.69	5.70
80	-12.8	19.52	19.53	18.94	18.94
	-8.3	16.87	16.88	16.35	16.35
	-5.0	14.92	14.94	14.45	14.45
	0	11.97	11.99	11.56	11.57
	5.0	9.02	9.04	8.68	8.69
	10.2	5.94	5.97	5.68	5.69
60	-12.8	19.24	19.24	18.87	18.87
	-8.3	16.62	16.63	16.28	16.28
	-5.0	14.70	14.71	14.39	14.39
	0	11.79	11.80	11.51	11.52
	5.0	8.88	8.89	8.64	8.65
	10.2	5.85	5.86	5.66	5.66
40	-12.8	18.96	18.96	18.80	18.80
	-8.3	16.37	16.37	16.21	16.21
	-5.0	14.48	14.28	14.33	14.33
	0	11.61	11.62	11.46	11.47
	5.0	8.74	8.76	8.60	8.61
	10.2	5.76	5.75	5.64	5.63

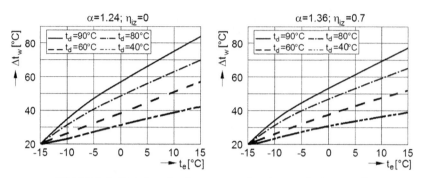

FIGURE 6.4   Variation of real temperature-drop for quantitative control.

control. If the flow rate decreases by quantitative control to assure the radiator thermal power, then Δt_w increases as the outdoor temperature increases due to greater heat losses through the pipe walls.

In Fig. 6.5, the percent energy saving e_s is given for a hot-water supply temperature t_d lower than 90°C and noninsulated or well-insulated pipes. It can be observed that the energy saving increases insignificantly for higher values of the radiator exponent α. The energy saving also increases for lower values of the hot-water temperature t_d, reaching approximately 7.5% and 5% for t_d values of 40°C and 45°C, respectively. The energy saving decreases when the pipe thermal insulation efficiency η_{iz} is higher reaching approximately 5.2% and 3.7% for t_d values of 40°C and 45°C, respectively.

Fig. 6.6 shows the percent energy saving e_s with well-insulated pipes ($\eta_{iz} = 0.7$) depending on the control method for different values of the hot-water temperature t_d. It can be observed that the energy saving decreases from 8.6 to 5−6% for lower values of the hot-water temperature in the case of qualitative control and remains approximately constant (5%) in the case of quantitative control. Both the energy-saving decrease at higher values of the radiator exponent α in the case of quantitative control and the energy-saving increase with the radiator exponent in the case of qualitative control occur insignificantly.

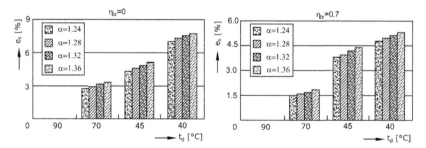

FIGURE 6.5   Energy saving depending on the supply temperature.

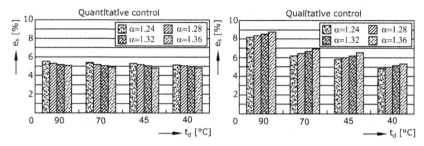

FIGURE 6.6   Energy saving by pipe insulation depending on system control.

6.3 RADIANT HEATING SYSTEMS

6.3.1 Preliminaries

In low-energy buildings, the low-temperature heating system usually works with a supply water temperature below 45°C [10]. Embedded radiant systems are used in all types of buildings. Radiant heating systems supply heat directly to the floor or to panels in the wall or ceiling of a house. Hydronic (liquid-based) systems can use a wide variety of energy sources to heat the distribution liquid, including solar water heaters, standard gas- or oil-fired boilers, wood-fired boilers, or a combination of these sources.

Radiant heating application is classified as panel heating if the panel surface temperature is below 150°C [11]. In thermal radiation, heat is transferred by electromagnetic waves that travel in straight lines and can be reflected. The water temperatures are operated at very close to room temperature and, depending on the position of the piping, the system can take advantage of the thermal storage capacity of the building structure.

Fig. 6.7 shows the available types of embedded hydronic radiant systems [12]. These systems are usually insulated from the main building structure (floor, wall, and ceiling), and the actual operation mode (heating/cooling) of the systems depends on the heat transfer between the water and the space.

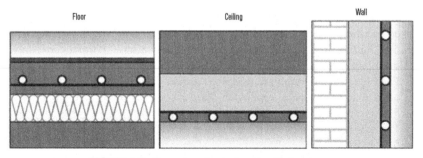

FIGURE 6.7   Examples of water-based radiant systems.

Panel heating provides a comfortable environment by controlling surface temperatures and minimizing air motion within a space. A radiant system is a sensible heating system that provides more than 50% of the total heat flux by thermal radiation. The controlled temperature surfaces may be in the floor, walls, or ceiling, with the temperature maintained by circulation of water or air.

The radiant heat transfer is, in all cases, 5.5 W/(m^{2}K). The convective heat transfer then varies between 0.5 and 5.5 W/(m^{2}K), depending on the surface type and on heating or cooling mode. This shows that the radiant heat transfer varies between 50% and 90% of the total heat transfer [13].

The low-temperature radiant systems are very complex because they involve different mechanisms of heat transfer: heat conduction through the walls, convection between the heating panel and the indoor air, heat radiation between the heating panel and the surrounding areas, and the heat conduction between the floor and the ground. The main goal of low-temperature radiant systems is to provide adequate thermal comfort at significantly lower temperatures.

Radiant panel heating is characterized by the fact that heating is associated with a yielding of heat with low temperature because of physiological reasons. Thus, at the radiant floor panels, the temperature must not exceed +29°C, and at the radiant ceiling panels, the temperature will not exceed 35−40°C, depending on the position of the occupier (in feet) and the occupier distance to the panels, in accordance with thermal comfort criteria established by ISO Standard 7730 [14]. A vertical air temperature difference between the head and feet of less than 3°C is recommended.

In a well-insulated building, the selected floor surface material is of crucial importance with regard to how warm the floor feels. For example, oak parquet at a temperature of 21°C and stone floor at a temperature of 26°C feel neutral and roughly the same under a bare foot according ISO/TS 13,732-2 [15]. However, this is not always the case, the percent dissatisfied (PD) in % has a relation with floor surface temperature as follows [16]:

$$PD = 100 - 94 \exp\left(-1.387 + 0.118t_f - 0.0025t_f^2\right), \qquad (6.4)$$

where t_f is the floor surface temperature.

The vertical profile of the air temperature for two types of radiant heating panels is illustrated in Fig. 6.8. The radiant part is lower (70%) at the floor heating than the ceiling in terms of heating (85%) because thermal convection is developed more in the case of floor heating panels [17].

The higher mean radiant temperature in radiantly heated space means that the air temperature can be kept lower than in convectively heated space. This has the advantage that the relative humidity in winter may be a little higher.

The heat transfer between the water and surface is different for each system configuration. Therefore, the estimation of heating/cooling capacity of systems is very important for the proper system design. Two calculation methods

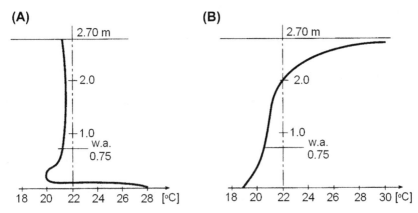

FIGURE 6.8 Temperature profile for low-temperature radiant heating: (A) floor heating; (B) ceiling heating.

included in ISO 11,855 are simplified calculation methods depending on the type of system, and finite element method or finite difference method.

The simplified calculation methods are specific for the given system types within the boundary conditions. Based on the calculated average surface temperature at given distribution fluid temperature and operative temperature in the space, it is possible to determine the steady-state heating and cooling capacity. Thus, the heating/cooling capacity of the floor, wall, and ceiling heating systems is [18]:

- floor heating and ceiling cooling:

$$q = 8.92|t_o - t_{S,m}|^{1.1} \tag{6.5}$$

- wall heating and wall cooling:

$$q = 8.0|t_o - t_{S,m}| \tag{6.6}$$

- ceiling heating:

$$q = 6.0|t_o - t_{S,m}| \tag{6.7}$$

- floor cooling:

$$q = 7.0|t_o - t_{S,m}|, \tag{6.8}$$

where q is the heating capacity, in W/m^2; t_o is the operative (comfort) temperature in the space, in °C; and $t_{S,m}$ is the average surface temperature, in °C.

The heating capacity for floor and ceiling is up to 100 W/m^2 and 40 W/m^2, respectively. The sensible cooling capacity for floor and ceiling is 40 W/m^2 and up to 100 W/m^2, respectively.

Control of the heating system needs to be able to maintain the indoor air temperatures within the comfort range under the varying internal loads and external climates. To maintain a stable thermal environment, the control system needs to maintain the balance between the heat gain of the building and the supplied energy from the system.

6.3.2 Radiant Floor Heating

6.3.2.1 Description of the System

There are three types of radiant floor heating: radiant air floors (air is the distribution fluid), electric radiant floors, and hot-water (hydronic) radiant floors.

Hydronic systems are the most popular and cost-effective radiant heating systems for heating-dominated climates. Hydronic radiant floor systems pump heated water from a boiler through tubing laid in a pattern under the floor. The tubing can be embedded in a thick concrete foundation slab or in a thin layer of concrete, gypsum, or other materials installed on top of a subfloor. Thick concrete slabs are ideal for storing heat from solar energy systems, which have a fluctuating heat output. The downside of thick slabs is their slow thermal response time, which makes strategies such as night or daytime setbacks difficult, if not impossible. Most experts recommend maintaining a constant temperature in homes with these heating systems.

In some systems, controlling the flow of hot water through each tubing loop by using zoning valves or pumps and thermostats regulates room temperatures. Ceramic tile is the most common and effective floor covering for radiant floor heating, because it conducts heat well and adds thermal storage. Common floor coverings like vinyl and linoleum sheet goods, carpeting, or wood can also be used, but any covering that insulates the floor from the room will decrease the efficiency of the system.

The cost of installing a hydronic radiant floor varies by location and depends on the size of the home, the type of installation, the floor covering, and the cost of labor.

6.3.2.2 Numerical Modeling of Thermal Emission at Radiant Floor

Floor heating construction corresponds with the mostly used system that is plastic tubing (PEX) embedded in the concrete slab.

Mathematical model: For heat exchange modeling between the radiant floor and environment, the "virtual tube" method is used [19]. This allows the calculation of temperature at a point P around a tube T of radius ρ placed at a distance b from the surface S, maintained at constant temperature of $0°C$.

The temperature at point P is given by the following equation:

$$t_P = \frac{t_{hc}}{\ln \frac{\rho}{2b}} \ln \frac{r}{r'}, \tag{6.9}$$

where t_{hc} is the mean temperature of the heat carrier; r is the distance from point P to the center of the tube cross-section; and r' is the distance from point P to the center of the virtual tube cross-section.

The virtual tube method was adapted to model thermal emission of radiant floors, considering that the distance b is the equivalent of the sum between the thermal resistances of floor layers above tube (R) and the superficial heat transfer resistance (R_i). In this case, the surface S of constant temperature is represented by the environment (Fig. 6.9).

Eq. (6.9) to calculate the temperature at point P located on the floor surface thus becomes the following:

$$t_P = t_i + \frac{t_{hc} - t_i}{\ln \frac{p}{2b}} \ln \frac{r}{r'} \tag{6.10}$$

in which

$$r = \sqrt{R^2 + x^2}; \quad r' = \sqrt{(R + 2R_i)^2 + x^2} \tag{6.11}$$

$$R = \sum_{j=1}^{N} \frac{\delta_j}{\lambda_j}; \quad R_i = \frac{1}{\alpha_i}, \tag{6.12}$$

where t_i is the indoor air temperature; R is the thermal diffusion resistance of compound floor layers above the tube, in $m^2 K/W$; R_i is the superficial heat transfer resistance at the internal surface, in $m^2 K/W$; δ_j and λ_j are the thicknesses, in m, and thermal conductivity of layer j, in $W/(m \cdot K)$, respectively; and α_i is the superficial heat transfer coefficient of the floor surface, in $W/(m^2 \cdot K)$.

Therefore, the modeling allows, for different floor structures, the determination of floor surface temperature at any point on its surface. In addition, the temperature at any point within the floor can be calculated. Thermal emission is proportional to the difference between the surface temperature and indoor air temperature. However, the surface temperature is not uniform. This

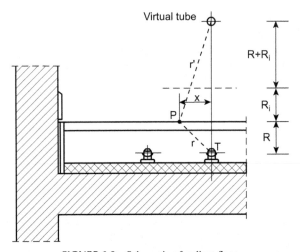

FIGURE 6.9 Schematic of radiant floor.

parameter has a nonlinear variation in relation to the distance from the vertical of tube section, according to Eqs. (6.9) and (6.10).

The mean floor surface temperature is given by the following:

$$t_{f,m} = \frac{1}{\frac{s}{2}} \int_0^{s/2} \left[t_i + \frac{t_{hc} - t_i}{\ln \frac{\rho}{2(R+R_i)}} \ln \frac{r}{r'} \right] dx. \tag{6.13}$$

Integrating Eq. (6.13) obtains the following:

$$t_{f,m} = \left[t_i + \frac{t_{hc} - t_i}{\ln \frac{\rho}{2(R+R_i)}} \right] \frac{A - B}{s} \tag{6.14}$$

in which

$$A = \left[\frac{s}{2} \ln \left(\frac{s^2}{4} + R^2 \right) \right] - \left[s - \left(2R \, \text{arctg} \frac{s}{2R} \right) \right] \tag{6.15}$$

$$B = \left\{ \frac{s}{2} \ln \left[\frac{s^2}{4} + (R + 2R_i)^2 \right] \right\} - \left\{ s - \left[2(R + 2R_i) \text{arctg} \frac{s}{2(R + 2R_i)} \right] \right\}, \tag{6.16}$$

where s is the arrangement step of the radiant floor tubes, and A, B are notations.

Superficial heat transfer: Radiative heat transfer between the floor and the room walls is calculated with the following relation:

$$q_r = \sigma \varepsilon_1 \varepsilon_2 \left(T_1^4 - T_2^4 \right) \tag{6.17}$$

in which q_r is the radiant flux, in W/m^2; $\sigma = 5.67 \times 10^{-8}$ W/(m$^2 \cdot$K^4) is the Stefan−Boltzmann constant; ε_1 and ε_2 are the thermal emittances of the floor surface and room walls, respectively (dimensionless); T_1 is the absolute temperature of radiant floor surface, in K; and T_2 is the weighted average absolute temperature of all room walls, in K [11].

The radiative heat transfer coefficient α_r can be calculated by [20]

$$\alpha_r = \frac{q_r}{T_1 - T_2}. \tag{6.18}$$

Convective heat transfer is determined with a criteria group expressed by dimensionless numbers as

$$\text{Nu} = \frac{\alpha_c L}{\lambda} = C (Gr \cdot Pr)^n \tag{6.19}$$

$$\text{Gr} = \frac{g \beta L^3 \Delta t_{f-a}}{\nu^2} \tag{6.20}$$

$$Pr = \frac{v}{a} \qquad (6.21)$$

in which Nu, Gr, and Pr are Nusselt, Grashoff, and Prandtl numbers, respectively; α_c is the convective heat transfer coefficient, in W/(m²·K); L is the characteristic dimension of the element surface, in m; λ is the thermal conductivity of air, in W/(m·K); g is the gravitational acceleration, in m/s²; C and n are the parameters depending on the $Gr·Pr$ product; β is the volumetric expansion coefficient of air, in K^{-1}; Δt_{f-a} is the temperature difference between the floor surface and air, in K; v is the kinematic viscosity of air, in m²/s; and a is the thermal diffusivity of air, in m²/s.

Thermal emission at floor surface: Thermal emission of radiant floor is computed by taking into account the weighted mean temperature of walls and indoor air temperature.

The operative (comfort) temperature t_o may be defined as the average of the mean radiant temperature t_r and indoor air temperature t_i weighted by their respective heat transfer coefficients as [21]

$$t_o = \frac{\alpha_r t_2 + \alpha_c t_i}{\alpha_r + \alpha_c} \qquad (6.22)$$

in which α_r and α_c are the radiative and the convective heat transfer coefficient between body and environment, in W/(m²K), respectively. The mean radiant temperature is approximated with t_2.

The mean superficial heat flux q is given by the product of the total heat transfer coefficient $(\alpha_r + \alpha_c)$ and the difference between mean floor surface temperature $t_{f,m}$ and operative temperature t_o as

$$q = (\alpha_r + \alpha_c)(t_{f,m} - t_o). \qquad (6.23)$$

The analytical model described above allows the determination of the maximum heat carrier temperature for any type of radiant floor structure. Exceeding this temperature leads to temperature values at the intersection point between the floor surface and the vertical of tube section beyond the maximum allowed 29°C.

Computation example: The radiant floor under consideration is used for an office room with geometrical dimensions of 6.7 × 3.3 × 3.45 m (Fig. 6.10) located in Timisoara, Romania. The latitude and longitude of this city are 45°47′ N and 21°17′ E, respectively. The radiant floor has embedded plastic tubes (netlike polyethylene) with a diameter of 20 × 2 mm, arranged with a step $s = 20$ cm, through which circulates the heat carrier of 42/36°C. Above the tubes, there is a concrete layer with thickness $\delta_1 = 5$ cm and a wood layer with thickness $\delta_2 = 8$ mm. The indoor air temperature is 22°C, and the weighted average wall temperature measured by thermograph is 24°C. The thermal emittances are $\varepsilon_1 = 0.9$ and $\varepsilon_2 = 0.85$.

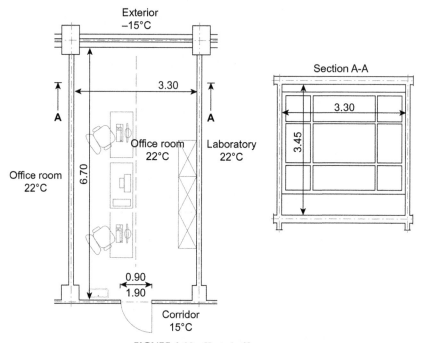

FIGURE 6.10 Heated office room.

The floor temperature $t_{f,m}$ at a point situated in the center of the distance between two consecutive tubes is determined, and finally, the operative temperature t_o and mean superficial heat flux q results are obtained. The numerical results obtained applying a previously developed computation model are summarized in Table 6.3.

Validation of mathematical model: Some statistical methods such as the root-mean square (RMS), the coefficient of variation (c_v), the coefficient of multiple determinations (R^2), and percentage difference may be used to compare simulated (computed) and actual values for model validation.

TABLE 6.3 Numerical Computation Results

R (m²K/W)	R_i (m²K/W)	B (m)	t_P (°C)	$t_{f,m}$ (°C)	q_r (W/m²)	α_r (W/m²K)	α_c (W/m²K)	t_o (°C)	Q (W/m²)
0.065	0.125	0.19	27.6	29.0	23.2	4.64	5.23	22.9	61.9

The simulation error can be estimated by the RMS defined as [22]

$$RMS = \sqrt{\frac{\sum_{i=1}^{n} \left(y_{sim,i} - y_{mea,i}\right)^2}{n}}. \tag{6.24}$$

In addition, the coefficient of variation c_v, in percent, and the coefficient of multiple determinations R^2 are defined as follows [22]:

$$c_v = \frac{RMS}{\left|\overline{y}_{mea,i}\right|} 100 \tag{6.25}$$

$$R^2 = 1 - \frac{\sum_{i=1}^{n} \left(y_{sim,i} - y_{mea,i}\right)^2}{\sum_{i=1}^{n} y_{mea,i}^2}, \tag{6.26}$$

where n is the number of measured data in the independent data set; $y_{mea,i}$ is the measured value of one data point i; $y_{sim,i}$ is the simulated value; and $\overline{y}_{mea,i}$ is the mean value of all measured data points.

The floor surface temperature values between two consecutive tubes, computed with the proposed model, are compared with measured values on a radiant floor heating system with the structure from the previous calculation example.

The temperatures measured at every point with a TESTO 350 instrument are represented by the average of three measurements. The maximum deviation between the extreme measured values was 0.15°C. The obtained results are plotted in Fig. 6.11, and they show a good agreement between the experimental temperature measurements and the computed values. Statistical values such as RMS, c_v, and R^2 are 0.12536, 0.00486, and 0.99998, respectively, which can be considered as very satisfactory. Thus, the computational model was validated by the experimental data.

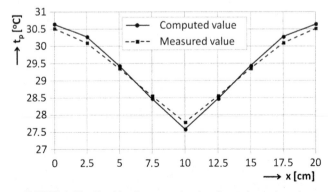

FIGURE 6.11 Graphic of computational and experimental results.

6.3.3 Comparative Analysis of Radiant Panel System Performance

An investigated residential building located in Timisoara, Romania, is shown in Fig. 6.3. The investigated building and its heating systems are modeled using TRNSYS software.

6.3.3.1 Description of Panel Systems

The panel heaters may be the floor heating panels, wall heating panels, ceiling heating panels, and floor-ceiling heating panels. The floor heating panels have a total surface area of 160 m². The wall heating panel is located at the exterior wall and has a total surface area of 177 m². The ceiling heating panel is located at the ceiling of the first and second storeys of the house and has total surface area of 160 m². The floor-ceiling heating panel operates as a ceiling heating of the lower story and as a floor heating of the upper story. Its total surface area is 80 m².

The main component of the heating panels is the pipe where the hot water flows. The hot water inlet temperature has the same value of 37°C for all heating systems.

For all heating panels, classic boilers were used to generate heat by using natural gas. The water circulation pump uses electricity to operate.

Four water-based radiant systems are analyzed: (1) floor heating, (2) wall panel heating, (3) ceiling heating, and (4) floor-ceiling heating.

6.3.3.2 Primary Energy Consumption of Heating System

The primary energy consumption E per heating season of the analyzed building is calculated by using the following equation:

$$E = E_g + \frac{E_{el}}{\eta_{el}}, \tag{6.27}$$

where E_g is the consumption of natural gas per heating season; E_{el} is the consumption of electrical energy per heating season; and $\eta_{el} = 0.4$ is the electricity generation efficiency, defined as ratio between finally produced electric energy and the primary energy consumption for electricity generation. If the boilers utilize solar energy, then a significant primary energy saving is possible.

6.3.3.3 Operating Cost

Total operating cost C_T to run the heating system is calculated by using the following equation:

$$C_T = c_{el}E_{el} + kc_g E_g, \tag{6.28}$$

where c_{el} is the specific cost of electricity; c_g is the specific cost of natural gas with energy value of 33,338 kJ/m³; and k is the correction coefficient of the natural gas consumption. The specific conditions of Romania can be considered: $c_{el} = 0.11$ €/kWh, $c_g = 0.29$ €/m^3, and $k = 1.05$ [23].

6.3.3.4 Carbon Dioxide (CO$_2$) Emission

The CO_2 emission of the heating system during its operation is calculated with following equation:

$$CO_2 = g_g E_g + g_{el} E_{el}, \tag{6.29}$$

where g_g is the specific CO_2 emission factor for natural gas and g_{el} is the specific CO_2 emission factor for electricity (Table 6.4) [24].

6.3.3.5 Results and Discussion

In these investigations, the four analyzed panel systems are simulated by using TRNSYS software during their operation at the heating season.

For each of the four analyzed panel systems, the energy consumption, the operating cost of building heating, the CO_2 emission due to the building heating, and temperatures of rooms were calculated.

The primary energy consumption during the heating season is illustrated in Fig. 6.12. The system with the floor-ceiling heating panel has the lowest energy consumption (7005 kWh/year) and the system with a ceiling heating panel has the highest energy consumption (9630 kWh/year). Their difference

TABLE 6.4 CO$_2$ Emission Factors g_{el} and g_g

Type of Fuel	Emission Factor (kg CO$_2$/kWh)
Electric energy	0.547
Natural gas	0.205

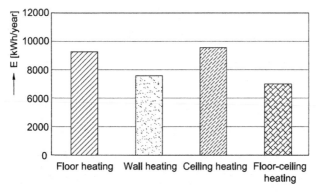

FIGURE 6.12   Energy consumption of heating systems.

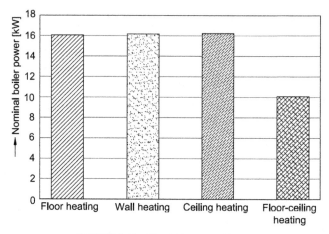

FIGURE 6.13   Nominal power of boiler.

is approximately 27.2%. The system with a floor-ceiling heating panel has a 10% lower energy consumption lower than that of the system with the wall heating panel and 22% lower than that of a system with the floor heating panel.

Fig. 6.13 shows the value of the boiler nominal power for all four systems. The minimum nominal power of the boiler is required for system with the floor-ceiling heating panel (8.5 kW), and the highest is required for the system with the ceiling heating panel (13.7 kW). The nominal power outputs of the other two systems are similar to that of system with the ceiling heating panel.

In Fig. 6.14, the operation cost of heating is shown. These data indicate that the largest heating cost is for the heating system with a ceiling heating panel (500 €), and the lowest cost is for the heating system with the floor heating panel (380 €). The use of the system with the floor heating panel instead of a

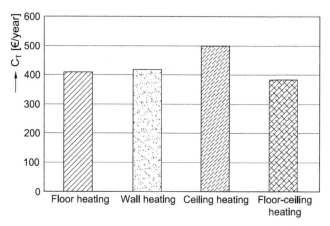

FIGURE 6.14   Operating cost during heating season.

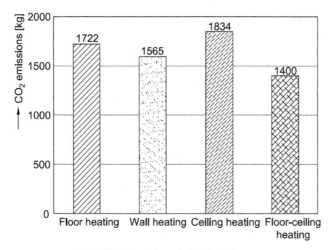

FIGURE 6.15 CO_2 emission for February.

system with the ceiling heating panel would yield a total financial savings of 120 € per heating season.

As seen in Fig. 6.15, the CO_2 emission of the heating systems calculated for February, month with lowest winter temperatures, is the lowest for the system with the floor-ceiling heating panel (1400 kg CO_2) and the highest for the system with the ceiling heating panel (1834 kg CO_2), a difference of 23.7%.

In terms of checking proper operation of all the four systems, Fig. 6.16 illustrates the mean operative indoor air temperature of rooms and the desired temperature of rooms in February. It is found that for all heating systems, the mean indoor air temperatures do not significantly deviate from the set-point operative temperature. The largest deviation (approximately 1°C) is found with the floor-ceiling panel system in room BR3 on the first storey of building, such that the mean temperature in this room is 19.1°C instead of the desired 20°C.

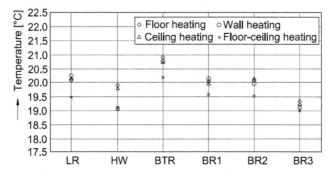

FIGURE 6.16   Mean operative temperature of building rooms.

6.3.3.6 Conclusions

This study showed that radiant heating panels works better than radiator heating. In well-insulated buildings, the energy consumption of the radiator heating system is up to 10% greater than that of the floor heating system, which also provided good thermal comfort.

The proposed analytical model of thermal emission for radiant floors was validated by the measured values, and a reasonable agreement prevailed. This model permits the determination of the floor surface temperature at any point on its surface and the emitted mean heat flux.

The floor-ceiling heating system has the best performance in terms of the lowest energy consumption, operation cost, CO_2 emission, and the nominal boiler power. In addition, it is important to note that the next best performing radiant system is the system with the wall heating panel. The classical ceiling heating system displays the worst performances in terms of the listed parameters.

The comparison of the room operative air temperatures and the set-point operative air temperature indicates that all radiant panel systems provide satisfactory results without significant deviations.

New investigations should be performed to examine other low-temperature heating systems and their combinations to be integrated in solar systems.

6.4 ROOM AIR HEATERS

Air collectors can be installed on a roof or an exterior (south-facing) wall for heating one or more rooms. Although factory-built collectors for on-site installation are available, do-it-yourselves may choose to build and install their own air collector.

The collector has an airtight and insulated metal frame and a black metal plate for absorbing heat with glazing in front of it. Solar radiation heats the plate that, in turn, heats the air in the collector. An electric fan or blower pulls air from the room through the collector, and blows it back into the room. Roof-mounted collectors require ducts to carry air between the room and the collector. Wall-mounted collectors are placed directly on a south-facing wall, and holes are cut through the wall for the collector air inlet and outlets.

Simple "window box collectors" fit in an existing window opening. They can be active (using a fan) or passive. In passive types, air enters the bottom of the collector, rises as it is heated, and enters the room. A baffle or damper keeps the room air from flowing back into the panel (reverse thermosiphoning) when the sun is not shining. These systems provide only a small amount of heat, because the collector area is relatively small.

6.5 CONTROL OF HEATING SYSTEMS

The control of heating and cooling system can be classified as central control, zone control and individual room control [13]. Fig. 6.17 is a diagram on the principles of control [18].

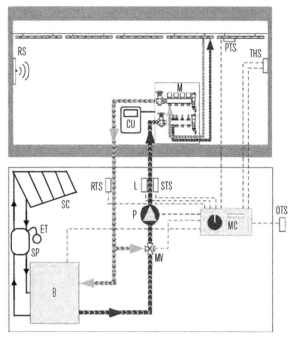

FIGURE 6.17   Diagram of the principles of control exemplified by a radiant floor system. *B*, boiler; *CU*, control unit; *ET*, expansion tank; *MC*, main controller; *M*, manifold; *MV*, mixing valve; *L*, limiter; *OTS*, outdoor temperature sensor; *PTS*, panel temperature sensor; *P*, pump; *RTS*, return medium temperature sensor; *RS*, room sensor; *SC*, solar collector; *SP*, solar pump; *STS*, supply medium temperature sensor; *THS*, temperaturehumidity sensor.

The central control controls the supply water temperature for the heating system based on the outdoor air temperature. The room control then controls the water flow rate or water temperature for each room according to the room set-point temperature.

Instead of controlling the supply water temperature, it is recommended to control the average water temperature (mean value of supply and return water temperature) according to outdoor and/or indoor temperatures. During the heating period, as the internal load increases, the heat output from the heating system will decrease and the return temperature will rise. If the control system controls the average water temperature, the supply water temperature will automatically decrease due to the increased return-water temperature. This will result in a faster and more accurate control of the thermal output to the space and will give better energy performance than controlling the supply water temperature. Radiant surface cooling systems need controls to avoid condensation.

Larger buildings should be divided in several different thermal zones to optimize energy and control performance. Each zone can be controlled with reference to a temperature sensor in a representative space of the zone.

For the improved comfort and further energy savings, use an individual room temperature control. Each valve on the manifold is controlled by each room thermostat. An apartment or one-family house normally was regarded as one zone, but installing thermostats for each room is becoming popular. For better thermal comfort, it is preferable to control the room temperature as a function of the operative temperature.

The heat capacity of surfaces with embedded pipes plays a significant role for the thermodynamic properties of the heating system and, hence, for the control strategy. An obvious consequence of the response time of a conventional floor structures is that the instant control of the heating power is not necessary. The temperature of heat carrier, the time response, and the thermal capacity of systems depend on the thickness of the surface layer where the pipes are embedded.

For a low-temperature heating and high-temperature cooling system, a significant effect is the "self-regulating" control [13]. This self-regulating effect depends partially on the temperature difference between room and heated surface, and partly on the difference between room and the average water temperature in the embedded pipes. This impact is bigger for systems with surface temperatures close to a room temperature because the small temperature change represents a higher percentage compared to the same temperature change at a high-temperature difference. The self-regulating effect supports the control equipment in maintaining a stable thermal environment, and providing comfort to the occupants in the room.

6.6 EFFICIENCY OF HEATING SYSTEMS

Efficiency η of a heating system is expressed as a function of the partial efficiencies of technological chain components (generation, distribution, emission, and control) as

$$\eta = \eta_G \eta_D \eta_E \eta_C, \tag{6.30}$$

where η_G is the generation efficiency (equal to collector efficiency); η_D is the distribution efficiency; η_E is the emission efficiency of the heaters; and η_C is the control efficiency.

The generation efficiency defines the generator (solar collector) capacity to convert the primary energy into thermal energy. The solar collector efficiency can be calculated using Eq. (3.21).

The distribution efficiency η_D characterizes the recovery level of heat losses in distribution pipes for heat carrier and DHW. For preliminary calculations, the η_D values can be adopted as follows [25]:

- in a central heating system, for the transport networks between source and buildings: 0.50−0.65 (noninsulated pipes) and 0.82−0.90 (insulated pipes);
- in a local central heating system, for the distribution pipes in buildings: 0.80−0.90 (noninsulated pipes) and 0.85−0.95 (insulated pipes);

- in a local heating system, for the pipes between generator and heaters: 0.92–0.98 (noninsulated pipes).

The emission efficiency η_E characterizes the heat emission in heated space to compensate the heat losses and to assure the thermal comfort. The following η_E values are recommended [25]: 0.95 for radiators and radiant panels, and 1.00 for radiant floors.

The control efficiency η_C represents the control systems capacity to ensure ever-changing heat demand in a space. This depends of the heating system, heater type, and control methods applied. For preliminary calculations, the following η_C values can be considered [25]:

- for a central heating system: 0.75–0.78 (without control), 0.78–0.81 (the control depending on the outdoor air temperature), 0.82–0.84 (the control depending on the outdoor and indoor air temperatures);
- for a local central heating system: 0.84–0.88 (without control), 0.87–0.95 (the control depending on the outdoor and indoor air temperatures).

The determination of the real deficiencies can be done only by monitoring the building and the heating system based on measurements during functional conclusive cycles.

6.7 ENERGY ANALYSIS OF SOLAR HEATING SYSTEMS

Neglecting the collector efficiency factor F in Eq. (3.20) to allow use of the average HTF temperature [21], the collector efficiency η_c can be approximated as

$$\eta_c = \eta_0 - \frac{U_L \Delta t}{I_T} \qquad (6.31)$$

where η_0 is the optical collector efficiency; U_L is the overall heat loss coefficient, in kW/(m^{2}K); I_T is the total radiation on solar collector surface, in kW/m^2; and Δt is the difference between average fluid temperature and ambient temperature, in K.

Based on Eq. (6.31), the operation regime of the solar collector can be analyzed:

- if $I_T > U_L \Delta t / \eta_0$, the circulating pump is in operation and HTF temperature in solar collector will increase;
- if $I_T = U_L \Delta t / \eta_0$, the circulating pump switches off;
- if $I_T < U_L \Delta t / \eta_0$, the HTF does not circulate in solar system.

The solar energy collected on collector surface can be written as

$$I_T = I_D + I_d, \qquad (6.32)$$

where I_T is the total radiation; I_D is the direct radiation; and I_d is the diffuse radiation.

The possible collected energy is given by [26]

$$E = \int_{n} I_D(\tau)d\tau + \int_{n} I_d(\tau)d\tau + \int_{N-n} I_d(\tau)d\tau, \qquad (6.33)$$

where n is the effective hours with solar radiation and N is the maximum possible hours with solar radiation.

Introducing the ratio $f = n/N$, Eq. (6.33) becomes:

$$E = \int_{fN} [I_D(\tau) + I_d(\tau)]d\tau + \int_{1+fN} I_d(\tau)d\tau. \qquad (6.34)$$

The maximum collected energy, depending on the collector efficiency, is given by

$$E_{max} = \int_{N} [\eta_0 E(\tau) - U_L \Delta t]d\tau \qquad (6.35)$$

Eq. (6.35) has two solutions, τ_1 and τ_2, when $E = U_L \Delta t / \eta_0$ ($\eta_c = 0$). Analyzing the variation of daily maximum solar energy (Fig. 6.18) results that:

- above the line $\eta_c = 0$, solar energy can be used to heat up the fluid in collector, and the marked area placed between the $E(\tau)$ curve and the line $\eta_c = 0$ represents the maximum value of solar energy E_{max} in a day. Thus, it is obtained as

$$E_{max} = \int_{\tau_1}^{\tau_2} \eta_0 E(\tau)d\tau - U_L \Delta t(\tau_2 - \tau_1). \qquad (6.36)$$

- under the line $\eta_c = 0$, solar energy cannot be used because it is lower than the energy demand for distribution fluid heating.

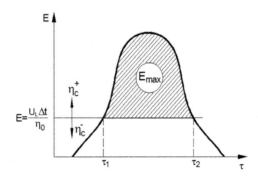

FIGURE 6.18   Daily maximum solar energy.

Because during a day there are periods with and without sun, the effective solar energy E_{ef} is different from the maximum solar energy E_{max}, and it is given by

$$E_{ef} = \int \{\eta_0[I_D(\tau) + I_d(\tau)] - U_L\Delta t\}d\tau + \int [\eta_0 I_d(\tau) - U_L\Delta t]d\tau. \quad (6.37)$$

Using the notations τ_1 and τ_2 for the solutions of equation,

$$\eta_0[I_D(\tau) + I_d(\tau)] - U_L\Delta t = 0 \quad (6.38)$$

and τ'_1, τ'_2 for the solutions of equation

$$\eta_0 I_D(\tau) - U_L\Delta t = 0. \quad (6.39)$$

Eq. (6.37) can be written as

$$E_{ef} = \int_\xi \{\eta_0[I_D(\tau) + I_d(\tau)] - U_L\Delta t\}d\tau + \int_{\xi'} [\eta_0 I_d(\tau) - U_L\Delta t]d\tau, \quad (6.40)$$

where ξ is the number of hours between τ_1 and τ_2 with sun and ξ' is the number of hours between τ'_1 and τ'_2 without sun.

The integral Eq. (6.40) can be solved when the meteorological data are known. At the same time, one considered the day like a sum of $(f \times N)$ hours with the sun and $(1 - f \times N)$ hours without the sun. Taking into consideration this assumption to calculate the effective solar energy, the following cases should be analyzed (Fig. 6.19):

- if τ_1 and τ_2 do not exist, that means: $U_L\Delta t/\eta_0 > I_D + I_d$, then $E_{ef} = E_{max} = 0$, and the solar energy cannot be used.
- if τ'_1 and τ'_2 do not exist, that means: $U_L\Delta t/\eta_0 > I_D$ and $U_L\Delta t/\eta_0 < I_D + I_d$, then the effective solar energy is given by equation as

$$E_{ef} = f \int_{\tau_1}^{\tau_2} \{\eta_0[I_D(\tau) + I_d(\tau)] - U_L\Delta t\}d\tau \quad (6.41)$$

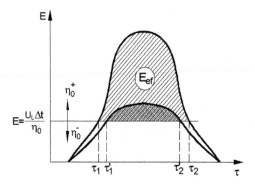

FIGURE 6.19 Daily maximum effective solar energy.

which is equivalent with

$$E_{ef} = fE_{max} \tag{6.42}$$

- if $U_L\Delta t/\eta_0 < I_d < I_D + I_d$, then the effective solar energy is given by

$$E_{ef} = f \int_{\tau_1}^{\tau_2} \{\eta_0[I_D(\tau) + I_d(\tau)] - U_L\Delta t\}d\tau + (1-f) \int_{\tau_1}^{\tau_2} [\eta_0 I_d(\tau) - U_L\Delta t]d\tau \tag{6.43}$$

or

$$E_{ef} = fE_{max} + \eta_0(1-f) \int_{\tau_1'}^{\tau_2'} I_d(\tau)d\tau + (1-f)U_L\Delta t(\tau_2' - \tau_1') \tag{6.44}$$

Substituting notation,

$$E_d = \int_{\tau_1'}^{\tau_2'} I_d(\tau)d\tau - \frac{U_L\Delta t}{\eta_0}(\tau_2' - \tau_1'). \tag{6.45}$$

Eq. (6.45) can be written as

$$E_{ef} = fE_{max} + (1-f)\eta_0 E_d. \tag{6.46}$$

Neglecting the heat losses between collector and storage tank, the average temperature of fluid in collector $((t_{ic} + t_{oc})/2$, where t_{ic} is the fluid inlet temperature of solar collector and t_{oc} is the fluid outlet temperature of solar collector) will be approximately equal to the average temperature in the storage tank (t_{hw}). The average energy delivered by storage tank E_{hw} is calculated as

$$E_{hw} = 24U'(t_{hw} - t_i), \tag{6.47}$$

where U' is the conventional heat transfer coefficient between storage and heated spae, in W/(m² K) and t_i is the indoor air temperature, in K.

The energy conservation law of the radiator can be written as

$$G_{hw}c_p(t_d - t_r) = U_R(t_R - t_i), \tag{6.48}$$

where G_{hw} is the distribution fluid (hot water) flow rate, in kg/s; c_p is the specific heat of hot water, in J/(kg K); t_d, t_r are the inlet and outlet temperatures of hot water in the radiator, in K; U_R is the heat transfer coefficient of radiator, in W/(m²·K); and t_R is the average temperature of the radiator surface, in K.

Assuming the average temperature of the radiator surface equals to the average temperature of hot water in radiator, Eq. (6.48) becomes

$$G_{hw}c_p(t_d - t_r) = U_R\left(\frac{t_d + t_r}{2} - t_i\right). \tag{6.49}$$

Using Eqs. (6.48) and (6.49),the following is obtained:

$$t_R = \frac{2G_{hw}c_p t_d + U_R t_i}{U_R + 2G_{hw}c_p}.$$

(6.50)

Neglecting the heat losses between storage tank and radiators ($t_d = t_{hw}$), the heat provided by radiator will be equal to the heat delivered by storage tank as

$$U_R(t_R - t_i) = U'(t_{hw} - t_i).$$

(6.51)

Combining Eqs. (6.48) and (6.51) gives

$$U' = \frac{2G_{hw}c_p U_R}{U_R + 2G_{hw}c_p}.$$

(6.52)

The daily energy requirement for heating is given by

$$E_{req} = 24q_0 V(t_i - t_a),$$

(6.53)

where q_0 is the specific heat loss of the space, in $W/(m^3\ K)$; V is the heated space volume, in m^3; t_i is the indoor air temperature, in K; and t_a is the outdoor air temperature, in K.

Representing the curves $E_{ef} = f(t)$, $E_{hw} = f(t)$, $E_{req} = f(t)$ the optimal collector surface could be established (Fig. 6.20). In Fig. 6.20A, the intersection point A between E_{ef} and E_{hw} curves is under the E_{req} line. In this case the stored energy E_{hw} is lower than the energy demand E_{req} of the consumer. Thus an auxiliary heat source is necessary. Abscise of the point A represents the average temperature in storage tank, t_{hw}.

When the intersection point A′ is above the E_{req} line (Fig. 6.20B), the stored energy E_{hw} is higher than the energy demand E_{req} of the consumer, which can lead to lower collector efficiency.

The optimal surface of solar collector is obtained when the collected and stored energy is equal to the consumer energy demand (Fig. 6.20C).

The efficiency of solar heating system increases in the case of buildings with low-energy demand for heating.

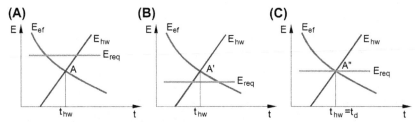

FIGURE 6.20 Daily variation of E_{ef}, E_{hw} and E_{req} with storage fluid temperature, (A) $E_{hw} < E_{req}$; (B) $E_{hw} > E_{req}$; (C) $E_{hw} = E_{req}$.

6.8 ECONOMIC ANALYSIS INDICATORS

In the economic analysis of a system, different methods could be used to evaluate the systems. Some of them are: the present value (PV) method, the net present cost (NPC), the future value (FV) method, the total annual cost (TAC) method, the total updated cost (TUC) method, the annual life cycle cost (ALCC), and other methods [27,28].

- Present value analysis or discounting is often used to address the time value of money in project planning. With an appropriate discount rate, PV analysis translates a series of annual costs over a design life time for a project, of say 50 or 100 years, from the future to the present, enabling effects occurring at different times to be compared [29]. The PV of a future payment can be calculated using the equation

$$PV = \frac{C}{(1+i)^{\tau}},\qquad(6.54)$$

where C is the payment/cost on a given future date; τ is the number of periods to that future date; and i is the discount (interest) rate. Therefore, PV is the present value of a future payment that occurs at the end of the τ-th period.

Similarly, the PV of a stream of costs with a specified number of fixed periodic payments can be expressed as

$$PV = C \sum_{n=1}^{\tau} \left[\frac{1}{(1+i)^{n}} \right],\qquad(6.55)$$

where C is the periodic payment that occurs at the end of each period; n is the number of periods (years).

The following equality can be demonstrated rather easily as

$$\sum_{n=1}^{\tau} \frac{1}{(1+i)^{n}} = \frac{(1+i)^{\tau} - 1}{i(1+i)^{\tau}}\qquad(6.56)$$

and is defined update rate u_r (or present value factor, PF)

$$u_r = \frac{(1+i)^{\tau} - 1}{i(1+i)^{\tau}} = \frac{1}{CRF},\qquad(6.57)$$

where CRF is the capital recovery factor.

Taking into account Eqs. (6.56) and (6.57), Eq. (6.55) yields the following equation:

$$PV = u_r C = \frac{C}{CRF}.\qquad(6.58)$$

The discount rate i has a significant impact on the present value of a future cost. Generally, the higher the discount rate, the more reduction that occurs when a future value is converted into a PV.

The interest rate i can be calculated using the following equation [27]:

$$i = \left[\left(1 + \frac{P}{I_0} \right)^{1/q} - 1 \right]^q - 1, \tag{6.59}$$

where $q = \log[1 + (1/N)]/\log 2$; N is the number of payment; P is the payment amount; and I_0 is the initial investment cost.

- Net present cost of the project over its entire lifespan of operation includes expenses such as components, component replacements, operation and maintenance costs, and initial investment costs. The NPC can be computed using equation

$$NPC = \frac{TAC}{CRF}, \tag{6.60}$$

where TAC is the total annual cost (sum of all annual costs of each system component).

- Another economic indicator is total updated cost, shown as

$$TUC = I_0 + \sum_{n=1}^{\tau} \frac{C}{(1+i)^n}, \tag{6.61}$$

where I_0 is the initial investment cost in the operation, beginning date of the system inception; C is annual operation and maintenance cost of the system; i is the discount (inflation) rate; and τ is the number of years for which an update is made (20 years).

Taking into account Eqs. (6.55) and (6.56), Eq. (6.61) yields the following equation:

$$TUC = I_0 + u_r C. \tag{6.62}$$

- Usually the SHS achieves a fuel economy ΔC (operating costs) comparatively of the classical system with thermal station (TS). On the other hand, the SHS involves an additional investment I_{SHS} from the classical system I_{TS}, which produces the same amount of heat.

Thus, it can be determined the recovery time RT, in years, to increase investment, $\Delta I = I_{SHS} - I_{TS}$, taking into account the operation saving achieved through low fuel consumption $\Delta C = C_{TS} - C_{SHS}$:

$$RT = \frac{\Delta I}{\Delta C} \le RT_n, \tag{6.63}$$

where RT_n is normal recovery time.

It is estimated that for RT_n a number 8–10 years is acceptable, but this limit varies depending on the country's energy policy and environmental requirements.

- The TAC method can be used to compare the cost effectiveness of the SHS over other heating systems. The annual life cycle cost is given by

$$ALCC = CRF \cdot LCC, \tag{6.64}$$

where LCC is the life cycle cost which represents the cumulative cost of purchasing and running the system over its useful life and considers the influence of cost escalation on the annual operation and maintenance costs of the system. The LCC is given as

$$\text{LCC} = I_0 + \sum_{j=1}^{M} \frac{(1+e)^z}{(1+i)^z}\left(\frac{Q_{req}p_f}{\text{SPF}} + \text{MC}\right), \qquad (6.65)$$

where e is the annual fuel price escalation rate (%); i is the discount rate (%); z is the year after purchase of heating system (year); M is the total number of temperature bins (bin); Q_{req} is the heating demand (kW); p_f is the base year fuel price (€/year); SPF is the seasonal performance factor [Btu/(Wh)]; and MC is the annual maintenance cost (€/year).

Considering both the interest and fuel cost escalation rates, the payback period of an SHS compared with conventional heating sources would be the year that the PV of additional investment over that of the conventional source would equal the PV of the saving ΔC based on the first year of operation as

$$\text{PV} = \Delta C \frac{\left(\frac{1+e}{1+i}\right)^b - 1}{1 - \frac{1+e}{1+i}}, \qquad (6.66)$$

where b is the year to payback [28].

The discounted payback (DPB) is the time required to repay the investments (I) on the solar heating/cooling system when compared to the traditional system from the extra economic savings (s). The mathematical representation of DPB is given in Eq. (6.67) [30] as

$$\text{DPB} = \frac{\log[s/(s - 0.03I)]}{\log(1 + 0.03)}. \qquad (6.67)$$

REFERENCES

[1]   Andersen N. End users dictate the potential for low temperature district heating. Energy and Environment Journal 1999;4:30−1.

[2]   Sarbu I, Bancea O, Cinca M. Influence of forward temperature on energy consumption in central heating systems. WSEAS Transaction on Heat and Mass Transfer 2009;4(3):45−54.

[3]   Sarbu I. Energetically analysis of unbalanced central heating systems. In: Proceedings of the 5th IASME/WSEAS international conference on energy and environment. Cambridge, UK; 2010. p. 112−7.

[4]   Buderus. Handbuch fur Heizung-stechnik. Berlin: Beuth Verlag; 1994.

[5]   Ilina M, Burchiu S. Influence of heating systems on microclimate from living rooms. Fitter, Romania6; 1996. p. 24−9.

[6]   Berglund LG, Fobelets A. A subjective human response to low level air currents and asymmetric radiation. ASHRAE Transactions 1987;93(1):497−523.

[7]   Sarbu I, Sebarchievici C. Aspects of indoor environmental quality assessment in buildings. Energy and Buildings 2013;60(5):410−9.

[8] Sarbu I. Energy efficiency of low temperature central heating systems. In: Proceedings of the 4th WSEAS International conference on energy planning, energy saving and, environmental education, Kantaoui, Sousse, Tunisia; 2010. p. 30−5.

[9] Demidovitch B, Maron I. Elements of numerical computation. Moscow: Mir; 1979.

[10] Hesaraki A, Holmberg S. Energy performance of low temperature heating systems in five new-built Swedish dwellings: a case study using simulations and on-site measurements. Building and Environment 2013;64:85−93.

[11] ASHRAE handbook. HVAC systems and equipment. Atlanta (USA): American Society of Heating, Refrigerating and Air Conditioning Engineers; 2016.

[12] REHVA. Guidebook no 7: low temperature heating and high temperature cooling. 2007.

[13] Kim KW, Olesen BW. Radiant heating and cooling systems. ASHRAE Journal 2015; 57(2,3):28−37. 34−42.

[14] ISO 7730. Moderate thermal environment—determination of the PMV and PPD indices and specification of the conditions for thermal comfort. Geneva: International Organization for Standardization; 2005.

[15] ISO/TS 13732-2. Ergonomics of the thermal environment. Methods for the assessment of human responses to contact with surface (Part 2): human contact with surfaces at moderate temperature. Geneva: International Organization for Standardization; 2001.

[16] ASHRAE Standard 55. Thermal environmental conditions for human occupancy. Atlanta (USA): American Society of Heating, Refrigerating and Air-conditioning Engineers; 2010.

[17] Sarbu I, Sebarchievici C. A study of the performances of low-temperature heating systems. Energy Efficiency 2015;8(3):609−27.

[18] ISO 11855. Building environment design—design, dimensioning, installation and control of the embedded radiant heating and cooling systems. Geneva: International Organization for Standardization; 2012.

[19] Roumajon J. Modélisation numerique des émissions thermiques. Chaud, Froid and Plomberie, 579; 1996. p. 55−8 (4).

[20] ASHRAE handbook. Fundamentals. Atlanta (GA): American Society of Heating, Refrigerating and Air-Conditioning; 2013.

[21] ASHRAE handbook. HVAC applications. Atlanta (GA): American Society of Heating, Refrigerating and Air−Conditioning Engineers; 2015.

[22] Bechthler H, Browne MW, Bansal PK, Kecman V. New approach to dynamic modelling of vapour-compression liquid chillers: artificial neural networks. Applied Thermal Engineering 2001;21(9):941−53.

[23] ANRE. National Authority of Energy Settlement. 2012. http://www.anre.ro/energie-electrica/legislatie/preturi-si-tarife-ee/energia-electrica-2010-2012.

[24] IEE. Intelligent Energy Europe. 2013. http://ec.europa.eu/energy/environment.

[25] Sarbu I, Kalmar F, Cinca M. Thermal building equipments—energy optimization and modernization. Timisoara: Polytechnic Publishing House; 2007 [in Romanian].

[26] Sarbu I, Adam M. Application of solar energy for domestic hot-water and buildings heating/cooling. International Journal of Energy 2011;5(2):34−42.

[27] Thuesen GJ, Fabrycky WJ. Engineering economy. Prentice-Hall International Editions; 1989.

[28] Tassou SA, Maequand CJ, Wilson DR. Energy and economic comparisons of domestic heat pumps and conventional heating systems in the British Climate. Applied Energy 1986;34(2):127−38.

[29] Kaen FR. Corporate finance: concepts and policies. USA: Blackwell Business; 1995.

[30] Noro M, Lazzarin R, Busato F. Solar cooling and heating plants: an energy and economic analysis of liquid sensible vs phase change material (PCM) heat storage. International Journal of Refrigeration 2014;39:104−16.

Chapter 7

Solar Thermal-Driven Cooling Systems

7.1 GENERALITIES

Usual vapor compression-based cycles are electrically powered, consuming large amount of high quality energy, which significantly increases the fossil fuel consumption. The *International Institute of Refrigeration* in Paris estimated that approximately 15% of all the electrical energy produced worldwide is employed for air-conditioning (A/C) and refrigeration processes [1]. In recent years, these sectors have witnessed a manifold growth and become essential not only for human comfort but also for a variety of applications. Moreover, electricity peak demands during summer are becoming more and more frequent due to the general increase in A/C and refrigeration equipment usage. Providing cooling by utilizing renewable energy such as solar energy is a key solution to the energy and environmental issues. Solar cooling serves the cold storage needs in industries as varied as hospitality, pharmaceuticals, chemicals, dairy, and food processing, besides serving the residential and office A/C needs.

Solar cooling depends primarily on solar energy, either by hot water production through solar collectors (SCs) or electricity production through photovoltaic (PV) panels. In comparison with conventional electrically driven compression systems, substantial primary energy savings can be expected from solar cooling, thus aiding in conserving energy and preserving the environment. Another advantage of using solar energy is the coincidence of the peak of the cooling demand and the availability of solar radiation.

Solar electrical- and thermal-powered refrigeration systems can be used to produce cooling [2]. The first is a PV-based solar energy system, in which solar energy is initially converted into electrical energy and then utilized for producing the cooling, similar to conventional methods [3] or by thermoelectric processes [4,5]. The second one utilizes solar thermal energy to power the generator of a sorption cooling system or converts the thermal energy to mechanical energy, which is utilized to produce the cooling effect. Thermal-powered cooling systems are classified into two categories: sorption system (absorption, adsorption, and desiccant system) and thermomechanical system (ejector system). Since

Solar Heating and Cooling Systems. http://dx.doi.org/10.1016/B978-0-12-811662-3.00007-4

solar energy is time-dependent, the successful utilization of all these cooling systems is to a very large degree dependent on the thermal ST employed. Table 7.1 shows the stages and options in solar cooling techniques.

Fig. 7.1 depicts the main alternative routes from solar energy into cooling effect using thermodynamic cycles. The main options are solar "thermal" or "electric" PV. Since the first decades after the energy crisis and till very recently, the PV option was excluded for the high cost of the modules. So, solar thermal collectors have been widely developed in the last decades in order to improve efficiency and durability and decrease cost.

Fig. 7.2 illustrates a schematic diagram of a solar thermal cooling system. The solar collection and storage system consists of an SC connected through pipes to the thermal storage tank (ST). SCs transform solar radiation into heat and transfer that heat to the heat transfer fluid in the collector. The fluid is then stored in a thermal ST to be subsequently utilized for various applications. The thermal A/C unit is run by the hot refrigerant coming from the ST, and the refrigerant circulates through the entire system.

The performance of cooling systems is determined based on energy indicators of these systems. The coefficient of performance (COP) of a cooling system can be calculated as follows:

$$COP = \frac{E_u}{E_c} \tag{7.1}$$

where E_u is the cooling usable energy, in watt-hour (Wh), and E_c is the consumed energy by system, in Wh.

The solar COP is defined in Eq. (7.2) as

$$COP_{sol} = \frac{E_c}{E_s} \tag{7.2}$$

where E_c is the usable energy produced by SCs, in Wh, and E_s is the energy received by SCs, in Wh.

The overall COP of a solar thermal cooling system (COP_{sys}) is given by combination of the two COPs in Eqs. (7.1) and (7.2):

$$COP_{sys} = COP_{sol} \times COP = \frac{E_u}{E_s} \tag{7.3}$$

Additionally, energy efficiency ratio (EER), in British thermal unit (Btu) per Wh, is defined by equation as

$$EER = 3.412 \ COP \tag{7.4}$$

where 3.412 is the transformation factor from W in Btu/h.

This chapter provides a detailed review of different solar thermal-driven refrigeration and cooling systems. Theoretical basis and practical applications for cooling systems within various working fluids assisted by solar energy and their recent advances are presented. The first aim of this chapter is

TABLE 7.1 Stages and Options in Solar Cooling Techniques

Conversion	Thermal Storage (Hot Energy)	Production of Cool Energy	Thermal Storage (Cool Energy)	Applications
Solar thermal **1.** Flat-plate collector **2.** Evacuated-tube collector **3.** Concentrated collector	**1.** Sensible **2.** Latent **3.** Thermochemical	**1.** Absorption 　**a.** Single-effect 　**b.** Half-effect 　**c.** Double-effect 　**d.** Triple-effect **2.** Adsorption **3.** Desiccant **4.** Ejector	**1.** Sensible **2.** Latent **3.** Thermochemical	**1.** Air-conditioning 　**a.** Office 　**b.** Building 　**c.** Hotel 　**d.** Laboratory **2.** Process industries 　**a.** Dairy 　**b.** Pharmaceutical 　**c.** Chemical **3.** Food preservation 　**a.** Vegetables 　**b.** Fruits 　**c.** Meat and Fish
Solar PV (electric)		Vapor compression Thermoelectric		

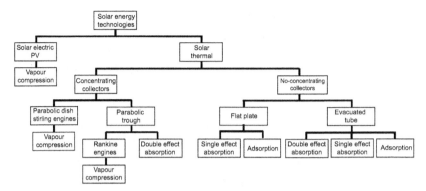

FIGURE 7.1   Alternative routes from solar energy into cooling effect.

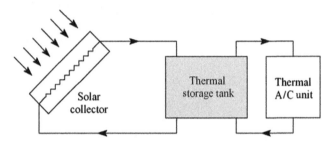

FIGURE 7.2   Schematic of a solar thermal cooling system.

to give an overview of the state-of-the-art of the sorption and thermo-mechanical technologies that are available to deliver cooling from solar energy. The second aim is to compare the potential of these technologies in delivering competitive sustainable solutions. The topics approached in present chapter are similar to that of the other works but they are focused on solar closed-sorption refrigeration systems providing useful information updated and more extensive on their principles, development history, applications, and recent advances. The application areas of these technologies are categorized by their cooling temperature demands (air-conditioning, refrigeration, ice making). Additionally, the thermodynamic properties of most common working fluids and the use of ternary mixtures in solar-powered absorption systems are reviewed. A mini-type solar-absorption cooling system using both fan coils and the radiant ceiling as terminals was designed and installed. The system performance as well as the indoor thermal comfort is analyzed. Finally, some information on design, control, and operation of hybrid cooling and heating systems are included. The study also refers to a comparison of various solar thermal-powered cooling systems and, to some use, suggestions of these systems.

7.2 SOLAR-POWERED SORPTION COOLING SYSTEMS

Sorption technology is utilized in thermal cooling techniques. This technology can be classified as either open sorption systems or closed sorption systems [6]. Desiccant cycles represent the open systems. Absorption and adsorption technologies represent closed systems.

Sorption refrigeration uses physical or chemical attraction between a pair of substances to produce the refrigeration effect. A sorption system has the unique capability of transforming thermal energy directly into cooling power. Between a pair of substances, the substance with the lower boiling temperature is called the refrigerant (sorbate) and the other is called the sorbent.

7.2.1 Desiccant Cooling Systems

Open sorption cooling is more commonly called desiccant cooling because sorbent is used to dehumidify air. Basically, desiccant systems transfer moisture from one air stream to another by using two processes: sorption and desorption (regeneration) processes [2]. Various desiccants are available in liquid or solid phases. Basically all water-absorbing sorbents can be used as a desiccant.

In the *sorption process* the desiccant system transfers moisture from the air into a desiccant material by using the difference in the water vapor pressure of the humid air and the desiccant. If the desiccant material is dry and cold, then its surface vapor pressure is lower than that of the moist air, and moisture in the air is attracted and absorbed to the desiccant material. In *regeneration process*, the captured moisture is released to the airstream by increasing the desiccant temperature. After regeneration, the desiccant material is cooled down by the cold airstream. Then it is ready to absorb the moisture again. When these processes are cycled, the desiccant system can transfer the moisture continuously by changing the desiccant surface vapor pressures, as illustrated in Fig. 7.3 [2]. To drive this cycle, thermal energy is needed during the desorption process. The difference between solid and liquid desiccants is their reaction to moisture.

7.2.1.1 Liquid Desiccant System

Materials typically used in liquid desiccant systems are lithium chloride (LiCl), calcium chloride ($CaCl_2$), and lithium bromide (LiBr). In a liquid desiccant cooling system, the liquid desiccant circulates between an absorber and a regenerator in the same way as in an absorption system. Main difference is that the equilibrium temperature of a liquid desiccant is determined not by the total pressure but by the partial pressure of water in the humid air to which the solution is exposed to. A typical liquid desiccant system is shown in Fig. 7.4 [7].

In the dehumidifier, a concentrated solution is sprayed at point A over the cooling coil at point B while ambient or return air at point 1 is blown across

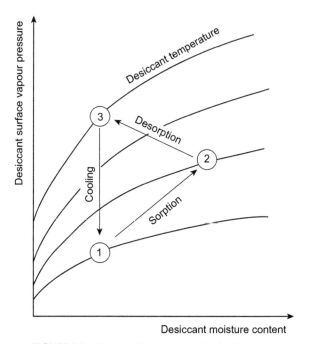

FIGURE 7.3   Process of moisture transfer by desiccant.

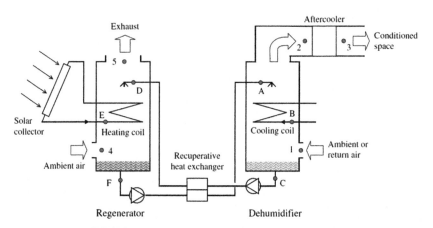

FIGURE 7.4   A liquid desiccant cooling system with SC.

the stream. The solution absorbs moisture from the air and is simultaneously cooled down by the cooling coil. The results of this process are the cool dry air at point 2 and the diluted solution at point C. Eventually, an after-cooler cools this air stream further down. In the regenerator, the diluted solution from the dehumidifier is sprayed over the heating coil at point E that is connected to

SCs, and the ambient air at point 4 is blown across the solution stream. Some water is taken away from the diluted solution by the air while the solution is being heated by the heating coil. The resulting concentrated solution is collected at point F, and hot humid air is rejected to the ambient at point 5. A recuperative heat exchanger preheats the cool diluted solution from the dehumidifier using the waste heat of the hot concentrated solution from the regenerator, resulting in a higher COP.

Since Lof [8] investigated liquid desiccant solar cooling, most of the research on liquid desiccant solar cooling began in the early 1990s. Moreover, the latest developments are focused on liquid sorption applications since the liquid sorption materials have advantages of higher air dehumidification at the same driving temperature, as well as the possibility of high energy storage by means of hygroscopic solutions.

Ameel et al. [9] compared the performance of various absorbents, including LiCl, CaCl, and LiBr. They concluded that LiBr outperformed the other absorbents.

Gommed and Grossman [10] developed the prototype of the liquid desiccant cooling system assisted by the flat SCs using $LiCl/H_2O$ as its working fluid. Through the parametric study, they demonstrated that conditions of the ambient air are the major parameters considerably affecting the dehumidification process in the liquid desiccant system. They reported that the system provided 16 kW of dehumidification capacity with a thermal COP of 0.8.

In efforts to reduce a building's energy consumption, designers have successfully integrated liquid desiccant equipment with standard absorption chillers [11]. In a more general approach, the absorption chiller is modified so that rejected heat from its absorber can be used to help regenerate liquid desiccants.

7.2.1.2 Solid Desiccant System

The solid desiccant system is constructed by placing a thin layer of desiccant material, such as silica gel, on a support structure [6]. Fig. 7.5 shows an example of a solar-driven solid desiccant cooling system. The system has two slowly revolving wheels and several other components between the two air streams from and to a conditioned space. The return air from the conditioned space first goes through a direct evaporative cooler and enters the heat exchange wheel with a reduced temperature (A → B). It cools down a segment of the heat exchange wheel when it passes through (B → C). This resulting warm and humid air stream is further heated to an elevated temperature by the solar heat in the heating coil (C → D). The resulting hot and humid air regenerates the desiccant wheel and is rejected to ambient (D → E). On the other side, fresh air from ambient enters the regenerated part of desiccant wheel (1 → 2). Dry and hot air comes out of the wheel as the result of dehumidification. This air is cooled down by the heat exchange wheel to a certain temperature (2 → 3).

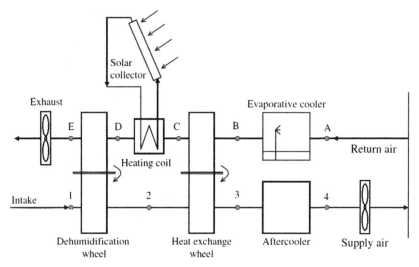

FIGURE 7.5   A solid desiccant cooling system with SC.

Depending on the temperature level, it is directly supplied to the conditioned space or further cooled in an after-cooler ($3 \rightarrow 4$). If no after-cooler is used, cooling effect is created only by the heat exchange wheel, which was previously cooled by the humid return air at point B on the other side. Temperature t_3 at point 3 cannot be lower than t_B, which in turn is a function of the return air condition at point A.

In principle, the COP of an open desiccant system is similar to its closed counterpart. For example, COP of 0.7 was said achievable with a solid desiccant cooling system under "normal" operating conditions [12].

Henning et al. [13] installed a solar-assisted desiccant cooling system with a 20 m² flat-plate SC and a 2 m³ hot-water ST. They reported that a solar fraction of the cooling between the solar heat and auxiliary heat provided was 76%, with an overall collector efficiency of 54% and a cooling COP of 0.6 during typical summer conditions. In addition, they proposed a combination of a solar-assisted solid desiccant cooling system with a conventional vapor compression chiller for warm and humid climates and claimed up to 50% of primary energy savings.

A desiccant cooling system is actually a complete system that has ventilation, humidity, and temperature-control devices in a ductwork. Therefore, it is inappropriate to compare a desiccant cooling system with components such as chillers. Desiccant dehumidification offers a more efficient humidity control than the other technologies. When there is a large ventilation or dehumidification demand, solar-driven desiccant dehumidification can be a very good option.

7.2.1.3 Desiccant-Based Evaporative Cooling Systems

A well suitable alternative of mechanical vapor compression system is evaporative cooling system that can be efficiently used for A/C applications with less power requirements, that is, one-fourth of the mechanical vapor-compression. It is an energy-saving, cost-effective, simple, and environment-friendly A/C technique. Many researchers have investigated different types of evaporative coolers such as direct, indirect, and modified coolers. Evaporative cooling systems are suitable for dry and high-temperature climatic conditions [14].

In the indirect evaporative system, the process air stream does not interact directly with the cooling fluid stream rather it is cooled sensibly. The cooling process inside an indirect evaporative cooler is represented on a psychrometric chart shown in Fig. 7.6 [15]. The temperature of air is lowered using some type of heat exchange arrangement in which primary air is cooled sensibly using a secondary air stream. The secondary air is cooled using water. In the indirect evaporative cooling system, both dry and wet bulb temperatures of the air are lowered.

The indirect evaporative cooling has an efficiency of 60−70%. The schematic and flow arrangement inside the indirect evaporative cooler are shown in Figs. 7.7 and 7.8, respectively [15].

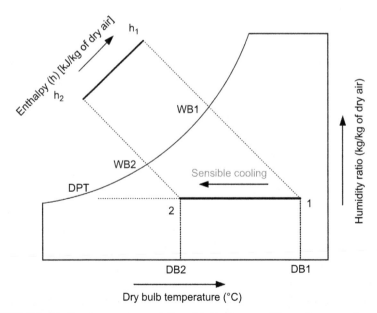

FIGURE 7.6 Cooling process representation of indirect evaporative cooler on psychrometric chart. *DB*, dry bulb temperature; *DPT*, dew point temperature; *WB*, wet bulb temperature.

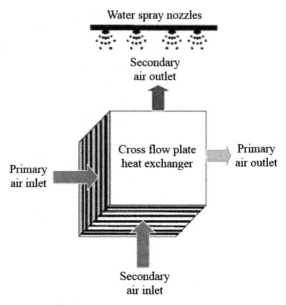

FIGURE 7.7   Schematic of the indirect evaporative cooler.

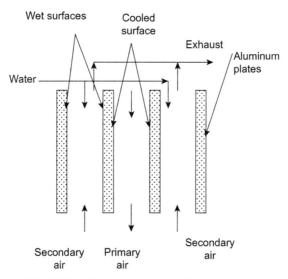

FIGURE 7.8   Flow arrangement inside indirect cooler.

In direct evaporative system, moisture is also added to the cooled air stream because process air stream comes in direct contact with the cooling water. The temperature of the process air is lowered because of the high moisture content in the air, so it is an adiabatic process, which is only suitable for hot and dry climates, and for hot and humid climates indirect evaporative cooler is preferred. In the direct evaporative cooling, dry bulb temperature of the air is lowered, and wet bulb temperature remains unchanged. The wet bulb temperature is an important parameter for the performance of direct evaporative cooler. The efficiency of a well-made direct evaporative cooler reaches an efficiency of approximately 85% [16]. Both the schematic and psychrometric processes of the direct evaporative cooler are shown in Fig. 7.9. The ambient air comes in direct contact with the sprayed water that decreases the temperature of the supply air and adds moisture content to it as shown on the psychrometric chart.

The process of indirect evaporative cooling needs input energy only for the water pump and fan, and that is why this system has a high COP.

The evaporative desiccant cooling system consists of a desiccant dehumidifier, a regenerator, and a cooling unit. The basic working principle of a solar activated evaporative desiccant cooling system is illustrated in Fig. 7.10 [17]. The air is dehumidified using desiccant dehumidifier, and its temperature is lowered using evaporative cooler or some other cooling device. For continuous operation of the system, the desiccant dehumidifier is regenerated using heat energy provided by SCs as shown in Fig. 7.10. Some heat recovery units are also utilized to make the system more efficient.

In desiccant-based evaporative cooling technique, latent and sensible loads are separately removed using desiccant dehumidification system and cooling unit, respectively. The type of cooling units used to reduce the temperature of dehumidified air, mainly, defines the type of hybrid desiccant cooling system. The selection of the cooling unit depends on operating conditions, that is, humidity and temperature of the air. The most commonly used cycles for desiccant-based evaporative cooling systems are recirculation

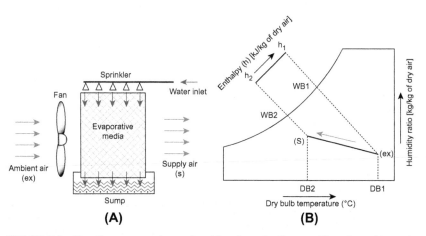

FIGURE 7.9 The direct evaporative cooler (A) schematic diagram (B) and psychrometric process. *DB*, dry bulb temperature; *WB*, wet bulb temperature.

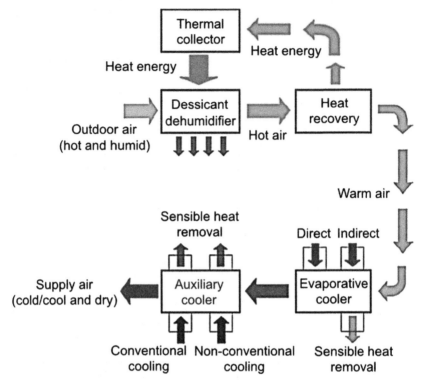

FIGURE 7.10   Schematic of solar-driven evaporative desiccant cooling systems.

and ventilation. The schematic of a typical desiccant dehumidification system in conjunction with evaporative cooler, shown in Fig. 7.5, is operated on ventilation mode.

Some advantages of the desiccant-aided evaporative cooling systems:

- They can be used for hot and humid climates because evaporative cooling alone is not feasible for such conditions.
- A lot of energy is saved as compared to vapor compression system because no preheating is required.
- Environment-friendly system because of no use of refrigerant, which affects the ozone layer.
- Separate and better control of sensible and latent loads; the desiccant wheel is controlling the latent part and the evaporative cooler is controlling the sensible one.
- The overall system has low maintenance cost because it operates at almost atmospheric conditions.
- Low grade energy such as solar, biomass, etc., can be effectively used to drive the system.

7.2.1.4 Solar Liquid Desiccant Regeneration Methods

Currently, solar desiccant regeneration systems are mostly driven by solar thermal energy. Solar electrodialysis regeneration is also being investigated as means of new solar regeneration method.

A common type of *solar thermal regeneration system* for liquid-desiccant air conditioner (LDAC) is shown in Fig. 7.11 [18]. Weak desiccant solution flows from the dehumidifier into the SC and causes an increase in the temperature. Then, hot but still dilute desiccant solution is discharged into the regenerator and gets contacted with the air passing upward. Water molecules in the desiccant solution are absorbed by the air stream, and the dilute solution becomes regenerated. The concentration of the desiccant solution is adjusted, and strong desiccant solution is obtained as a result of this process. The strong solution is introduced to the dehumidifier from the strong solution storage to carry on the dehumidification cycle.

A recent solar method for the liquid desiccant regeneration is being studied. *Electrodialysis* (ED) is a technology-based method that transports ions through the selective membranes under the influence of an electrical field [19]. Cation- and anion-exchange membranes are alternately set in between a cathode and an anode within an electrodialyser. Under the electrical field, anions and cations inside the electrodialyser cells move toward the anode and cathode. During this process, anions and cations pass through anion-exchange membranes and cation-exchange membranes, respectively. This flow causes a rise in the ions concentration in the concentrate compartments and falls into the dilute compartment. By this way, both the concentrated desiccant solution

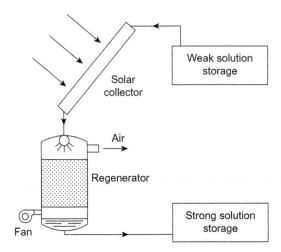

FIGURE 7.11   Schematic of a solar thermal regeneration system for liquid-desiccant air conditioner.

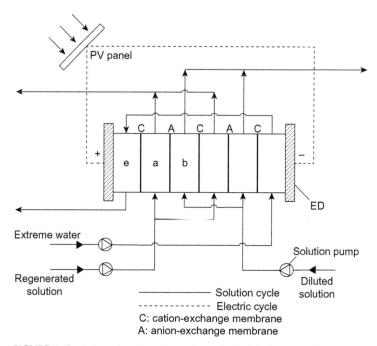

FIGURE 7.12 Schematic of the photovoltaic—electrodialysis regeneration system.

and pure water can be acquired. A schematic diagram of the PVED regeneration method is presented in Fig. 7.12.

In a PVED regeneration process, the dilute desiccant solution is drained from the dehumidifier to the regenerator. The PV panels-driven regenerator is made of ED stacks formed from a mass of cells located in parallel between two electrodes. As shown in a schematic of an ED regenerator, the cells a, b, and e are the concentrate, dilute, and electrode rinse cells, respectively. The weak solution is introduced to the dilute cell and regenerated solution is introduced to the concentrate cell. After this process, the dilute solution becomes regenerated that is sent to the dehumidifier unit of the LDAC system [20].

New developments of solar-assisted liquid desiccant evaporative cooling system with small capacity will open up new market segments, alike solar combisystems.

7.2.2 Principles of Closed Sorption Systems

In closed sorption technology, there are two basic methods: absorption refrigeration and adsorption refrigeration.

Fig. 7.13 shows a schematic diagram of a closed sorption system. The component where sorption takes place is denoted as absorber Ab, and the one where desorption takes place is denoted as generator G. The generator receives

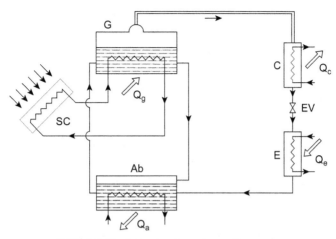

FIGURE 7.13   Solar closed-sorption cooling system.

heat Q_g from the SC to regenerate the sorbent that has absorbed the refrigerant in the absorber. The refrigerant vapor generated in this process condenses in the condenser C, rejecting the condensation heat Q_c to the ambient. The regenerated sorbent from the generator is sent back to the absorber, where the sorbent absorbs the refrigerant vapor from the evaporator E, rejecting the sorption heat Q_a to ambient. In the evaporator, the liquefied refrigerant from the condenser evaporates, removing the heat Q_e from the cooling load.

In an adsorption system, each of the adsorbent beds functions alternatively as the generator and absorber due to the difficulty of transporting solid sorbent from one to another.

For sorption cooling cycle, Eq. (7.1) can be rewritten in Eq. (7.5) as

$$\text{COP} = \frac{Q_e}{Q_g + P_{el}} \tag{7.5}$$

where COP is the coefficient of performance of the sorption system; Q_e is the cooling power of evaporator, in W; Q_g is the thermal power consumed by generator, in W; and P_{el} is the electrical power consumed in system, in W.

Absorption refers to a sorption process where a liquid or solid sorbent absorbs refrigerant molecules into its interior and changes physically and/or chemically in the process. Adsorption involves a solid sorbent that attracts refrigerant molecules onto its surface by physical or chemical force and does not change its form in the process.

7.2.3 Absorption Systems

Absorption refrigeration has been the most frequently adopted for solar cooling. It requires very low or no electric input, and, for the same capacity, adsorption systems are larger than absorption systems due to the low specific cooling power of the adsorbent.

Absorption is the process by which a substance changes from one state into a different state. These two states create a strong attraction to make a strong solution or mixture. The absorption system is one of the oldest refrigeration technologies. The first evolution of an absorption system began in the 1700s. It was observed that in the presence of H_2SO_4 (sulfuric acid), ice can be made by evaporating pure H_2O within an evacuated container. In 1859 Ferdinand Carre designed an installation that used a working fluid pair of ammonia/water (NH_3/H_2O). In 1950, a new system was introduced with a water/lithium bromide ($H_2O/LiBr$) pairing as working fluids for commercial purposes [21].

The absorption cooling technology consists of a generator, a pump, and an absorber that are collectively capable of compressing the refrigerant vapor. The evaporator draws the vapor refrigerant by absorption into the absorber. The extra thermal energy separates the refrigerant vapor from the rich solution. The refrigerant is condensed by rejecting the heat in a condenser, and then the cooled liquid refrigerant is expanded by the evaporator, and the cycle is completed.

The refrigerant side of the absorption system essentially works under the same principle as the vapor compression system. However, the mechanical compressor used in the vapor compression cycle is replaced by a thermal compressor in the absorption system. The thermal compressor consists of the absorber, the generator, the solution pump, and the expansion valve. The attractive feature of the absorption system is that any type of heat source, including solar heat and waste heat, can be utilized in the desorber.

NH_3/H_2O and $H_2O/LiBr$ are typical refrigerant/absorbent pairs used in absorption systems. Each working pair has its advantages and disadvantages, as shown in Table 7.2. The NH_3/H_2O systems are often used for refrigeration and in industrial applications, whereas the $H_2O/LiBr$ systems are more suitable for A/C purposes. The operation of the $H_2O/LiBr$-based absorption system is limited in terms of the evaporating temperature and the absorber temperature due to the freezing of the water and the solidification of the LiBr-rich solution, respectively. The operation of the NH_3/H_2O-based absorption system is not limited in terms of either the evaporation temperature or the absorption temperature. However, ammonia is toxic, and its usage is limited to large-capacity systems.

The most suitable refrigerant/absorbent working pair alternative to NH_3/H_2O and $H_2O/LiBr$ can be ammonia/lithium nitrate ($NH_3/LiNO_3$), lithium chloride/water ($LiCl/H_2O$), ammonia/calcium chloride ($NH_3/CaCl_2$), ammonia/sodium thiocyanate ($NH_3/NaSCN$), methanol/TEGDME (tetraethylene glycol dimethyl ether), and trifluoroethanol (TFE)/TEGDME. Abdulateef et al. [22] performed a comparative study among the performances of NH_3/H_2O, $NH_3/LiNO_3$, and $NH_3/NaSCN$ mixtures for absorption systems. Their results indicated that $NH_3/LiNO_3$ and $NH_3/NaSCN$ mixtures gives better performance

TABLE 7.2 Comparison Between the Absorption System With NH_3/H_2O and $H_2O/LiBr$

Working Pair	Advantages	Disadvantages
NH_3/H_2O	Evaporative at the temperatures below 0°C	Toxic and dangerous for health (NH_3)
		In need of a column of rectifier
		Operation at high pressure
$H_2O/LiBr$	High COP	The risk of congelation, therefore a device anticrystallization is necessary
	Low operation pressures	
	Environmental friendly and innoxious	Relatively expensive (LiBr)
	Large latent heat of vaporisation	

compared to NH_3/H_2O mixture at temperatures below the freezing point of water. However, above the freezing point of water, NH_3/H_2O has comparatively better performance than the other two. They also stated that $NH_3/LiNO_3$ and $NH_3/NaSCN$ mixtures are simpler in operation because of no requirement of rectifier in their operation. Their results also indicated that $NH_3/NaSCN$ mixture cannot operate below $-10\ °C$ because of possibility of crystallization.

Experimental studies. Although NH_3/H_2O and $H_2O/LiBr$ pairs have been used throughout the world, researchers continue to search for new pairs. In 1994, Erhard and Hahne [23] developed a solar-powered absorption refrigeration system with NH_3 and $SrCl_2$ as the working pair. The overall COP of the cooling system has been calculated to be 0.49. In 1995, a better overall COP (0.45−0.82) was attained. Moreno-Quintanar et al. [24] analyzed the use of a ternary mixture consisting of a binary absorbent solution ($LiNO_3H_2O$) and a refrigerant (NH_3) in comparison with the binary working pair $NH_3/LiNO_3$ for producing 8 kg of ice. Compound parabolic concentrators (CPCs) were used in the experimental investigation. They reported an increase of 24% in the COP using the ternary mixture compared to the binary mixture.

Simulation studies. Medrano et al. [25] discussed the potential of using the organic fluid mixtures TFE/TEGDME and methanol/TEGDME as working pairs in a double lift absorption system. Their simulation results indicated that the COP using TFE/TEGDME and methanol/TEGDME is 0.45, which is almost 15% higher than that for NH_3/H_2O. Pilatowsky et al. [26] analyzed the monomethyl-amine/H_2O working pair for an absorption cooling system powered by flat-plate SCs. Their simulation results indicated that a COP of 0.72 can be reached while operating the system at a generator temperature of 60 °C and producing a cooling effect of 10 °C. Fong et al. [27] optimized a solar thermal cooling system employing $LiCl/H_2O$ mixtures. The optimization results indicated a reduction in the primary energy consumption of 12.2% for $LiCl/H_2O$ systems.

Rivera and Rivera [28] simulated the performance of a solar intermittent adsorption refrigeration chiller with a $H_2O/LiBr$ pair for Mexico. A compound parabolic concentrator with a glass cover is used for powering its system by solar energy. They reported that 11.8 kg of ice could be produced at a COP ranging from 0.15 to 0.40 when operating at a generation temperature of 120 °C. One recent trend in the field of solar-powered absorption systems is the use of a ternary mixture as the working fluid for such systems.

Based on the thermodynamic cycle of operation and solution regeneration, the absorption systems can be divided into three categories: single-, half-, and multieffect (double-effect and triple-effect) solar absorption cycles. The single-effect and half-effect chillers require relatively lower hot water temperatures with respect to multieffect systems [29].

Grossman [30] provided typical performances of the single- and multieffect absorption systems, as shown in Table 7.3. Typical cooling COPs of the single-, double-, and triple-effect absorption systems are 0.7, 1.2, and 1.7, respectively.

As shown in Table 7.3, a flat-plate SC can be used for the single-effect cycle. However, the multieffect absorption cycles require high temperatures above 85 °C, which can be delivered by evacuated tube or concentrating-type collectors. According to the *collector catalog* [31], a 40%-efficient evacuated tube collector (ETC) working at 150 °C costs 600–700 €/m² (gross area). For less expensive collectors, working at approximately 90 °C, a single-effect $LiBr/H_2O$ or NH_3/H_2O absorption system with a COP between 0.6 and 0.8 can be considered [7]. The price of a SC varies widely in this temperature range. The price of a 50%-efficient collector at 90 °C ranges between 300 and 600 €/m². It must be noted that the SC efficiencies listed above are only indicative, and the actual efficiencies will depend on the ambient air temperature and solar radiation.

7.2.3.1 Continuous Operation Systems

The continuous operation systems belong to a specific classification of absorption cooling systems, in which both the generation and absorption processes take place simultaneously. Such systems operate with a cyclic behavior with a cycle time of less than a day (24 h). The schematic diagram of the continuous operation-based solar-powered absorption system is shown in Fig. 7.14.

The basic components of a continuous operation absorption refrigeration system are the generator G, absorber Ab, condenser C, evaporator E, expansion valve EV, and a solution pump P. The generator is powered by SCs in the case of a solar-powered absorption system. To ensure continuous operation and reliability of the system, a hot-water ST is used.

TABLE 7.3 Typical Performance of Absorption Cycles

No.	Absorption System	COP (−)	EER (Btu/ (Wh))	Heat-Source Temperature (°C)	Type of Solar Collectors Matched
1	Single-effect	0.7	2.39	85	Flat-plate
2	Double-effect	1.2	4.10	130	Flat-plate/Compound parabolic concentrator
3	Triple-effect	1.7	5.80	220	Evacuated-tube/ Concentrating collector

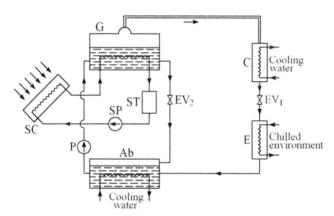

FIGURE 7.14 Continuous operation-based solar absorption system.

In the absorber, the absorbent-rich solution is diluted with the refrigerant. During this process, the absorber is cooled to keep its pressure at a low level. Then, the solution pump increases the pressure of the absorbent/refrigerant mixture to the high pressure level. The solution pump's electrical power requirement is much less than that of the compressors in the vapor-compression systems. Since the refrigerant is more volatile than the absorbent, it is separated from the solution when adequate heat is added in the generator (desorber).

Along with the basic components, certain heat-recovery components are added to the continuous-operation absorption system to increase its COP. These heat-recovery components are a solution heat exchanger (SHX) and a refrigerant precooler (PC). The performance of the absorption system falls by 3–5% without the PC, and the usage of the SHX in the system does not increase the system performance as expected [32]. Also, additional refrigerant rectification equipment such as a rectifier and a dephlegmator are added in the design to rectify the refrigerant vapor in the case of a volatile absorbent. Rectification equipment is added to restrict the volatile absorbent (water) within the generator and absorber, thus preventing it from entering into the evaporator.

The rectification equipment is normally constructed in the form of a column and divided into two subsections [33]. Fernandez-Seara et al. [34] used a helical coil rectifier in an absorption refrigeration system and analyzed the influence of the heat- and mass-transfer coefficients on the rectifier performance. The construction of the rectifier plays a very important role in the refrigerant-purification process. In this regard, Sieres and Fernandez-Seara [35] experimentally determined that there are appreciable differences between the experimental results and the available correlations that describe the volumetric mass-transfer coefficient within the rectification column.

7.2.3.2 Intermittent Operation Systems

The intermittent operation systems comprise a specific class of absorption refrigeration systems in which the generation and absorption processes do not take place simultaneously but rather follow each other in an intermittent manner. Because of the intermittent nature, it is possible to utilize the NH_3/H_2O vessel to behave as a generator G during the daytime and as an absorber Ab during the night (Fig. 7.15). Such systems operate cyclically with a cycle time of one complete day (24 h). The pressurization process in the intermittent operation systems is carried out by isochoric heating of the NH_3/H_2O solution in the generator. In this way, electrical energy is not required at all in the operation of intermittent absorption systems. The intermittent solar-absorption cooling system has two configurations: the first is a single stage and the second is a two-stage configuration. Here "two stages" distinctly refer to stages of generation, namely, high-pressure generation and low-pressure generation. The overall COP of the two-stage system operating at this temperature is 0.105, which is twice that of a single-stage system operating at 120 °C. Thus a two-stage system operating at a low-generation temperature is better than a

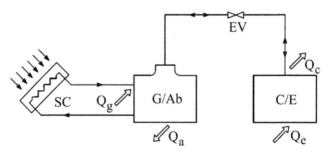

FIGURE 7.15 Intermittent operation—based solar absorption system.

single-stage system, even when the single-stage system is operating at a high temperature [36]. Some of the recent developments regarding these systems will be reviewed in this chapter.

7.2.3.3 Single-Effect Solar Absorption Cycle

Most absorption cooling systems use a single-effect absorption cycle operating with an $H_2O/LiBr$ working pair and a solar flat-plate collector (FPC) or an ETC with hot water to drive these systems.

The single-effect absorption cooling system is based on the basic absorption cycle that contains a single absorber and generator, as shown in Fig. 7.14. In the generator G, the refrigerant is separated from the absorbent by the heat provided by the SC. The vapor-refrigerant is condensed in condenser C, then laminated in expansion valve EV_1, and evaporated at a low pressure and temperature in the evaporator E. The cooled refrigerant is absorbed in the absorber Ab by a weak solution that returns from the generator after the lamination in the expansion valve EV_2. The rich mixture created in the absorber is pumped by pump P and returned in G. A typical SHX can be used to improve the cycle efficiency. A 60% higher COP can be achieved by using the SHX [37]. Because the absorption is exothermic, the absorber is chilled by cooling water.

For low-temperature heat sources, the degassing zone is unacceptably small, the release of the refrigerant-vapor in the generator is slowed down, and the operation of the system becomes unstable or impossible. To improve the COP and use lower temperatures in generator, a solar resorption cooling system can be used (Fig. 7.16) [38].

In this case, in the generator, the refrigerant is also separated from the absorbent by the heat provided by the SC, but the vapor-refrigerant is reabsorbed by a weak solution in the resorber Rb, and the system operates with

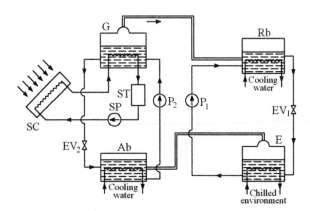

FIGURE 7.16 Solar resorption cooling system.

a cycle similar to the one above. System pressures may be allowed to be as close to atmospheric pressure as possible, which simplifies sealing problems, the manufacture of the pumps, and reduces the temperature in generator.

The single-effect system is the simplest type of these systems. Design of a single-effect absorption cooling system depends on the working fluid types. The system shows better performance with a nonvolatile working pair, such as H_2O/LiBr. An extra rectifier should be used before the condenser to provide pure refrigerant if the system operates with a volatile working pair, such as NH_3/H_2O [21].

A low-cost nonconcentrating flat-plate or evacuated-tube SC is sufficient to obtain the required temperature for the generator. Though economical, its COP is lower. For obtaining a higher COP, multieffect systems such as double-effect and triple-effect absorption chillers are used, which are run by steam produced from the concentrating SCs.

Experimental studies. Nakahara et al. [39] developed a single-effect H_2O/LiBr absorption chiller of 7 kW nominal cooling capacity assisted by a 32.2-m^2 array of flat-plate SCs. In their system, thermal energy produced by the SC was stored in a 2.5 m^3 hot-water ST. Their experimental results during the summer period showed that the cooling capacity was 6.5 kW. The measured COP of the absorption system was in the range of 0.4−0.8 at a generator temperature of 70 °C−100 °C. Another investigation on a H_2O/LiBr absorption system consisting of 49.9 m^2 of FPC was performed by Syed et al. [40]. The system performs cooling with generation temperatures of 65 °C−90 °C while maintaining a capacity of 35 kW. They achieved an average collector efficiency of ∼50%. Darkwa et al. [41] performed an experimental analysis of a H_2O/LiBr absorption system powered by evacuated-tube and flat-plate SCs and integrated with four hot-water STs. Their experimental results indicated that a COP of 0.69 could be achieved when supplied with the heated water at 96.3 °C from the SCs.

De-Francisco et al. [42] tested a prototype of a 2-kW solar-powered NH_3/H_2O absorption chiller that utilizes concentrating collectors and uses a transfer tank instead of a pump. Their experimentation resulted in a COP of 0.05 when the collectors operated at temperatures greater than 150 °C. They suggested the inefficient operation of the transfer tank was responsible for the low COP of their system. Brendel et al. [43] developed an experimental setup for a small-scale solar-powered 10-kW NH_3/H_2O absorption chiller. They used plate heat exchangers in the system, except for the generator, which is fabricated as a helical-coil heat exchanger inside a cylindrical shell. They reported a COP of 0.58−0.74 for the absorption system operating at a generator temperature ranging from 80 °C to 120 °C. Rosiek and Batles [44] experimentally investigated the performance of a 70-kW solar absorption chiller operating under two modes of heat rejection from the absorber and condenser of the absorption cycle. The two operation modes were compared in terms of energy consumption, water consumption, and CO_2 savings. For one cooling period, the shallow geothermal system reduced the electricity by 31%, reduced water consumption by 116 m^3, and led to a CO_2 savings of 833 kg.

Simulation studies. El-Shaarawi and Ramadan [45] investigated the performance of a solar-powered intermittent NH_3/H_2O absorption system for varying condensation temperatures. They reported that decreasing the condenser temperature at any fixed initial solution concentration and temperature results in an increase in the COP of the system. A theoretical simulation has also been conducted by Said et al. [46] of an intermittent solar absorption system designed to provide 120 kWh of cooling effect throughout the day and night operations. Their simulation results indicated that the intermittent absorption system can achieve a COP of 0.23 when operating at the generation temperature of 120 °C while producing a cooling effect at a temperature of −9 °C.

Recently, El-Shaarawi et al. [47] developed a simplified correlation for an NH_3/H_2O intermittent solar-powered absorption refrigeration system. They developed a set of correlations of polynomial form for directly estimating the performance of intermittent absorption systems as a function of the generator, absorber, condenser, and evaporator temperatures. They reported that their developed correlation estimates the design parameters of intermittent absorption refrigeration systems with an accuracy of greater than 97%.

To improve the unsteady nature of the solar heat from the SC provided to the absorption system, Chen and Hihara [48] proposed a new type of absorption cycle that was codriven both by solar energy and electricity. In their proposed system, the total energy delivered to the generator could be controlled by adjusting the mass flow rate through the compressor. Their numerical simulation model results showed the steady COP value of 0.8 for the new cycle, which was higher than the conventional cycle. Chinnappa et al. [49] proposed a conventional vapor compression A/C system cascaded with a solar-assisted NH_3/H_2O absorption system. They concluded that by reducing the R22 condensation temperature to 27 °C, the hybrid system achieved a COP of 5, which is higher than that of the vapor compression cycle at 2.55.

Recently, the approach of direct air cooling of a solar-powered NH_3/H_2O chiller has been simulated by Lin et al. [50] for a two-stage absorption system. They tested several arrangements for the series connection of condenser, low-pressure absorber, and medium-pressure absorber. The simulation results for the best arrangement indicated a thermal COP of 0.34 under typical summer conditions.

7.2.3.4 Half-Effect Solar Absorption Cycle

The half-effect absorption cycle, also called two-stage or double-lift cycle, can provide cold with a relatively low driving temperature. The name "half effect" arises from the value of the COP, which is almost half than that of the single-effect cycle. A schematic diagram for this cycle is shown in Fig. 7.17.

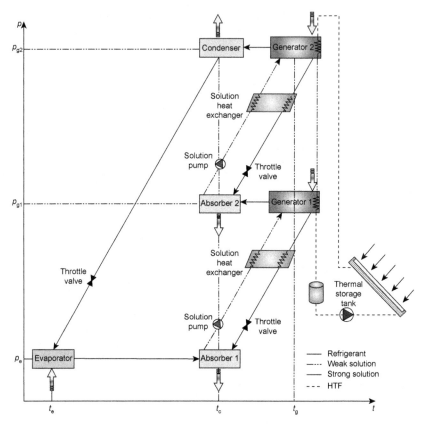

FIGURE 7.17   Schematic of the half-effect solar-absorption cooling system.

Sumathy et al. [51] proposed a two-stage H$_2$O/LiBr chiller for cooling purposes in southern China. Test results have proved that the two-stage chiller could be driven by low-temperature hot water ranging from 60 °C to 75 °C, which can be easily provided by conventional solar hot-water systems. Based on the successes of this system, they integrated the solar cooling and heating systems with two-stage absorption chiller and cooling capacity of 100 kW. Operating results from the system indicated that this type of system could be efficient and cost-effective, comparing to the conventional cooling system with single-stage chiller. The proposed system with a two-stage chiller could achieve the same total COP as of the conventional system but with a cost reduction of about 50%.

Izquierdo et al. [52] designed a solar double-stage absorption plant with H$_2$O/LiBr that contained FPCs to feed the generator. The results show that a generation temperature of ~80 °C was required in the absorption system when the condensation temperature reached 50 °C, and they obtained a COP of

0.38 without crystallization problems. They also performed an exergetic analysis of this system and concluded that the irreversibility generated by the double-stage thermal compressor will tend to increase with the absorption temperature up to 45 °C. The conclusions show that the double-stage half-effect system has ~22% less exergetic efficiency than the single-effect system and 32% less exergetic efficiency than the double-effect one. The entropy generated and the exergy destroyed by the air-cooled system are higher than those by the water-cooled one. For an absorption temperature equal to 50 °C, the air-cooled mode generates 14% more entropy and destroys 14% more exergy than the water-cooled one.

Arivazhagan et al. [53] performed an investigation with a two-stage half-effect absorption system operating with an R134a/DMAC working pair. They obtained an evaporation temperature of −7 °C for generation temperatures of 55−75 °C. They concluded that a COP of approximately 0.36 could be achieved within the optimum temperature range (65−70 °C).

7.2.3.5 Double-Effect Solar Absorption Cycle

Double-effect absorption cooling technology increases the system performance using a heat source at higher temperatures. Fig. 7.18 illustrates a double-effect absorption system with a $H_2O/LiBr$ pair.

The cycle begins with generator G-I providing heat to generator G-II. The condenser C rejects the heat and passes the working fluid toward the evaporator E, in this step, the required refrigeration occurs. Then, the fluids pass through the heat exchangers HX-I and HX-II from the absorber Ab to G-I by means of a pump P. Through this process, HX-II can pass the fluids to G-II, and then G-II passes them to HX-I. The complete cycle includes three different pressure levels: high, medium, and low.

The combination of two single-effect systems effectively comprises a double-effect absorption cooling system. Therefore, the COP of a double-effect system is almost twice as that of a single-effect absorption system. For

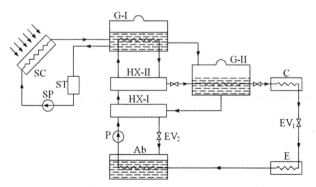

FIGURE 7.18   Solar-assisted double-effect $H_2O/LiBr$ absorption system.

example, an analysis performed by Srikhirin et al. [21] indicates that the COP of a double-effect system is 0.96, whereas the single-effect system has a COP of only 0.6. In the past decade, the COP of double-effect absorption systems has reached values of 1.1—1.2 by using gas-fired absorption technology [29].

Experimental studies. Tierney [54] performed a comparison among four systems with different chiller—collector combinations, at four different latitudes. He concluded that the double-effect chiller with a trough collector had the highest potential savings (86%) among the four systems to handle the demand for a 50-kW load. Bermejo et al. [55] developed a 174-kW gas/solar-powered double-effect $H_2O/LiBr$ absorption chiller in Spain using linear concentrating Fresnel collectors with a total area of 352 m^2. Their experimental results indicated the solar energy contributed 75% of the total heat input to the generator producing a COP of 1.10—1.25 when operated at the evaporator temperature of 8 °C.

Simulation studies. Several simulation studies have been conducted using simulation software such as FORTRAN, MATLAB, ASPEN, TRNSYS, Engineering Equation Solver (EES), etc. One recent simulation study was performed by Gomri [56] using mathematical code developed in the FORTRAN programming language. He compared the performance of single-effect and double-effect absorption systems. His results indicated that the double-effect systems have a COP approximately double than that of single-effect systems. His case study indicated a COP of 1.22—1.42 for the double-effect systems, whereas the single-effect systems could only reach a COP of 0.73—0.79 while operating under the same conditions.

Somers et al. [57] compared the results of simulations using ASPEN and EES software for the performance of a solar-powered $H_2O/LiBr$ system. They successfully modeled the single-effect and double-effect $H_2O/LiBr$ systems using the ASPEN program, and their results indicated errors of less than 3% and 5% respectively, compared to the EES results.

Hong et al. [58] proposed an evaporator—absorber-exchange absorption cycle that can utilize the heat of condensation of the vapor generated from the high-temperature generator. Their simulation results indicated that the COP of such a system is 40% higher than that of the conventional single-effect cycle but lower than that of the conventional double-effect cycle while operating at generator temperatures between 127.5 °C and 150 °C.

7.2.3.6 Triple-Effect Solar Absorption Cycle

Triple-effect absorption cooling can be classified as having either single-loop or dual-loop cycles. Single-loop triple-effect cycles are basically double-effect cycles with an additional generator and condenser. The resulting system, with three generators and three condensers, operates similar to the double-effect system. Primary heat concentrates the absorbent solution in a first-stage generator at approximately 200—230 °C. A fluid pair other than $H_2O/LiBr$ must be used for the high-temperature cycle. The refrigerant vapor produced is then used to concentrate additional absorbent solution in a second-stage generator at ~ 150 °C. Finally, the refrigerant vapor produced in the second-stage generator concentrates additional absorbent solution in a third-stage generator at ~ 93 °C. The typical SHXs can be used to improve the cycle efficiency. Theoretically, these triple-effect cycles can obtain COPs of approximately 1.7 [11].

A double-loop triple-effect cycle consists of two cascaded single-effect cycles. One cycle operates at normal single-effect operating temperatures and the other at higher temperatures. The smaller high-temperature topping cycle has a generator temperature of approximately 200−230 °C. A fluid pair other than $H_2O/LiBr$ must be used for the high-temperature cycle. Heat is rejected from the high-temperature cycle at 93 °C and is used as the energy input for the conventional single-effect bottoming cycle. Theoretically, this triple-effect cycle can obtain an overall COP of approximately 1.8 [11].

Multieffect cycles are costlier but more energy-efficient. Double- and triple-effect chillers employ an additional generator and heat exchanger to liberate the refrigerant from the absorbent solution with less heat input. Multieffect absorption cycles require high temperatures above 200 °C that can limit choices in materials and working pairs. In order to handle the absorption liquid of such high temperature and high pressure, a high-temperature generator must be manufactured as a boiler, and corrosion suppressive measures are needed. The available solar intensity, cooling capacity requirements, overall performance, and cost determine the selection of a particular configuration.

A combined cooling concept arose due to integrate different pairs or systems for obtaining better cooling performance. Combined solar-absorption cooling system refers to the integration of three individual cooling technologies: radiant cooling, desiccant cooling, and absorption cooling.

Table 7.4 summarizes the above-mentioned absorption cooling systems. Solar-absorption cooling systems are used in A/C applications, for food preservation, and in ice production.

7.2.3.7 Thermodynamic Properties of Working Pairs

The thermodynamic properties of the refrigerant/absorbent working pair are the most important factors in analyzing and optimizing the performance of the absorption systems. NH_3/H_2O is one of the most widely used refrigerant/absorbent working pairs in absorption systems. Thus several correlations for determining the vapor−liquid-equilibrium (VLE) thermodynamic properties of the NH_3/H_2O working pair have been proposed by researchers using different methodologies.

Park and Sonntag [59] used a generalized equation of state based on a four-parameter corresponding states principle to determine the thermodynamic properties of NH_3/H_2O mixtures. Later, Ibrahim and Klein [60] used the Gibbs

TABLE 7.4 The Characteristics of Working Fluids Found From Various Absorption Refrigeration Technologies

Absorption Cooling Systems	Working Fluids	Results/References
Single-effect	H_2O/LiBr, NH_3/H_2O	• A rectifier is needed to purify the refrigerant if the pair is volatile/[36]. • Approximately 60% more COP can be achieved by using a solution heat exchanger/[37]. • A system capacity of 70 kW can be achieved by using a vacuum tubular collector (108 m^2) with flat-plate collectors (FPCs) (9124 m^2)/[29]. • COP can be increased by 15% using a partitioned hot-water tank with a FPC (38 m^2) and chillers (4.7 kW)/[2].
Half-effect	H_2O/LiBr	• Within the optimum temperature range of 65−70°C, the COP = .36 and the evaporation temperature is −7°C/[53] • The pair is capable of providing the same COP as a conventional cooling system with reducing the cost by half/[51] • The system has 22% lower exergetic efficiency compared to the single-effect system/[52]
Double-effect	H_2O/LiBr	• The system has almost double (0.96) the COP compared to the single-effect system/[21] • The double-effect chillers with trough collectors show the maximum potential savings (86%)/[54]
Hybrid	Combination of mentioned pairs	• The types of systems are widely implemented for the cooling of larger places, such as offices, markets or auditorium/[2]

excess energy to estimate the VLE thermodynamic properties of NH_3/H_2O mixtures. Tillner-Roth and Friend [61] used the approach of the Helmholtz free energy for developing the correlations for estimating the VLE thermodynamic properties of NH_3/H_2O mixtures. Recently, El-Shaarawi et al. [62] used an EES program to develop polynomial forms of explicitly defined thermodynamic property correlations that can be used by any simulation software.

TABLE 7.5 Summary of Operating Temperature and Pressure Ranges

Methodology Used	Year	Operating Range		References
		Temperature [°C]	Pressure [bar]	
Generalized equation of state	1990	≤377	≤200	[60]
Gibbs excess energy	1993	≤327	≤110	[61]
Simple functional form	1995	≤180	≤20	[38]
Helmholtz free energy	1998	–	≤400	[62]
Polynomial form of equations	2013	–	≤100	[62]

The working temperature and pressure range of the VLE thermodynamic properties of NH_3/H_2O working mixtures is summarized in Table 7.5.

Patek and Klomfar [63] give a fast calculation of thermodynamic properties. As an example, the bubble point and dew point temperatures of the NH_3/H_2O mixture are found from the correlations in Eqs. (7.6) and (7.7), developed by Patek and Klomfar [63]:

$$t_b(p,x) = t_o \sum_i a_i (1-x)^{m_i} \left(\ln\frac{p_o}{p} \right)^{n_i} \tag{7.6}$$

$$t_d(p,y) = t_o \sum_i a_i (1-y)^{m_i/4} \left(\ln\frac{p_o}{p} \right)^{n_i} \tag{7.7}$$

where t_b is the bubble point temperature, in K; t_d is the dew point temperature, in K; p is the pressure, in Pa; x is the NH_3 mole fraction in liquid phase; y is the NH_3 mole fraction in vapor phase; a_i, m_i, and n_i are the coefficients; and subscript o represents the ideal gas state.

7.2.3.8 Design, Control, and Operation Guidelines

Absorption is the most popular heat-driven cooling technology. Proper calculations for collector and storage size depend on the solar cooling technology used. Hot water storage may be integrated between the SCs and the heat-driven chiller to dampen fluctuations in the return temperature of the hot water from the chiller. Storage size depends on the application—when cooling loads mainly occur during the day, a smaller storage is necessary than when the loads peak in the evening. The storage exists only to store excess heat of the solar circuit and make it available when sufficient solar heat is not available.

The single-effect absorption system gives best results with a heat supply temperature of 80−100 °C. Double- and triple-effect systems require higher supply temperatures. Therefore, these systems require a higher temperature collector.

Most large-scale applications (≥ 300 kW) use $H_2O/LiBr$ and produce chilled water at about 6−7 °C, and the COP is relatively higher than NH_3/H_2O. However, LiBr systems must be water cooled and thus usually require a cooling tower, whereas NH_3 systems can have an air-cooled condenser. Because of the large vapor volume of the water refrigerant, $H_2O/LiBr$ chillers usually have large physical dimensions. For small cooling loads and applications where it is not possible to use water cooling, an NH_3/H_2O system is preferred.

In hot and sunny climates, the required SC area is approximately 3−4 m^2 per kW cooling. Higher heat-supply temperature for multieffect chillers require higher cost evacuated tube or concentrating collectors, and may need high-temperature storage.

In $H_2O/LiBr$ systems, the refrigerant freezes at 0°C, so care is necessary while the system is idle, especially in winter. Another potential problem is crystallization of the LiBr solution at high concentrations, which may result from high generator temperatures or from inadequate temperature control at other parts of the system. Thus the heat-supply temperature from the SCs or heat storage must be adequately controlled. The cooling-water temperature, particularly to the absorber, must also be monitored. Chiller capacity may be controlled by increasing the heat-supply temperature or decreasing the cooling-water temperature.

A viable design solution for a single-effect absorption cycle is to incorporate an auxiliary desorber powered by the backup, whereas the original desorber is powered by the solar energy. The weak solution goes first to the solar-powered desorber, where it is concentrated as much as possible with the available solar heat, and then proceeds to the auxiliary desorber, where it is concentrated further using heat from the backup source. Vapor from both desorbers are then supplied to the condenser [64].

7.2.4 Experimental Investigation on a Solar-Powered Absorption Radiant Cooling System

7.2.4.1 System Configuration

In this study, a mini-type solar-absorption cooling system using both fan coils and the radiant ceiling as terminals was designed and installed [65]. The solar-powered absorption cooling system mainly consists of evacuate collectors with area of 96 m^2, a hot-water ST with the capacity of 3 m^3, an 8 kW LiBr/H$_2$O absorption chiller, a cooling tower, and the terminals of the A/C system. In summer, the system was used to supply cooling to meet the cooling requirement of a test room with the area of 50 m^2. Two different terminals of the A/C system including the fan coils and the radiant cooling ceiling were fixed in the test room. Besides, a fresh-air handling unit (f-AHU) was installed to supply fresh air and remove latent heat away from the test room.

- SC *array*. The SCs were installed on the building roof and designed to be tilted 30 degrees° horizontal for better performance. The SC array was divided into four rows. Each row was composed of four SC units, which were connected in series for the purpose of achieving hot water with a relatively high temperature.
- *Heat storage*. The hot-water storage system has been proved significant for the whole system. The system employed a hot-water ST with the capacity of 3 m^3 that supplied hot water for the chiller generator. The water tank and the SC array are connected through a plate heat exchanger with the heat exchange capacity of 30 kW.
- *Absorption chiller*. The LiBr/H$_2$O single-effect absorption chiller was used. Under the nominal operating mode, the inlet hot-water temperature is 70−95°C with the flow rate of 4 t/h. The chilled water at approximately 9°C with the flow rate of 1.5 t/h is produced. The chiller COP of 0.7 can be achieved under the nominal condition. The absorption chiller is driven by the hot water to generate cooling effect. Meanwhile, it releases heat to the environment through the cooling tower. A solution circulation pump is fixed inside the chiller to drive rich solution to flow from the absorber to the generator.
- *Test room and terminals*. The area of the test room is 50 m^2. The north wall is an outer envelope with 6 m^2 glass windows. The other three walls can be treated as inner envelopes. The cooling ceiling with the area of 30 m^2 and two fan coils were fixed in the test room. Besides, an independent fresh-air system was installed in the neighbor room to supply fresh air and meet the latent load of the room. The fresh air can be cooled by both solar cooling and electric cooling. So when solar radiation is not enough, the electric cooling device can be operated to meet the entire indoor cooling load.
- *System control and data acquisition*. The entire system was automatically controlled. Fig. 7.19 shows the schematic diagram of the experimental system. By means of the PT1000 sensors, the circulation pumps can be switched on or off, which, in turn, dominates the SC cycle, hot-water tank cycle, cooling tower cycle, and chilled water cycle. Besides, the solar radiation and the temperatures were all recorded by a data logger. In the morning, when the hot water temperature achieves 75°C, the system is turned on. The system is turned off when the hot water temperature is below 65°C or when the time is after 18:00 h. Several kinds of sensors have been employed in the pipeline of the whole system including thermometers, flow meters, irradiation sensors, hygrometers, and pressure gauges. The platinum resistance sensors were fixed on the important locations such as the inlet and outlet of the water tank, chiller, and cooling tower. The

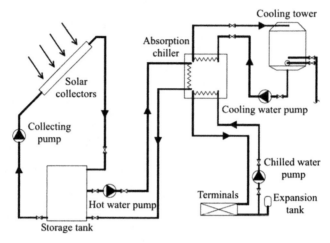

FIGURE 7.19   Schematic of the experimental system.

water flow-rate meters collect the data of flow rate and transmit them to a computer. There is a data acquisition logger of Keithley 2700, which connects to the computer and records the data every 15 s. The measuring accuracy is 1 W/m² for the irradiation sensor, 0.01°C for the platinum resistance sensor, and 0.1 t/h for the water flow meter.

7.2.4.2 Experimental Investigation

- SC *array*. Fig. 7.20 shows the variations of the solar collecting efficiency η_c of the SC array and the solar radiation intensity I_T in a typical summer day. It was obvious that I_T had a maximum value at 11:20. Meanwhile, the collector efficiency gently rose from the morning and then kept nearly constant from 8:30 to 12:30. The collector efficiency reduced with the decrease of I_T. From 12:30 to 15:30, the η_c decreased slowly although the solar radiation intensity remained at high level with above 400 W/m². When the solar radiation intensity was less than 300 W/m², the collector efficiency had a dramatic reduction because the heat loss to the surrounding was more than heat collected by the SC array. The heat storage water tank received heat from SC array by means of a flat-plate heat exchanger. The daily average solar collecting efficiency of the collector array was concluded to be 0.46 [65].
- *Operation of the solar absorption chiller.* Fig. 7.21 shows the variations of heat consumption, cooling power, and COP of the chiller. The cooling power was estimated through the flow rate of the chilled water and the temperature difference of the chilled water between the inlet and outlet of the absorption chiller. It is seen that 4.5-kW cooling power can be obtained from 10:00 to 18:00. The chiller COP varied from 0.25 to 0.38 with an average value of

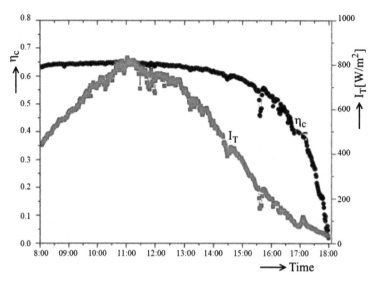

FIGURE 7.20   Solar radiation and the collector efficiency.

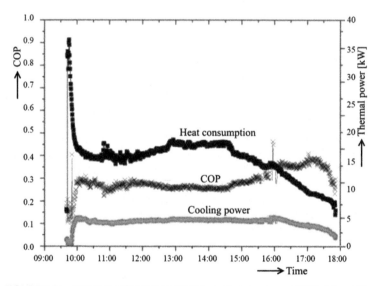

FIGURE 7.21   Variations of heat consumption, cooling power, and COP of the chiller.

0.32 [65]. After 16:00, the hot-water temperature began to decrease, which resulted in the decrease of heat consumption. The COP went up from 16:00 because the decrease of heat consumption overwhelmed the decrease of cooling output. In this experiment, because the chiller worked at partial load, a low efficiency from solar to cooling was obtained. Therefore, it is

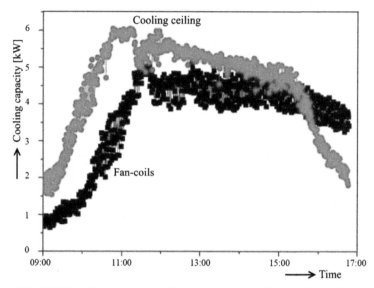

FIGURE 7.22 Comparison of cooling capacity by fan coils and radiant ceiling.

important to select the adequate size of the absorption system to make it work at nominal operating conditions, so that the system COP reaches the nominal value and the solar fraction, consequently, is higher.

- *Comparison of different terminals.* The fan coils and the cooling ceiling have been employed as terminals in this system, respectively. The performance of the solar-driven absorption chiller combined with fan coils was compared with that of combined with radiant cooling panel, as shown in Fig. 7.22. The average cooling power (capacity) of the former was 3.62 kW and, correspondingly, 4.47 kW for the latter based upon the experimental results under similar weather conditions. By means of the radiant ceiling, the cooling power of the absorption chiller increased by 23.5% [65]. That is due to the higher evaporation temperature of the chiller when the cooling radiant process was applied.

7.2.4.3 Conclusions

The main conclusions of this study can be summarized as follows:

1. The daily-average solar collecting efficiency of the collector array was 0.46.
2. The cooling capacity of 4.5 kW can be obtained during the continuous operation of 8 h. The chiller COP was 0.32.
3. Under similar weather conditions, the cooling power by means of radiant cooling panels increased by 23.5% compared to the operation mode of fan coils.

7.2.5 Adsorption Systems

Adsorption technology was first used in refrigeration and heat pumps in the early 1990s. The adsorption phenomenon results from the interaction between a solid and a fluid (refrigerant) based on a physical or chemical reaction. Physical adsorption (physisorption) occurs when the molecules of the refrigerant (adsorbate) fix themselves to the surface of a porous solid element (adsorbent) due to Van der Waals forces and electrostatic forces, and the molecules are adsorbed onto the surface. By applying heat, the refrigerant molecules can be released (desorption), whereby this is a reversible process. Physical adsorption is nonspecific and occurs for any refrigerant—adsorbent system. In turn, the chemical adsorption (chemisorption) results from the ionic or covalent bonds formed between the refrigerant molecules and the solid substance. The bonding forces are much greater than those of physical adsorption, and thus more heat is released. However, the process cannot be easily reversed [66]. The chemisorption process is specific and occurs between a certain gas and a certain corresponding adsorbent solid.

When fixed adsorbent beds are employed, these cycles can be operated without any moving parts. The use of fixed beds results in silence, simplicity, high reliability, and long lifetime. On the other hand, it leads to intermittent cycle operation with adsorbent beds changing between adsorption and desorption stages, which decreases the system COP. Thus, when a continuous cooling effect is required, two or more adsorbent beds must be operating out of phase [67].

As in the case of absorption, both physisorption and chemisorption, the process of adsorption is exothermic and is accompanied by the evolution of heat, whereas the process of desorption is endothermic and is accompanied by absorption of heat. These characteristics are used to produce the cooling effect in refrigeration and A/C applications.

7.2.5.1 Basic Solar Adsorption Refrigeration Cycle

Solar energy is the energy source of most adsorption systems operating with the basic cycle. A solar adsorption cooling system based on the basic adsorption refrigeration cycle does not require any mechanical or electrical energy, just thermal energy, and it operates intermittently according to the daily cycle. This refrigerator is a closed system consisting of a SC containing the adsorbent bed, a condenser C, a receiver R equipped with a two-way valve V, and an evaporator E (Fig. 7.23). In this case, the compressor is an adsorber powered by thermal energy (SC), and the cooling effect is achieved by the evaporation of a refrigerant, while the vapor produced is adsorbed by the adsorbent layer in the adsorber.

The basic adsorption cycle consists of four stages (two isobaric and two isosteric lines), which can be represented in the Clapeyron diagram (Fig. 7.24). The process starts at point 1 when the adsorbent is at adsorption temperature t_a

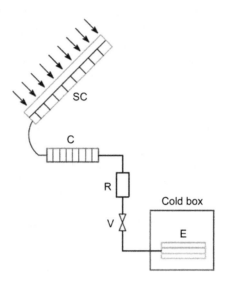

FIGURE 7.23   Schematic of a solar adsorption cooling device.

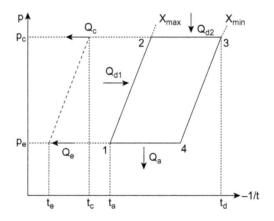

FIGURE 7.24   Basic adsorption cycle.

and at a low evaporation pressure p_e, and the content of the refrigerant is at its maximum value X_{max}.

The control valve V is closed, and as heat Q_{d1} is applied to the adsorbent, both the temperature and pressure increase along the isosteric line 1—2, while the mass of the refrigerant remains constant at the maximum value.

Upon reaching the condensation pressure p_c at point 2, the desorption process starts (isobaric line 2—3), in which the progressive heating Q_{d2} of the adsorbent from point 2 to point 3 causes the adsorbent to release the refrigerant vapor, which is liquefied in the condenser (releasing the condensation heat Q_c

at the condensation temperature t_c) and is then collected in a receiver. This stage ends when the adsorbent reaches its maximum regeneration temperature t_d and the refrigerant content decreases to the minimum value X_{min} (point 3). Then, the adsorbent cools along the isosteric line 3−4, while the refrigerant content remains constant at the minimum value. During this phase, the control valve V is opened, allowing for the refrigerant to flow into the evaporator, and the system pressure decreases until it reaches the evaporator pressure p_e, thus equalizing the refrigerant vaporization pressure (point 4).

Afterward, the adsorption−evaporation phase (isobaric line 4−1) from points 4 to point 1 occurs producing the cooling effect Q_e in the evaporator at evaporation temperature t_e. At this stage, the vaporized refrigerant in the evaporator flows to the adsorber, where it is adsorbed until the maximum content X_{max} is reached at point 1. During this phase, the adsorbent cools until it reaches the temperature t_a by rejecting the sensible heat and the heat of adsorption Q_a. At the end of this phase, the control valve V is closed, and the cycle restarts [66]. In the solar adsorption refrigeration cycle, stages 1−3 correspond to the daytime period, and the stages 3−1 correspond to the nighttime period. The adsorption cycle achieves a COP of 0.3−07, depending upon the driving heat temperature of 60−95°C [68].

The thermal COP of the adsorption cycle can be calculated using following equation:

$$COP = \frac{Q_e}{Q_d} \tag{7.8}$$

where Q_e is the cooling power, $Q_d = Q_{d1} + Q_{d2}$ is the heat transferred to the adsorber to promote its regeneration/desorption, and Q_{d1} and Q_{d2} are as described above in the equation.

The overall COP of a solar adsorption cooling system is defined in Eq. (7.9):

$$COP_{sys} = \frac{Q_e}{Q_s} \tag{7.9}$$

where Q_s is the solar-power received by the SC surface.

The specific cooling power (SCP), in W/kg, is defined in Eq. (7.10):

$$SCP = \frac{Q_e}{m_a} \tag{7.10}$$

where Q_e is the cooling power, in W and m_a is the mass of adsorbent, in kg.

Adsorption refrigeration technology has been used for many specific applications, such as purification, separation, and thermal refrigeration technologies.

Researchers worldwide are working to improve the performance of adsorption cooling systems in order to overcome its current technical and economic issues. There are many ways: improving working pairs, heat, and mass enhancement in adsorbent bed during adsorption and desorption process, optimization of the adsorbent bed structure and refrigeration cycle. The majority of the current research works are related with the necessity to improve the adsorber–collector operating parameters and to develop new or enhanced working pairs. On the other hand, several projects and research works have been undertaken to overcome these limitations by continuously developing operation systems with higher performances, recurring to multiple adsorbent beds, mass or heat recovery systems, or multistage systems. However, such systems usually require a continuous heat source to operate uninterruptedly, and they are also more complex and expensive.

7.2.5.2 Working Pairs

In an adsorption refrigeration technique, the adsorbent/refrigerant working pair plays a vital role in the optimal performance of the system. For any cooling application, the adsorbent must have high adsorptive capacity at ambient temperatures and low pressures but less adsorptive capacity at high temperatures and high pressures. The adsorbents are characterized by surface properties such as surface area and polarity. A large specific surface area is preferable for providing large adsorption capacity, but the creation of a large internal surface area in a limited volume inevitably gives rise to large numbers of small-sized pores between adsorption surfaces. The pore size distribution of micropores that determines the accessibility of adsorbate molecules to the internal adsorption surface is important for characterizing adsorptivity of adsorbents.

The choice of the adsorbent will depend mainly on the following factors:

- high adsorption and desorption capacity to attain high cooling effect;
- good thermal conductivity to shorten the cycle time;
- low specific heat;
- chemically compatible with the chosen refrigerant;
- low cost; and
- wide availability.

The selected refrigerant must have most of the following desirable thermodynamics and heat transfer properties:

- evaporation temperature below $0°C$;
- high latent heat per unit volume;
- high thermal conductivity;
- low viscosity;
- low specific heat;
- nontoxic, noninflammable, and noncorrosive nature; and
- chemically stableness in the working temperature range.

The most commonly used adsorbents are activated carbon, zeolite, and silica gel. Activated carbon (AC) offers a good compromise between high adsorption and desorption capacities. In turn, natural zeolites need to be present in large quantities because only a small amount of refrigerant is desorbed during the temperature increase. Silica-gel is expensive and may not be available in most countries. The most commonly used refrigerants are NH_3 (NH_3), methanol, and water (H_2O), which have relatively high latent-heat values (1368, 1160, and 2258 kJ/kg, respectively) and low specific volumes (of the order of 10^{-3} m^3/kg). NH_3 is toxic and corrosive, while H_2O and methanol are not, but the latter are flammable.

The most commonly used working pairs are: silica-gel/H_2O; activated-carbon/methanol; activated-carbon/NH_3; zeolite/H_2O; activated-carbon/granular and fiber adsorbent; metal chloride/NH_3; and composite adsorbent/ NH_3. The most common working pairs used in hybrid adsorption technology are silica-gel/chloride-water and chlorides/porous media/ NH_3.

Silica-gel/water is ideal for solar energy applications due to its low regeneration temperature, thus only requiring low-grade heat sources, commonly below 85°C. Moreover, water has the advantage of having a greater latent heat than the other conventional refrigerants. However, this pair has a low adsorption capacity as well as a low vapor pressure, which can hinder mass transfer. Furthermore, this working pair requires vacuum conditions in the system [29]. Main applications for this working pairs are refrigeration, water cooling, and ice making.

Activated-carbon/methanol is one of the most common working pairs in adsorption refrigeration systems. It also operates at low regeneration temperatures, while its adsorption–evaporation temperature lift is limited to 40°C. This pair is also characterized by its large cyclic adsorption capacity, low freezing point, low adsorption heat, and the high evaporation latent heat of methanol. However, AC has a low thermal conductivity, which decreases the system's COP. Additionally, methanol has a high toxicity and flammability [66]. Main applications for this working pairs are ice making, water cooling, and ventilation.

The *activated-carbon*/NH_3 pair requires regeneration temperatures that can exceed 150°C. Its adsorption heat is similar to that of activated-carbon/ methanol, but it requires higher operating pressures, which prevents the infiltration of air into the system and reduces the cycle time. These factors help to increase the specific cooling capacity of the system. However, the AC has a lower adsorption capacity when paired with NH_3 than methanol [29,66]. Main applications for this working pairs are ice making and refrigeration.

The *zeolite/water* pair requires regeneration temperatures that exceed 200°C, with an adsorption–evaporation temperature lift up to 70°C or more. This pair remains stable at high temperatures, and the latent heat of water is much higher than those of methanol or other classical refrigerants. However, a system operating with the zeolite/H_2O pair is more suitable for A/C applications due to the solidification temperature of water, which restrains the freezing process. The specific cooling capacity of these systems is not very high [29]. Main applications for this working pairs are refrigeration, ice making, and water cooling.

In addition to the most common working pairs, other pairs have also been investigated. In Germany, Bansal et al. [69] developed a solar refrigerator with the capacity to produce a daily 4.4-MJ cooling effect operating with strontium chloride as the adsorbent and NH_3 as the refrigerant. The maximum solar COP obtained was 0.08 with a daily solar radiation of 26 MJ/m^2.

Anyanwu and Ogueke [70] evaluated the thermodynamic performance of different working pairs when designing a solar adsorption system. It was concluded that zeolite/H_2O is the best pair for A/C applications, whereas the activated-carbon/NH_3 pair is preferred for ice making, deep freezing, and food conservation applications. The maximum solar COP was found to be 0.3, 0.19, and 0.16 for zeolite/H_2O, activated-carbon/ NH_3, and activated-carbon/methanol, respectively, when a conventional flat-plate SC was used. The activated-carbon/methanol pair is also suitable for ice production and freezing applications [68].

More working pairs have been evaluated in the recent years. For example, Jribi et al. [71] simulated the dynamic behavior of a four-bed adsorption cycle using AC and low global warming potential R1234ze. Table 7.6 provides the summary of a comparison between various solid adsorbent pairs.

The main direction for future research work on adsorption working pairs is related to advance adsorbent technology with high heat and mass transfer performance. Promising results have been obtained with the composite adsorbents and consolidated adsorbents. However, improvements in heat transfer often result in deployment of the mass transfer performance, thus research should be focused in producing adsorbent where both mass and heat transfer can be satisfactory. Generally, consolidated adsorbent with high density and short-mass transfer path has both good heat and mass transfer performance. However, the increase of the volume occupied by the mass transfer channels reduces the amount of the adsorbent that can be placed in a specific space.

Another direction for future research work is the search for working pairs that could be powered by low temperature and used to obtain temperatures below $-10°C$.

7.2.5.3 Physical Adsorption

Adsorbents such as silica gel, AC, and zeolite are physical adsorbents having highly porous structures with surface-to-volume ratios on the order of several hundred that can selectively catch and hold refrigerants. When saturated, they can be regenerated simply by being heated.

The employed adsorbent/refrigerant working pairs include silica-gel/H_2O [72] and AC with methanol or NH_3 [73]. Several small-capacity silica-gel/ H_2O adsorption chillers have been developed for solar A/C [74]. Cooling capacities were reported between 3.2 and 3.6 kW. The COPs ranged from 0.2

TABLE 7.6 Comparison Between Various Solid Adsorbent Pairs

Adsorbent	Refrigerant	Adsorption Heat [kJ/kg]	Density of Refrigerant [kg/m³]	Considerations
Activated alumina	H_2O	3000	1000	Water is applicable except for very low operating pressure
Zeolite	H_2O	3300–4200	1000	Natural zeolite has lower values than synthetic zeolite
	NH_3	4000–6000	681	
	CH_3OH	2300–2600	791	
Silica gel	CH_3OH	1000–1500	791	Suitable for temperature less than 200°C
Silica gel	H_2O	2800	1000	Used mostly for descent cooling
Calcium chloride	CH_3OH	1800–2000	791	Used for cooling
Metal hydrides	Hydrogen	2300–2600	1000	Used for air-conditioning
Complex compounds	Salts and NH_3 or H_2O	2000–2700	681	Used for refrigeration

to 0.6 with heating temperatures from 55°C to 95°C. Current solar adsorption technology can provide a daily ice production of 4–7 kg/m² of SC with a cooling COP between 0.10 and 0.15 [75]. In 2012, Omisanya et al. [76] presented a solar adsorption cooling system working with zeolite/H_2O and comprising two CPC SCs (1 m² of total area), a condenser, and an evaporator inside a cold box. The experiment resulted in an average temperature of 11°C in the evaporator throughout the daytime period and a maximum temperature of 110°C in the adsorber. The average daily-hourly mean cycle COP ranged between 0.8 and 1.5 with an average daily solar radiation of 14.7 MJ/m².

7.2.5.4 Chemical Adsorption

Chemical adsorption is characterized by the strong chemical bond between the refrigerant and the absorbent. Therefore, it is more difficult to reverse and thus requires more energy to remove the adsorbed molecules than physical adsorption does.

The most commonly used chemical adsorbent in solar cooling applications has been $CaCl_2$. $CaCl_2$ adsorbs NH_3 to produce $CaCl_2 \cdot 8NH_3$ and water to produce $CaCl_2 \cdot 6H_2O$ as a product. Erhard et al. [77] used the strontium-chloride/NH_3 pair in a demonstration adsorption refrigeration device constructed in Germany. However, a global COP of only 0.08 was reached with the evaporation temperature reaching $-5°C$ during the cycle. Maggio et al. [78] developed a mathematical model to evaluate the performance of a solar adsorption refrigerator working with LiCl in the pores of silica gel (composite material) as the adsorbent and methanol as the refrigerant. The maximum solar COP was 0.33 with a maximum daily production of 20 kg of ice per m^2 of collector.

Metal hydride refrigeration uses hydrogen as a refrigerant. Metal hydride refrigeration systems are of interest for integration into hydrogen-fuelled systems. The research issues for metal hydride refrigeration are basically the same as the other adsorption technologies, including the enhancement of the specific cooling capacity and heat transfer in the beds. The driving temperature of a single-stage system is as low as $80°C$ depending on the hydride and the heat rejection temperature. The COPs of single-stage systems are approximately 0.5 [79].

7.2.5.5 Study of Solar Adsorption Cooling Systems

Solar energy can be easily used in adsorption cooling systems. The performance of solar adsorption cooling systems was reported by several researchers.

Experimental studies. Some researchers [8083] reported COP values of 0.10−0.12 for solar-assisted adsorption systems using zeolite/H_2O, and Critoph [83] reported the COP value of 0.05 using activated-carbon/NH_3. Wang et al. [73] developed a prototype solar adsorption cooling system with a 2-m^2 SC using the activated-carbon/water pair. They concluded that the prototype system was capable of producing 10 kg of ice per day and 60 kg of hot water at $90°C$.

Henning and Glase [84] designed a pilot adsorption cooling system using silica-gel/H_2O as the working pair. The system was powered with the solar heat produced by vacuum tube collectors having a surface area of 170 m^2. The reported COP varied between 0.2 and 0.3.

Sumathy et al. [85] provided literature reviews of the solar adsorption cooling technologies using various adsorption pairs and their performances.

Luo et al. [86] used a solar adsorption cooling system for low-temperature grain storage with silica-gel/H_2O. They concluded that a COP value ranging from 0.096 to 0.13 could be achieved.

González and Rodríguez [87] presented a solar refrigeration system comprising a 0.55 m^2 parabolic SC with four parallel tube receptors containing a bed of activated-carbon (Fig. 7.25). The cooling water flow is promoted by an electrically driven pump powered by a PV module. The evaporator consists

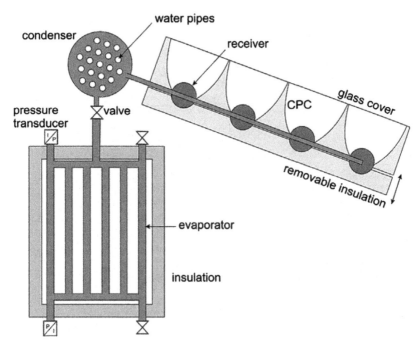

FIGURE 7.25   Solar refrigerator with parabolic collector.

of several vertical pipes, each surrounded by a small cylindrical tank containing the water to refrigerate. The test results led to a maximum solar COP of 0.10, while the evaporation temperature reached $-1.1°C$ for 19.5 MJ/ m^2 of daily solar radiation.

Sapienza et al. [88] developed a new composite sorbent named SWS-9V (LiNO$_3$/vermiculite) that was specially designed for low-temperature heat sources ($<70°C$). The experimental results showed that the system had an SCP of 230 W/kg and a COP of 0.66 at $t_e = 10°C$, $t_c = 35°C$, and $t_d = 90°C$.

Lu et al. [89] investigated a heat-pipe type adsorption refrigerator system, which can be powered by solar energy or waste heat of engine. The study assesses the performance of compound adsorbent (CaCl$_2$ and activated-carbon)/NH$_3$ adsorption refrigeration cycle with different orifice sets and different mass and heat recovery processes by experimental prototype machine. The SCP and COP were calculated with experimental data to analyze the influences of the operating condition. The results show that the jaw opening of the hand needle nozzle can obviously influence the adsorption performance and the thermostatic expansion valve is effective in the inter-mediate cycle time in the adsorption refrigeration system.

Recently, a solar adsorption refrigeration system operating with the silica-gel/H$_2$O pair was developed in Portugal (Fig. 7.26). The system consists of a

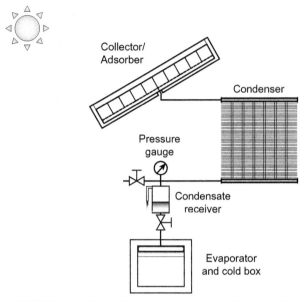

FIGURE 7.26   Solar refrigerator operating with silica-gel/H₂O.

1-m² FPC, a condenser, a condensate receiver, and an evaporator inside a cold box [66]. From the experimental results, it was found that the unit is capable of reaching a solar COP of 0.07, cooling a daily load of 6 kg of water and still produce a significant amount of ice inside the evaporator to maintain its temperature constant all the time (near 0°C).

Simulation studies. More recently, Li et al. [90] presented simulation results of a solar refrigerator in which the zeolite is placed inside the evacuated tubes of the SC. The adsorbent can reach 200°C, and the global system performance is relatively high compared to the previous solar adsorption refrigerators with the theoretical solar COP values reaching greater than 0.25.

For the summer climate in Morocco, El-Fadar et al. [91] simulated a solar adsorption refrigerator with no moving parts and with a parabolic SC connected to a cylindrical adsorber through a water heat pipe (Fig. 7.27). The influences of several parameters were analyzed, and it was determined that the COP increases with the adsorbent mass, up to a critical value of 14.5 kg, which corresponds to a 72.8 cm collector opening and a solar COP of 0.18.

Hassan et al. [92] presented more recently in Canada a theoretical simulation model of a tubular solar adsorption refrigeration system using the activated-carbon/methanol working pair. The 1-m² flat-plate SC consists of several steel pipes containing the adsorbent. The test results indicate that the solar COP value was 0.21 for a maximum solar radiation intensity of 900 W/m².

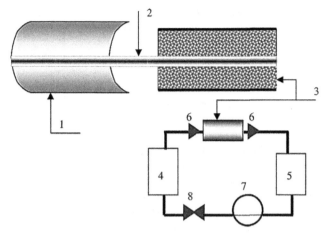

FIGURE 7.27 Solar refrigerator with cylindrical adsorber: 1-solar collector; 2-heat pipe; 3-adsorber; 4-evaporator; 5-condenser; 6-valves; 7-tank; and 8-expansion valve.

Table 7.7 summarizes the performance of the solar adsorption cooling systems using various adsorption pairs. Fig. 7.28 compares the performance of the adsorption cycles using different working pairs. One can notice that silica gel and AC are popular choice for adsorbents.

TABLE 7.7 Performance of the Solar Adsorption Refrigeration System

Working Pair	COP$_{sys}$	Solar Collector	System Conditions	References
Activated-carbon/methanol	0.10–0.12	FPC (A = 6 m^2)	$t_e = -3°C$, $t_c = 25°C$, $t_d = 110°C$	[80]
Activated-carbon/methanol	0.10–0.12	FPC (A = 6 m^2)	$t_e = -6°C$, $t_d = 70°C–78°C$	[84]
Zeolite/water	0.11	FPC (A = 20 m^2)	$t_e = 1°C$, $t_c = 30°C$, $t_d = 118°C$	[81]
Zeolite/water	0.10–0.12	FPC (A = 1.5 m^2)	–	[79,80]
Activated-carbon/ammonia	0.05	FPC (A = 1 m^2)	–	[82]
Activated-carbon/water	0.07	FPC (A = 2 m^2)	–	[72]
Silica-gel/water	0.20–0.30	VTC (A = 170 m^2)	–	[83]
Silica-gel/water	0.10–0.13	–	–	[85]

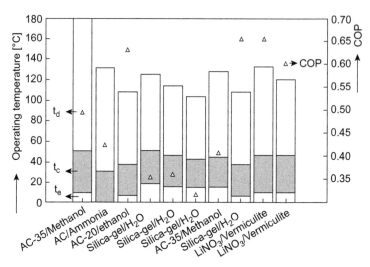

FIGURE 7.28   Comparison of the adsorption cycle performance using different working pairs.

7.2.6 Comprehensive Review of Solar Sorption Cooling System Applications

7.2.6.1 Air-Conditioning (A/C)

The main purpose of most buildings and A/C systems is to provide an acceptable environment that does not impair the health and performance of the occupants. Solar sorption refrigeration systems are suitable for A/C due to their low-installation costs and high-cooling capacities.

Experimental studies. In Japan, a solar heating and cooling system with FPCs and absorption refrigeration system was installed [39]. Yeung et al. [93] developed a solar-powered absorption A/C system on the campus of University of Hong Kong. The system consists of a FPC array with a surface area of 38.2 m^2, a 4.7-kW cooling capacity $H_2O/LiBr$ absorption chiller, a 2.75-m^3 hot-water ST, a cooling tower, a fan-coil unit, and an electrical auxiliary heater. It had an annual system COP of 7.8%. In China, solar-powered sorption systems are all adsorption refrigeration systems with activated-carbon/NH_3 or zeolite/H_2O as the working pair.

Syed et al. [40] studied a solar cooling system for typical Spanish houses in Madrid. A FPC array with a surface area of 49.9 m^2 was used to drive the system. A single-effect $H_2O/LiBr$ absorption chiller of 35 kW cooling capacity is used. This system operated within the generation and absorption temperature ranges of 57−67°C and 32−36°C, respectively. The measured maximum instantaneous, daily average, and period average COPs were 0.60, 0.42, and 0.34, respectively. The potential energy savings and limitations of solar thermal A/C in comparison to conventional technologies for Europe are illustrated and discussed by Balaras et al. [94].

Bujedo et al. [95] studied different control strategies for a solar-powered $H_2O/LiBr$ absorption cycle. A 77.5-m^2 collector area was used to drive a 35-kW chiller with two 2-m^3 hot-water STs and a 1-m^3 cold-storage tank to air condition 200-m^2 of offices under Spanish weather conditions. The new strategies had better results than the conventional one. An improvement of the yield of the solar field ranged from 7% to 12%, whereas the improvement in the total system COP was 44—48%.

Agyenim et al. [96] designed a solar-absorption cooling system with cold storage. A 4.5-kW chiller unit using $H_2O/LiBr$ was driven by 12 m^2 of vacuum tube collectors. The chiller was activated at a minimum temperature of 80°C and able to produce chilled water with a temperature in the range of 7—16°C. The solar cooling system was experimentally tested in Cardiff, UK. It was recommended that, to commercialize the system, the solar cooling system should be integrated with the domestic hot water and space-heating systems used during the winter season.

Lizarte et al. [97] introduced a directly air-cooled single-effect $H_2O/LiBr$ absorption chiller that used an innovative adiabatic flat-fan sheet absorber. The flat-fan sheet configuration was also investigated by Palacios et al. [98] and found to exhibit a mass transfer coefficient in an order of magnitude higher than that of the falling-film or the spray-type absorbers. The system was used to air-condition a 40-m^2 space during the summer season in Madrid. The volume of the absorption unit was 1 m, whereas the hot-water ST volume was 1.5 m^3. For the test period, the mean collector efficiency, system COP, and solar COP were 0.27, 0.53, and 0.062, respectively.

Simulation studies. Al-Alili et al. [99] studied the performance and economic and environmental benefits of a 10-kW NH_3/H_2O absorption chiller under Abu Dhabi's (United Arab Emirates) weather conditions. The solar A/C system had a specific collector area of 6 m^2/kW and a specific tank volume of 0.10 m^3/kW. The system was found to consume 47% less electricity than that of the widespread systems with vapor-compression cycles of the same cooling capacity. The economic analysis showed that the collector area was the key parameter in reducing the payback period of the initial investment. Balghouthi et al. [100] assessed the feasibility of using a solar-powered $H_2O/LiBr$ absorption system in Tunisia by simulation. TRNSYS and EES programs were used to model the system and the conditioned space. To air-condition a 150 m^2 building, an 11-kW absorption cycle using a 30-m^2 FPC and a 0.8-m^3 hot-water tank was used. They studied the effect of the generator inlet temperature and heat transfer coefficient on the COP and cooling capacity.

Table 7.8 provides more details regarding the solar absorption A/C cycles.

7.2.6.2 Refrigeration

The low-temperature applications can also utilize sorption systems. These systems are attractive for refrigeration applications in remote or rural areas of

TABLE 7.8 Performance of the Solar-Absorption Air-Conditioning Cycles

Working Fluid	COP_{sys}	Solar Collector	System Conditions	References
NH_3/H_2O	0.427	ETC ($A = 11$ m^2)	$t_e = -9°C$, $t_c = 37-45°C$, $t_g = 110°C$	[46]
$H_2O/LiBr$	0.620	ETC ($A = 94$ m^2)	$t_e = 15.5-25°C$, $t_c = 26.5$ $-37.7°C$, $t_g = 90-106°C$	[96]
$H_2O/LiBr$	0.660	ETC ($A = 2.7$ m^2)	$t_e = 10.1°C$, $t_c = 24°C$, $t_g = 77.1°C$	[95]
$H_2O/LiBr$	0.490	FPC&ETC ($A = 1.5$ m^2)	–	[111]
$H_2O/LiBr$	0.740	FPC ($A = 2.7$ m^2)	$t_e = 7.5°C$, $t_c = 28°C$, $t_g = 86°C$	[99]
NH_3/H_2O	0.550	ETC ($A = 6$ m^2)	$t_e = 6°C$, $t_c = 24°C$, $t_g = 85°C$	[98]

developing countries, where access to electricity is impossible. Various solar sorption refrigerators have been developed.

Experimental studies. Uppal et al. [101], in 1986, built a small-capacity (56 L) solar-driven NH_3/H_2O absorption refrigerator to store vaccines in remote locations.

Sierra et al. [102] used a solar-pond to power an intermittent absorption system with NH_3/H_2O working fluids. They reported that generation temperatures as high as 73°C and evaporation temperatures as low as $-2°C$ could be obtained. The system COP was in the range of 0.24−0.28.

De-Francisco et al. [42] tested, in Madrid, a prototype of a 2-kW solar-powered NH_3/H_2O absorption chiller for refrigeration in small rural operations that utilizes concentrating collectors and uses a transfer tank instead of a pump. Their experimental results showed inefficient operation of the equipment with a COP less than 0.05 when the collectors operated at temperature greater than 150°C.

Pilatowski et al. [103] proposed a monomethylamine/H_2O absorption refrigeration system driven by ETCs coupled with a conventional auxiliary heating system for milk cooling in the rural regions of Mexico. The results indicated that it is possible to obtain evaporation temperatures from -5 to 10°C, a low generation temperature from 60°C to 80°C, condensation temperatures of 25°C, and COP values from 0.15 to 0.7.

Lemmini and Errougani [104] presented experimental work to evaluate the performance of a solar adsorption system using the AC35/methanol pair in Rabat, Morocco. The system consists of a FPC, a condenser, and a cold-box evaporator. The results indicated that the solar COP ranges between 0.05 and 0.08 for a solar radiation between $12 \, MJ/m^2$ and $27 \, MJ/m^2$, a daily average temperature between 14°C and 18°C, and a lowest temperature achieved by the evaporator between −5°C and 8°C.

Simulation studies. Hammad and Habali [105] simulated a solar absorption refrigeration system with NH_3/H_2O as the working mixture to cool a vaccine cabinet in the Middle East. The simulation results indicated that the system had a COP between 0.50 and 0.65 at the generation temperature of 100−120°C and the cabinet indoor temperature of 0−8°C. More recently, Abu-Hamdeh et al. [106] developed a model of a solar adsorption refrigeration system operating with the olive-waste/methanol pair. The system comprises a $3.7\text{-}m^2$ parabolic SC, a heat-storage tank, an adsorber, a condenser, an evaporator, a throttling valve, and a circulating pump. From the simulated results, the lowest temperature attained in the refrigerated space was 4°C, with a solar COP of 0.03 for a solar radiation flux of $56.2 \, MJ/m^2$.

7.2.6.3 Ice Making

An absorption or adsorption chiller can also be used for freezing applications that require temperatures below 0°C.

Experimental studies. In 1991 Medini et al. [107] studied a nonvalve solar adsorption ice maker with a $0.8\text{-}m^2$ collection surface. The system employed an intermittent daily cycle with the AC35/methanol pair. The results indicated that it is possible to produce 4 kg of ice per day during the summer and to obtain a solar COP of 0.15 with a collection efficiency of 0.41 and a system COP of 0.40.

Sumathy and Li [108] designed a solar ice maker with the AC/methanol pair using a $0.92\text{-}m^2$ flat-plate SC. The system produced 4−5 kg of ice daily at an evaporation temperature of −6°C for a daily solar radiation between $17 \, MJ/m^2$ and $19 \, MJ/m^2$ and achieved solar COP values of 0.10−0.12.

In Switzerland, Hildbrand et al. [72] constructed and tested a new high-efficiency solar adsorption system. The working pair is silica-gel/H_2O. Cylindrical tubes function as both the adsorber and the solar FPC ($2\text{-}m^2$ double glazed). The condenser is air-cooled, and the evaporator contains 40 L of water that can freeze. This ice functions as a cold storage for the cabinet. This system has a solar COP of 0.16.

More recently, the German company Zeo-Tech GmbH patented a solar-thermal ice maker based on the adsorption principle and operating with the zeolite/H_2O pair [109].

Ji et al. [110] built a solar-powered solid adsorption refrigeration system with the finned-tube absorbent bed collector. Activated-carbon/methanol was utilized as the working pair for adsorption refrigeration in the experiments. The experiments achieved the maximum COP of 0.122 and the maximum daily ice making of 6.5 kg. The cooling efficiency of the solar-powered adsorption refrigeration system with a valve control in the adsorption/desorption process was significantly higher than that without a valve control.

Simulation study. In Italy, Vasta et al. [111] presented the numerical model of a solar adsorption refrigerator that simulates the different stages of the thermodynamic cycle and the processes occurring in the system components: a 1.5-m^2 SC, containing the adsorbent bed, and a condenser and a cold box containing an evaporator, and the water to freeze. It was found that for most of the year (February–October) the system has the ability to produce between 4 and 5 kg of ice per day. For the colder months (November–January), it is possible to produce 2–3.5 kg of ice per day. The average monthly solar COP ranged from 0.05 (July) to 0.11 (January), with a yearly average COP of 0.07.

7.3 SOLAR THERMOMECHANICAL COOLING SYSTEMS

7.3.1 Description of the System

In the thermomechanical solar cooling system, the thermal energy is converted to the mechanical energy. Then the mechanical energy is utilized to produce the cooling effect.

The steam ejector system represents the thermomechanical cooling technology. Fig. 7.29 illustrates the steam ejector system integrated with a parabolic SC. The steam produced by the SC is passing through the steam jet ejector E. During this process, the evaporator pressure is reduced, and water is vaporized in the evaporator V by absorbing the heat from the cold water.

The ejector is the key component in the cycle. The ejector mainly consists of a nozzle, a mixing chamber, and a diffuser, as shown in Fig. 7.30.

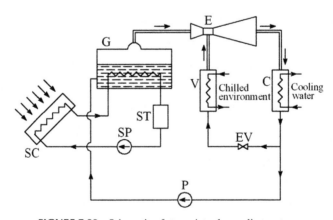

FIGURE 7.29   Schematic of steam jet solar cooling system.

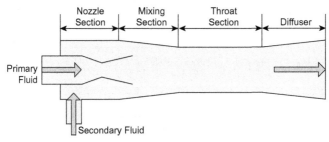

FIGURE 7.30   Schematic of ejector.

The working principle of the ejector is based on the expansion of a high-pressure steam jet in the converging/diverging nozzle section. The internal energy of the motive steam is converted to kinetic energy. The high-speed steam jet (primary fluid) entrains the low-pressure secondary steam jet (secondary fluid). The two steam jets enter the mixing section where momentum is transferred from the primary fluid to the secondary fluid causing acceleration of the secondary steam jet. Before exiting the ejector, a diffuser is used to convert the mixed steam's kinetic energy into internal energy in order to reach a pressure higher than the back pressure.

When cooling is not needed, the steam turbines can be used to produce electricity. Most of the steam ejector system requires steam at pressures in the range of 0.1−1.0 MPa and temperatures in the range of 120−180°C [112]. However, Loehrke [113] proposed and demonstrated that the steam ejector system could be operated using low-temperature solar heat by reducing the operating pressure under atmospheric pressure.

The pump does not determine a high growth in cost or electricity consumption (i.e., the required pump power consumption is ∼0.18% of the energy received from the SC). However, the pump requires more maintenance than other parts because it is the only moving part in the system. Hence to replace the pump, several solutions have been found [114]:

- gravitational/rotational ejector cooling system;
- bi-ejector cooling system; and
- heat pipe/ejector cooling system.

In a *gravitational ejector* cooling system, the heat exchangers are placed on different vertical positions, equalizing the pressure differences between them. The steam generator has the highest pressure, and the evaporator has the lowest pressure. There are also complex mechanisms of self-regulation of the generator, evaporator, and condenser. A major drawback of the system is the requirement of height differences (depending on the working fluid and the temperature differences) and the length of pipes (which causes high friction and heat losses).

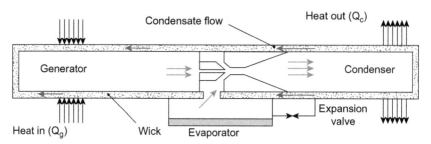

FIGURE 7.31   Heat pipe/ejector cycle.

In the *bi-ejector* cooling system, a second ejector, which replaces the pump, carries the liquid condensate to the generator. Therefore, the ejector is a vapor/liquid ejector.

An interesting technology is the coupling *heat pipe/ejector*. The coupling of the heat pipe and the ejector technology is interesting because it results in a system that is both compact and with high performance. This system is composed of a heat pipe, an ejector, an evaporator, and an expansion valve (Fig. 7.31); the working principles will not be described here because they are the same as those of other ejector cooling systems. A description can be found in the work of Smirnov and Kosov [115].

7.3.2 Performance Parameters

Several parameters are used to describe the performance of ejectors in cooling cycles, as provided below.

The entrainment ratio ω is the ratio between the secondary fluid mass flow rate m_s, in kg/s, and the primary fluid mass flow rate m_p, in kg/s:

$$\omega = \frac{m_s}{m_p} \qquad (7.11)$$

The compression ratio H_c is the static pressure at the exit of the diffuser p_c, in Pa, divided by the static pressure of the secondary fluid p_e, in Pa:

$$H_c = \frac{p_c}{p_e} \qquad (7.12)$$

The theoretical COP of the thermodynamic cycle is defined as the ratio between evaporation heat (cooling power) Q_e and the total incoming energy in the cycle $Q_g + P_{el}$:

$$COP = \frac{Q_e}{Q_g + P_{el}} \qquad (7.13)$$

where Q_g is the generator thermal power, in W, and P_{el} is the electrical power of pump, in W.

The ejector efficiency η_{ej} is defined as the ratio between the actual recovered compression energy and the available theoretical energy in the motive fluid [116]:

$$\eta_{ej} = \frac{(m_p + m_s)(h_{c,in} - h_{e,out})}{m_p(h_{g,out} - h_{e,out})} \tag{7.14}$$

where $h_{c,in}$ is the specific enthalpy at inlet of condenser, in kJ/kg; $h_{e,out}$ is the specific enthalpy at outlet of evaporator, in kJ/kg; and $h_{g,out}$ is the specific enthalpy at outlet of generator, in kJ/kg.

The real COP of the thermodynamic cycle (COP_r) is given by

$$COP_r = \eta_{ej}COP \tag{7.15}$$

The overall COP of a solar thermomechanical cooling system is given by the following equation:

$$COP_{sys} = \frac{Q_e}{Q_s} \tag{7.16}$$

where Q_s is the solar power received by the SC surface.

7.3.3 Working Fluid Selection

The selection of the appropriate refrigerant is of fundamental importance in the design of an ejector cooling system. In the past, the main principle for selection was the maximization of the performance; more recently, several factors (safety, cost, etc.) are considered, and the final choice depends on the compromise between the performance and the environmental impact. The working fluids can be classified based on the chemical compounds and can be classified into three main groups [117] (Table 7.9): (1) the halocarbon group [i.e., chlorofluorocarbons (CFCs), hydro-chlorofluorocarbons (HCFCs), hydro-fluorocarbons (HFCs), and hydro-fluoroolefin (HFO) and the hydrocarbon group (HC)], (2) organic compounds consisting of hydrogen and carbon (i.e., R290, R600, and R600a), and (3) other refrigerants, (i.e., water [R718b], NH_3 [R717], and carbon dioxide [R744]).

Generally, a suitable refrigerant for a refrigeration system should be able to guarantee high performance for the required operating conditions. Accordingly, working fluid thermophysical properties must be taken into account. Thermophysical properties should satisfy some constraints—they should have a large latent heat of vaporization and a large generator temperature range for limiting the circulation rate per unit of cooling capacity and the fluid should have a high critical temperature to compensate large variations in generator temperatures. The fluid pressure should not be too high in the generator for the design of the pressure vessel and limiting the pump energy consumption. Moreover, the viscosity, the thermal conductivity, and the other properties that influence the heat transfer should be favorable. A high molecular mass is

TABLE 7.9 Refrigerant Classification and Safety Characteristics

Group		Safety Group (Toxicity/Flammability)	Working Fluid
Halocarbon compounds	CFC	A1	R11, R12, R113, R114
	HCFC	A1-B1	R21, R22, R123, R141b, R142b, R500, R502
	HFC	A1-A2	R134a, R152a, R236fa, R245fa
	HFO	A2L	R1234yf
Hydrocarbon compounds	HC	A3	R290, R600, R600a
Other refrigerants		B1	CH_3OH
		B2L	R717
		A1	R718b, R744

desirable to increase ω and η_{ej}. Low environmental impact, as defined by the global warming potential and the ozone depletion potential, is also an important factor for consideration. The fluid should also be nonexplosive, nontoxic, noncorrosive, chemically stable, cheap, and available on the market.

The working fluid used in a solar ejector cooling system lead to different performances depending of operating conditions. In order to compare the performances of different used working fluids, in Table 7.10 the following values are presented: t_g—the generation temperature; t_c—the condensation temperature achieved in condenser C (37°C for cooling with cooling tower, 30°C for cooling with cold water); p_g—the pressure in the generator G (maximum pressure in the system); p_e—the pressure in the evaporator V (minimum pressure in the system); Q_{SC}—the heat needed to be supplied by SC in generator to achieve a cooling power of 1.16×10^4 W; A_c—the SC area, assuming a solar flux of 0.8 kW/m^2 and capture efficiency of 0.5 for achieving a cooling capacity of 1.16×10^4 W. Considering one FPC, the possible temperature that can easily provide solar heat is $t_g = 85°C$, and for a parabolic-cylinder concentrating collector can be adopted $t_g = 130°C$.

Analyzing the COP$_r$ values from this table results as the competitive working fluids: H_2O, R-11, and R-21, among which the best is R-11. H_2O and R-11 have comparable COP$_r$, but operating pressures in the system are very different. Thus, for the use of FPCs ($t_g = 85°C$), steam ejector cooling system works completely in depression (p_e and p_g is less than atmospheric pressure). So if H_2O is used as refrigerant, leakage problems are to be solved to avoid the air entering the system.

TABLE 7.10 Performances of Different Working Fluids Used in Solar Steam Ejector Systems

Working Fluid	t_g (°C)	t_c (°C)	p_g (kPa)	p_e (kPa)	Q_{SC} (W)	A_c (m²)	COP (−)	η_{ej} (−)	COP_r (−)
H₂O	85	37	392.2	0.88	69,130	173	0.913	0.184	0.168
		30	392.2	0.88	36,396	91	1.471	0.217	0.319
	130	37	475.5	0.88	24,717	62	2.076	0.226	0.469
		30	475.5	0.88	17,979	45	2.887	0.223	0.645
R-11	85	37	460.8	50.0	63,428	159	0.936	0.196	0.183
		30	460.8	50.1	26,356	66	1.947	0.226	0.440
	130	37	784.0	50.1	26,911	67	1.708	0.226	0.386
		30	784.0	50.1	17,175	43	3.121	0.216	0.675
R-21	85	37	754.9	90.2	85,630	215	0.790	0.172	0.135
		30	754.9	90.2	47,866	120	1.162	0.209	0.242
Propane	85	37	2745	539	298,969	750	0.496	0.078	0.039
		30	2745	539	56,585	142	1.038	0.198	0.209
Butane	85	37	882.3	147	302,475	758	0.423	0.091	0.038
		30	882.3	147	102,284	256	0.666	0.170	0.113
NH₃	85	37	2157	520	not possible solution				
	130	37	2157	520	2,130,150	5338	0.348	0.016	0.005

Various experimental studies [118,120] have examined the effect of the operation conditions such as the generator temperature, evaporator temperature and condenser temperature, the geometrical conditions, and the system conditions such as refrigerant and collector selections on the performance of the system. Other researchers [121,122] have presented numerical methods of simulating the ejector and studied the performance of system.

Nehad [118,119] compared the theoretical performance of the ejector system working with R-717, R-11, R-12, R-113, and R-114. Then he chose R-113 as the refrigerant for the experiment since it has a higher COP_r, a reasonable operating pressure, and is nontoxic.

Eames et al. [120] reported that the measured COP_r of the single-stage ejector system using H_2O as its working fluid ranged from 0.178 to 0.586 at a generating temperature t_g of 120—140°C, an evaporation temperature t_e of 5—10°C, and a condensation temperature t_c of 26.5—36.3°C.

Vidal et al. [121] analyzed the solar ejector system using R-141b as its refrigerant by using the TRNSYS and EES simulation software. The system was designed to deliver 10.5 kW of cooling with 80 m^2 of FPC tilted 22 degrees from the horizontal and a 4-m^3 hot-water ST. They reported the maximum COP_r of 0.22 at $t_g = 80$°C, $t_e = 8$°C, and $t_c = 32$°C. They also concluded that an efficient ejector system could work in a region only with decent solar radiation and where a sufficiently low condenser temperature could be kept.

Grazzini and Rocchetti [122] theoretically investigated the performance of the two-stage ejector system. They reported that the COP_r of the two-stage ejector system ranged from 0.13 to 0.53 at $t_g = 110$—120°C, $t_e = 7$—12°C, and $t_c = 30$—40°C.

Along with other researches, results [123] show that the low-ejection efficiency leads to values of COP for solar ejector cooling systems smaller than in the case of solar-absorption cooling systems. The performance of the ejector system depends on the mass-flow rate ratio through the motive nozzle and the suction nozzle. Fig. 7.32 shows the performance of ejector cycle using different refrigerants [124].

The ejector systems are mostly used in A/C applications, but they can be used in chemical and metallurgical industries for air cooling in areas with higher heat dissipation.

7.4 HYBRID COOLING AND HEATING SYSTEMS

In solar thermal systems with large capacity, both solar cooling and solar heating are provided synergistically, yielding a complete annual utilization. During the cold season, solar heat serves for space heating. During the warm season, solar heat is converted into useful cold by means of sorption cooling devices, avoiding overheating of the solar thermal system. There are already several hybrid systems using the basic closed sorption cycle for the heat and cold productions.

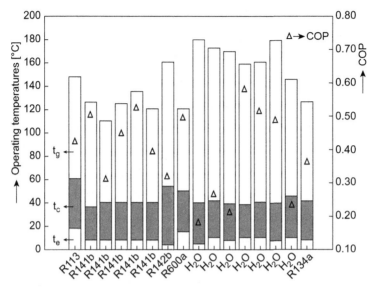

FIGURE 7.32 Comparison of the ejector cycle performance using different refrigerants.

Many applications of solar A/C will be done in conjunction with solar heating, with the same collector, storage, and auxiliary energy system serving both functions and supplying hot water. Fig. 7.33 shows a hybrid cooling and heating system using a H_2O/LiBr absorption cooler for A/C, in which solid flow lines are for cooling, dashed lines are for heating, and dotted lines are for air-conditioner coolant flow [125].

An important consideration in combined heating and cooling systems is the relative importance of the summer and winter loads. Either one may dictate the needed capacity of the collector and, consequently, its size and design. Climate is a major determining factor, and cooling requirements will dominate in warm climates. Commercial buildings are likely to have design fixed by cooling loads, even in cool climates. Also, building design features are important that can affect relative energy requirements for the two loads. These include fenestration, shading by overhangs, wing walls, and building orientation. Less obvious is the performance of the cooling and heating system; a poor absorption cooler would require a larger collector area than one with a high COP and thus could shift the determination of collector needs from winter heating to summer cooling.

The location of storage, whether inside or outside the building, will have an effect on heating or cooling loads. If heat is to be stored and if the storage unit is inside the structure, heat losses from storage become uncontrolled gains during the heating season and additional loads during the cooling season. If collectors are part of the envelope of the building, back losses from the collector will also become uncontrolled gains during heating and additional loads during cooling.

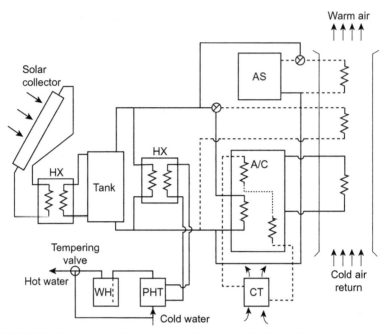

FIGURE 7.33   Schematic of a hybrid solar heating, air-conditioning, and hot-water system: *A/C*, air-conditioner; *AS*, auxiliary heat source; *CT*, cooling tower; *PHT*, preheat tank; *WH*, water heater.

Collector orientation may be affected by which load dominates; optimum orientation is approximately $\Sigma = \varphi + 15$ degrees for winter use, $\Sigma = \varphi - 15$ degrees for summer use, and $\Sigma = \varphi$ for all-year use. Heating loads are likely to be higher in the morning, suggesting that the surface azimuth angle ψ should be negative, while cooling loads peaking in the afternoon suggest that ψ should be positive.

As with solar heating alone, the major design problem is the determination of optimum collector area, with an underdesign leading to excessive use of auxiliary energy and overdesign leading to low use factors on the capital-intensive solar energy system. In climates where annual cooling loads are low, the use of absorption coolers will lead to higher cooling costs because of low use factors on the coolers.

Solar cooling systems predominantly cover cooling capacities in the range of 10–30 kW, requiring SCs of approximately 30–100 m2 surface area. Wet cooling towers designed for coolant supply/return temperatures of about 27/35°C are applied to transfer the heat rejected by absorber and condenser to the ambient. When a dry air cooler is to be used, cooling water temperatures have to be increased to 40/45°C. As a consequence of the increase of the cooling water temperature, the temperature level of the driving heat supplied to the generator of the absorption chiller has to be increased accordingly.

Both systems, wet cooling towers and dry air coolers allow significantly lower coolant temperatures during off-peak hours with moderate ambient air temperatures.

If latent heat storage is integrated in the coolant loop in addition to the dry air cooler, coolant cycle temperatures at design point operation can be reduced to about 40/32°C. Thus referring to the above standard design with 45/40°C coolant temperature, the same dry air cooler will be able to operate at reduced coolant temperatures of about 40/36°C at 32°C ambient temperature, providing about half of its nominal capacity specified for cooling water temperature of 45/40°C. The latent heat storage then would have to cool the brine from 36 to 32°C, which is reasonable for internal phase-change temperatures about 29°C. A further reduction of the coolant cycle temperatures during off-peak hours with moderate ambient air temperatures is possible for this configuration [126]. The impact of chilled water temperature and reject heat temperature on driving the heat temperature for the discussed system configuration are given in Fig. 7.34.

For chilled water 18/15°C, starting from a chiller design with 80/75°C, the required driving hot-water temperature rises with increasing cooling water temperature to 90/85°C and further to 105/100°C when a dry air cooler together with a latent heat storage or solely a dry air cooler is applied, respectively. Analogously, for chilled water temperature 12/6°C, the required driving hot water temperature increases from 90/85°C to 100/95°C for the system with dry air cooler assisted by latent heat storage. For this chilled water layout, that is, standard temperature 12/6°C, solely dry air cooling is not feasible [127].

Experimental studies. Helm et al. [126] have described the operation of an absorption cooling and heating hybrid system, that involves latent heat storage

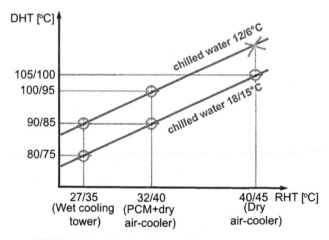

FIGURE 7.34 Variation of external heat-carrier parameters.

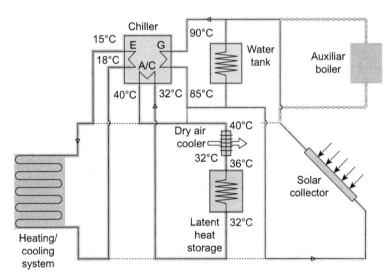

FIGURE 7.35   Solar cooling/heating system with an absorption chiller and latent-heat storage in cooling mode.

supporting the heat rejection of the absorption chiller, in conjunction with a dry cooling system as shown in Fig. 7.35. They have indicated low-temperature latent-heat storage together with a dry air cooler, as a promising alternative to the conventional wet cooling tower, as it substantially reduces the oversizing of the SC system.

Mammoli et al. [128] used FPCs and ETCs (232 m^2 of total area) to drive a 70 kW H_2O/LiBr absorption system. The system was used to provide heating and cooling to the mechanical engineering university building in New Mexico, U.S.A., wherein 34-m^3 hot-water ST and seven 50-m^3 cold water tanks were used. The system implemented four different control strategies for summer daytime, summer night-time, winter daytime, and winter night-time operations. The results indicated that the collector field had a daily average efficiency of 0.58, while the chiller had an average COP of 0.63. The hot storage tank was found to be the largest source of heat loss.

The first hybrid adsorption cooling and heating system was developed by Tchernev in the late 1970s [80]. The system operates with the zeolite/H_2O pair and is shown in Fig. 7.36. The condenser and the evaporator are combined in a single unit crossed by an external water loop. During the day, the condensation heat is rejected to the external water loop and can be used for providing domestic hot water (DHW) and space heating during the winter. During the night, the previously condensed water evaporates and is adsorbed by the zeolite, promoting the cooling effect on the condenser/evaporator. This will cool the water in the external loop, which can be used for A/C or stored for a later use.

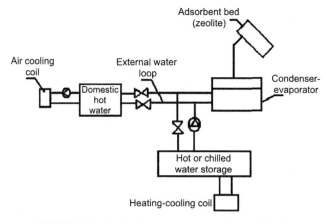

FIGURE 7.36   Zeolite/water system for cooling and heating.

In China, Wang et al. [73] developed an innovative solar hybrid system for water heating and ice production using the AC/methanol pair (Fig. 7.37). The adsorber bed is immersed in a water tank and is directly heated by a 2-m^2 evacuated tube SC. During the night, the hot water drained from the tank can be used for domestic purposes. During the adsorption stage, the sensible and adsorption heats from the adsorber are transferred to the water in the tank, producing useful heat. The authors estimated the following results: in the winter case, with a solar radiation of 24.6 MJ/day, the system produces 60 kg of hot water at 98°C and 10.5 kg of ice at −2.5°C with a system COP of 0.143 and a heating efficiency of 0.795; in the spring case, with a solar radiation of 22 MJ/day, the system produces 60 kg of hot water at 91.3°C and 10 kg of ice at −1.8°C with a system COP of 0.144 and a heating efficiency of 0.797.

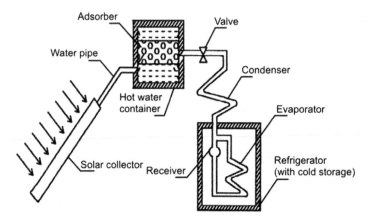

FIGURE 7.37   Solar hybrid system using the activated-carbon/methanol pair.

Simulation studies. Zhang and Wang [129] designed a continuous-operation hybrid solar adsorption system for heating and cooling purposes, working with the AC/methanol pair. The results were obtained by simulation. For a solar radiation of 21.6 MJ/day, the system is capable of heating 30 kg of water to 47.8°C with a heating COP of 0.34, while the evaporation temperature reaches 5°C with a solar COP of 0.18. The upper bed can reach 80−90°C, after which the collector is rotated by 180 degrees to replace the upper bed by the lower bed, which is now heated by solar energy, while the bed initially in the upper position is cooled by cold water coming from a water tank, which flows by natural convection.

A continuous combined solid adsorption-ejector cooling and heating system driven by solar energy, operating with zeolite/H_2O working pair, also was presented by Zhang and Wang [130]. The combined system consists of two parts: a heating cycle and a cooling cycle, as shown in Fig. 7.38.

The SC consists of a 2-m^2 CPC. During daytime, the adsorber receives solar energy and releases the water vapor from the zeolite. Then adsorber is connected to the ejector and disconnected from the evaporator. The water vapor is then ejected at a fast speed in the ejector, creating a low pressure area, into which water vapor from the evaporator is pulled. The mixed water vapor is compressed into the condenser by the diffuser the liquid enters the receiver, and finally returns to the evaporator by the throttle valve. In the afternoon, the adsorber is disconnected from the ejector and connected to the evaporator. Thus, the cold water in the tank circulates into the adsorber to be heated, thus cooling the adsorber. The hot water collected in the tank can then be used for

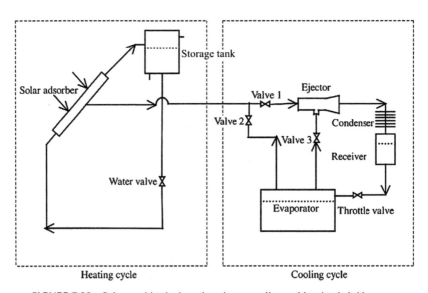

FIGURE 7.38   Solar combined adsorption-ejector cooling and heating hybrid system.

domestic applications in the evening. When the adsorber's temperature and pressure decrease to the evaporation state, evaporation starts in the evaporator. So, the system will refrigerate continuously. From the simulation results, the combined hybrid system had a total cooling capacity of 0.49 MJ per kg of zeolite with a COP of 0.33, and could deliver 290 kg of hot water at 45°C with a heating COP of 0.55.

More recently, Suleiman et al. [131] have performed a numerical model of a solar hybrid system for cooling and heating in Nigeria. The system comprises a 2-m² flat-plate SC and operates with the AC/methanol pair. Considering an evaporator temperature of 0°C and 25°C in the condenser, the results showed an average cooling capacity of 4815 kJ (regeneration temperature of 86°C) with a solar COP of 0.02 and a heating efficiency of 0.46.

7.5 COMPARISON OF VARIOUS SOLAR THERMAL COOLING SYSTEMS

Balaras et al. [94] described the main results of the European Union project SACE (Solar A/C in Europe), intended to assess the state-of-the-art, future needs, and overall prospects of solar cooling in Europe. For this purpose, they collected information on 54 solar-powered cooling projects conducted in various locations in Europe. They reported the thermal COP of different solar cooling technologies, as shown in Table 7.11. They concluded that the single-effect absorption systems haves a COP in the range of 0.50−0.73, adsorption systems haves a lower thermal COP of 0.59, a liquid desiccant system have a COP of 0.51, and a steam jet system have a relatively high COP of 0.85. Regarding the operating temperature of the systems, absorption systems operated at 60−165°C, adsorption systems operated at 53−82°C, a liquid desiccant system operated at 67°C, and a steam jet system operated at 118°C. For most of these systems operated below 100°C, flat-plate SCs could be used, while concentrating SCs had to be used to drive temperatures higher than 100°C. They also compared the annual EER, which is defined as the ratio of the annual cold production and the annual heat input. The $H_2O/LiBr$ absorption systems have the best annual performance, while the adsorption systems have low annual performance. This result reflects the fact that 70% of the systems employed absorption technology and also that 75% of the solar-assisted absorption systems used $H_2O/LiBr$ as their working fluid.

The dominant technology in the European solar refrigerating installation market is still absorption chillers. However, some newly developing trends currently observed are directed toward reducing the use of absorption chillers [132].

One reason for this situation is the possibility of taking advantage of alternative systems, such as adsorption, when the hot water is below 90°C [133]. However, the results of solar adsorption cooling systems show that their performance is still lower than that of absorption systems (Table 7.11) and

TABLE 7.11 Comparison of Solar Thermal Cooling Systems

Specification	Open		Closed			Thermomechanical
System	Liquid desiccant	Solid desiccant	Absorption	Adsorption		Ejector
Sorbent type	Liquid	Solid	Liquid	Solid		—
Working fluid (refrigerant/sorbent)	$H_2O/CaCl_2$, $H_2O/LiCl$	H_2O/silica-gel, $H_2O/LiCl$, cellulose	$H_2O/LiBr$, NH_3/H_2O	H_2O/silica-gel		Steam
COP (–)	0.74	0.51	0.50–0.73 (single-stage) <1.3 (two-stage)	0.59		0.85
EER (Btu/(Wh))	2.53	1.74	1.71–2.49 (single-stage) <4.44 (two-stage)	2.01		2.90
Operating temperature	67°C	45–95°C	60–110°C (single-stage) 130–165°C (two-stage)	53–82°C		118°C

needs improvement. Additional enhancement of adsorption systems' COP is needed through new material development, system loss reduction, or multistage approaches.

According to the works of different researchers reviewed in this chapter, the cooling capacity of absorption chillers could be situated in the range of 5–2300 kW for the temperatures $t_g = 35–211°C$, $t_a = 15–81°C$, $t_c = 10–108°C$, $t_e = -60–20°C$ and different working pairs, achieving a COP between 0.2 and 0.9. An adsorption chiller is more expensive than an absorption chiller. The low power density of an adsorbent tends to increase the price of an adsorption system by requiring bigger components for the same capacity.

Regarding the direction of future research (R) and development (D) in solar cooling, it would better be focused on low-temperature sorption systems. This is because first, the cost of an SC system tends to increase with working temperature more rapidly than the COP of a sorption machine does, and second, high temperature-driven chillers would not be compatible with the existing solar heating systems, which were originally designed to produce domestic hot water. Another important subject in the future R&D is the development of air-cooled machines. In addition, combined systems and domestic equipments using advanced micro-exchangers are also the trend of development.

7.6 CONCLUSIONS

As the world population is projected to increase and the supply of the fuel is projected to decrease, the increased supply of the renewable energy for the post-fossil fuel period is inevitable. Since the cooling demand has been increased associated with the recent climate change, cooling technologies based on solar energy are promising technologies for the future.

Thermally activated cooling systems are being used all over the world for domestic and industrial cooling purposes. Solar thermal cooling systems are more suitable than the conventional refrigeration systems because pollution-free working fluids (instead of chlorofluorocarbons) are used as refrigerants. Solar cooling systems can be used, either as stand-alone systems or with conventional A/C systems, to improve the indoor air quality of all building types (residential buildings, offices, schools, hotels, hospitals, and laboratories).

In this chapter, an extensive study of solar thermal-driven cooling systems was presented. The liquid desiccant system has a higher thermal COP than the solid desiccant system. The adsorption cycle needs a lower heat-source temperature than the absorption cycle. The ejector system has a higher COP, but needs a higher heat-source temperature than other systems. Based on the COP, the liquid desiccant system is preferred to the solid desiccant system and the absorption cooling systems are preferred to the adsorption cooling systems, and the higher temperature issues can be easily handled with solar adsorption systems. Solar thermal with single-effect absorption systems appear to be the best option closely followed by solar thermal with single-effect adsorption systems and by solar thermal with double-effect absorption system options at the same price levels.

The study indicates that much research has recently been performed on continuous operation absorption systems that use NH_3/H_2O and $H_2O/LiBr$ as the working fluid. However, there are also certain other working fluids that have theoretically shown good performance, such as TFE/TEGDME, methanol/TEGDME, monomethylamine/H_2O, and $LiCl/H_2O$.

Solar-powered adsorption refrigeration devices can meet, among things, the needs for refrigeration, A/C applications and ice making, with great potential for the conservation of various goods (medicines, food supplies) in remote areas. Nevertheless, the purpose of each system and the ambient conditions dictate its configuration (type of SC) and working pair, and therefore its performance.

Recently, many researchers have focused on the development of hybrid adsorption systems. Combined systems of solar adsorption cooling and heating improve the overall system performance by recovering the condensation heat and/or the heat of adsorption to produce hot water. Another possibility is to use adsorption systems as thermal energy storage devices, allowing for long-term heat or cold storage without losses. Moreover, the development of multi-functional systems that can provide space heating, water heating, and power in addition to providing cooling should be investigated to maximize the overall system efficiency and cost.

For solar sorption systems, considerable reduction in unit cost or significant improvements in its performances at present costs are still required to increase their competitiveness and commercialization potential.

The next several years will be decisive for the success of solar cooling systems, which depend on the encouragement and promotional schemes offered by policymakers and the efforts undertaken by manufacturers to improve the cost efficiency and develop better technologies. Additionally, the search for new working fluids that are environmentally friendly and require low operating temperatures is advised. Finally, research on the integration and control of various energy conversion systems for multiple uses may produce synergistic efficiency enhancements.

REFERENCES

[1] Santamouris M, Argiriou A. Renewable energies and energy conservation technologies for buildings in southern Europe. International Journal of Solar Energy 1994;15:69–79.
[2] Sarbu I, Sebarchievici C. Review of solar refrigeration and cooling systems. Energy and Buildings 2013;67(12):286–97.
[3] Saidur R, Masjuki H, Mahlia M, Tan C, Ooi J, et al. Performance investigation of a solar powered thermoelectric refrigerator. International Journal of Mechanical and Materials Engineering 2008;3:7–16.
[4] Vella GJ, Harris LB, Goldsmid HJ. A solar thermoelectric refrigerator. Solar Energy 1976;18(4):355–9.

[5]   Sarbu I. Refrigeration systems. Timisoara: Mirton Publishing House; 1998 [in Romanian].

[6]   Hwang Y, Randermacher R, Al-Alili A, Kubo I. Review of solar cooling technology. HVAC&R Research 2008;14(3):507−28.

[7]   Kim DS, Infante Ferreira CA. Solar refrigeration options − a state-of the-art review. International Journal of Refrigeration 2008;31:3−15.

[8]   Lof GO. Cooling with solar energy. In: Proceedings of the 1955 Congress on Solar Energy, Tucson, AZ; 1955. p. 171−89.

[9]   Ameel A, Gee KG, Wood BD. Performance predictions of alternative, low cost absorbents for open-cycle absorption solar cooling. Solar Energy 1995;54(2):65−73.

[10]   Gommed K, Grossman G. Experimental investigation of a liquid desiccant system for solar cooling and dehumidification. Solar Energy 2007;81(1):131−8.

[11]] ASHRAE handbook. HVAC systems and equipment. Atlanta, USA: American Society of Heating, Refrigerating and Air Conditioning Engineers; 2016.

[12]   Henning HM. Solar-assisted air-conditioning in buildings: a handbook for planners. Wien: Springer-Verlag; 2004.

[13]   Henning HM, Erpenbeck T, Hindenburg C, Santamaria IS. The potential of solar energy use in desiccant cooling cycles. International Journal of Refrigeration 2001;24(3):220−9.

[14]   Heidarinejad G, Bozorgmehr M, Delfani S, Esmaeelian J. Experimental investigation of two-stage indirect/direct evaporative cooling system in various climatic conditions. Buildings and Environment 2009;44(10):2073−9.

[15]   Rafique MM, Gandhidasan P, Rehman S, Al-Hadhrami LM. A review on desiccant based evaporative cooling systems. Renewable and Sustainable Energy Reviews 2015;45:145−59.

[16]   Bruno F. An indirect evaporative cooler for supplying air near the dew point. In: 48th AuSES Annual Conference, Canberra, ACT, Australia; 2010.

[17]   Enteria N, Mizutani K. The role of the thermally activated desiccant cooling technologies in the issue of energy and environment. Renewable and Sustainable Energy Reviews 2011;15(4):2095−122.

[18]   Cheng Q, Zhang X. Review of solar regeneration methods for liquid desiccant air-conditioning system. Energy and Buildings 2013;67:426−33.

[19]   Li X-W, Zhang X-S. Photovoltaic-electrodialysis regeneration method for liquid desiccant cooling system. Solar Energy 2009;83(12):2195−204.

[20]   Buker MS, Riffat SB. Recent developments in solar assisted liquid desiccant evaporative cooling technology − A review. Energy and Buildings 2015;96:95−108.

[21]   Srikhirin P, Aphornratana S, Chungpaibulpatana S. A review of absorption refrigeration technologies. Renewable and Sustainable Energy Reviews 2001;5:343−72.

[22]   Abdulateef JM, Sopian K, Alghoul MA. Optimum design for solar absorption refrigeration systems and comparison of the performances using ammonia-water, ammonia-lithium nitrate and ammonia-sodium thiocyanate solutions. International Journal of Mechanical and Materials Engineering 2008;3(1):17−24.

[23]   Erhard A, Hahne E. Test and simulation of a solar-powered absorption cooling machine. Solar Energy 1997;59:155−62.

[24]   Moreno-Quintanar G, Rivera W, Best R. Development of a solar intermittent refrigeration system for ice production. In: Proceedings of World Renewable Energy Congress, Linkoping, Sweden; 2011. p. 4033−44.

[25]   Medrano M, Bourouis M, Coronas A. Double-lift absorption refrigeration cycles driven by low-temperature heat sources using organic fluid mixtures as working pairs. Applied Energy 2001;68:173−85.

[26] Pilatowsky I, Rivera W, Romero RJ. Thermodynamic analysis of monomethylamine water solutions in a single-stage solar absorption refrigeration cycle at low generator temperatures. Solar Energy Materials and Solar Cells 2001;70:287–300.

[27] Fong KF, Lee CK, Chow CK, Yuen SY. Simulation–optimization of solar-thermal refrigeration systems for office use in subtropical Hong Kong. Energy 2001;36:6298–307.

[28] Rivera CO, Rivera W. Modeling of an intermittent solar absorption refrigeration system operating with ammonia/lithium nitrate mixture. Solar Energy Materials and Solar Cells 2003;76:417–27.

[29] Ullah KR, Saidur R, Ping HW, Akikur RK, Shuvo NH. A review of solar thermal refrigeration and cooling methods. Renewable and Sustainable Energy Reviews 2013;24:490–513.

[30] Grossman G. Solar–powered systems for cooling: dehumidification and air conditioning. Solar Energy 2002;72(1):53–62.

[31] Collector catalogue. Rapperswil, Switzerland: Institut fur Solartechnik SPF; Patent04.

[32] Sozen A. Effect of heat exchangers on performances of absorption refrigeration systems. Energy Conversion and Management 2001,42.1699–716.

[33] Fernandez-Seara J, Sieres J. The importance of the ammonia purification process in ammonia-water absorption systems. Energy Conversion and Management 2006;47:1975–87.

[34] Fernandez-Seara J, Sieres J, Vazquez M. Heat and mass transfer analysis of a helical coil rectifier in an ammonia-water absorption system. International Journal of Thermal Sciences 2003;42:783–94.

[35] Sieres J, Fernandez-Seara J. Mass transfer characteristics of a structured packing for ammonia rectification in ammonia-water absorption refrigeration systems. International Journal of Refrigeration 2007;30:58–67.

[36] Hassan H, Mohamad AA. A review on solar cold production through absorption technology. Renewable and Sustainable Energy Reviews 2012;16:5331–48.

[37] Erickson D. Solar icemakers in Maruata Mexico. Solar Today 1994;8:21–3.

[38] Sarbu I, Sebarchievici C. General review of solar-powered closed sorption refrigeration systems. Energy Conversion and Management 2015;105:403–22.

[39] Nakahara N, Miyakawa Y, Yamamoto M. Experimental study on house cooling and heating with solar energy using flat plate collector. Solar Energy 1977;19(6):657–62.

[40] Syed A, Izquierdo M, Rodriguez P, Maidmet G, Missenden J, Lecuona A, et al. A novel experimental investigation of a solar cooling system in Madrid. International Journal of Refrigeration 2005;28(6):859–71.

[41] Darkwa J, Fraser S, Chow DHC. Theoretical and practical analysis of an integrated solar hot water-powered absorption cooling system. Energy 2012;29:395–402.

[42] De-Francisco A, Illanes R, Torres JL, Castillo M, De-Blas M, Prieto E, et al. Development and testing of a prototype of low power water-ammonia absorption equipment for solar energy applications. Renewable Energy 2002;25:537–44.

[43] Brendel T, Zetzsche M, Muller-Steinhagen H. Development of small scale ammonia/water absorption chiller. In: Ninth IIR Gustav Lorentzen Conference, Sydney, vol. 1; 2010. p. 469–74.

[44] Rosiek S, Batlles FJ. Shallow geothermal energy applied to a solar-assisted air-conditioning system in southern Spain: two-year experience. Applied Energy 2012;100:267–76.

[45] El-Shaarawi MAI, Ramadan RA. Effect of condenser temperature on the performance of intermittent solar refrigerators. Energy Conversion and Management 1987;27(1):73–81.

[46] Said SAM, El-Shaarawi MAI, Siddiqui MU. Alternative designs for a 24-h operating solar-powered absorption refrigeration technology. International Journal of Refrigeration 2012;35(7):1967–77.

[47] El-Shaarawi MAI, Said SAM, Siddiqui MU. New simplified correlations for aqua-ammonia intermittent solar-powered absorption refrigeration systems. International Journal of Air-Conditioning and Refrigeration 2012;20(2):1250−8.

[48] Chen G, Hihara E. A new absorption refrigeration cycle using solar energy. Solar Energy 1999;66(6):479−82.

[49] Chinnappa JC, Crees MR, Murthy SS, Srinivasan K. Solar-assisted vapour compression/absorption cascaded air-conditioning systems. Solar Energy 1993;50(5):453−8.

[50] Lin P, Wang RZ, Xia ZZ. Numerical investigation of two-stage air-cooled absorption refrigeration system for solar cooling: cycle analysis and absorption cooling performances. Renewable Energy 2011;36:1401−12.

[51] Sumathy K, Huang Z, Li Z. Solar absorption cooling with low grade heat source − a strategy of development in South China. Solar Energy 2002;72:155−65.

[52] Izquierdo M, Venegas M, Rodriguez P, Lecuona A. Crystallization as a limit to develop solar air-cooled LiBr−H₂O absorption systems using low-grade heat. Solar Energy Materials and Solar Cells 2004;81:2005−16.

[53] Arivazhagan S, Saravanan R, Renganarayanan S. Experimental studies on HFC based two-stage half effect vapour absorption cooling system. Applied Thermal Engineering 2006;26:1455−62.

[54] Tierney MJ. Options for solar-assisted refrigeration trough collectors and double-effect chillers. Renewable Energy 2007;32:183−99.

[55] Bermejo P, Pino FJ, Rosa F. Solar absorption cooling plant in Seville. Solar Energy 2010;84:1503−12.

[56] Gomri R. Second law comparison of single effect and double effect vapour absorption refrigeration systems. Energy Conversion and Management 2009;50:1279−87.

[57] Somers C, Mortazavi A, Hwang Y, Radermacher R, Rodgers P, Al-Hashimi S. Modeling water/lithium bromide absorption chillers in ASPEN Plus. Applied Energy 2011;88:4197−205.

[58] Hong DL, Chen GM, Tang LM, He YJ. Simulation research on an EAX (evaporator-absorber-exchanger) absorption refrigeration cycle. Energy 2011;36:94−8.

[59] Park YM, Sonntag RE. Thermodynamic properties of ammonia-water mixtures: a generalized equation-of-state approach. ASHRAE Transactions 1990;97(1):150.

[60] Ibrahim OM, Klein SA. Thermodynamic properties of ammonia-water mixtures. ASHRAE Transactions 1993;99(1):1495.

[61] Tillner-Roth R, Friend DG. Survey and assessment of available measurements on thermodynamic properties of the mixture (water-ammonia). Journal of Physical Chemistry 1998;27(1):45.

[62] El-Shaarawi MAI, Said SAM, Siddiqui MU. New correlation equations for ammonia-water vapor-liquid equilibrium (VLE) thermodynamic properties. ASHRAE Transactions 2013;119(1) (DA-13-025).

[63] Patek J, Klomfar J. Simple functions for fast calculations of selected thermodynamic properties of the ammonia-water system. International Journal of Refrigeration 1995;18:228−34.

[64]] ASHRAE handbook. HVAC applications. Atlanta, GA: American Society of Heating, Refrigerating and Air Conditioning Engineers; 2015.

[65] Zhai X, Li Y, Cheng X, Wang R. Experimental investigation on a solar-powered absorption radiant cooling system. Energy Procedia 2015;70:552−9.

[66] Fernandes MS, Brites GJVN, Costa JJ, Gaspar AR, Costa VAF. Review and future trends of solar adsorption refrigeration systems. Renewable and Sustainable Energy Reviews 2014;39:102−23.

[67] Fan Y, Luo L, Souyri B. Review of solar sorption refrigeration technologies: development and applications. Renewable and Sustainable Energy Reviews 2007;11(8):1758−75.

[68] Sapienza A, Santamaria S, Frazzica A, Freni A. Influence of the management strategy and operating conditions on the performance of an adsorption chiller. Energy 2011;36:5532−8.

[69] Bansal NK, Blumenberg J, Kavasch HJ, Rorttinger T. Performance testing and evaluation of solid adsorption solar cooling unit. Solar Energy 1997;61(2):127−40.

[70] Anyanwu EE, Ogueke NV. Thermodynamic design procedure for solid adsorption solar refrigerator. Renewable Energy 2005;30(1):81−96.

[71] Jribi S, Saha BB, Koyama S, Chakraborty A, Ng KC. Study on activated carbon/HFO-1234ze(E) based adsorption cooling cycle. Applied Thermal Engineering 2013;50:1570−5.

[72] Wang D, Li YH, Li D, Xia YZ, Zhang JP. A review on adsorption refrigeration technology and adsorption deterioration in physical adsorption systems. Renewable and Sustainable Energy Reviews 2010;14(1):344−53.

[73] Wang RZ, Li M, Xu XX, Wu JY. An energy efficient hybrid system of solar powered water heater and adsorption ice maker. Solar Energy 2000;68(2):189−95.

[74] Liu YL, Wang RZ, Xia ZZ. Experimental study on a continuous adsorption water chiller with novel design. International Journal of Refrigeration 2005;28:218−30.

[75] Wang RZ, Oliveira RG. Adsorption refrigeration − an efficient way to make good use of waste heat and solar energy. In: Proceedings of International Sorption Heat Pump Conference, Denver, USA; 2005.

[76] Omisanya NO, Folayan CO, Aku SY, Adefila SS. Performance of a zeolite-water adsorption refrigerator. Advances in Applied Science Research 2012;3(6):3737−45.

[77] Erhard A, Spindler K, Hahne E. Test and simulation of a solar powered solid sorption cooling machine. International Journal of Refrigeration 1998;21(2):133−41.

[78] Maggio G, Gordeeva LG, Freni A, Aristov YI, Santori G, Polonara F, et al. Simulation of a solid sorption ice-maker based on the novel composite sorbent lithium chloride in silica gel pores. Applied Thermal Engineering 2009;29(8−9):1714−20.

[79] Gopal MR, Murthy SS. Performance a metal hydride cooling system. International Journal of Refrigeration 1995;18:413−20.

[80] Tchernev DI. Solar air conditioning and refrigeration systems utilizing zeolites. In: Proceedings of Meetings of Commissions E1−E2. Jerusalem, Israel: International Institute of Refrigeration; 1979. p. 209−15.

[81] Pons M, Guilleminot JJ. Experimental data on a solar powered ice maker using activated carbon and methanol adsorption pair. Journal of Solar Energy Engineering 1987;109(4):303−10.

[82] Grenier P, Guilleminot JJ, Meunier F, Pons M. Solar powered solid adsorption cold store. ASME Transaction. Journal of Solar Energy Engineering 1988;110:192−7.

[83] Critoph RE. Laboratory testing of van ammonia carbon solar refrigerator. Budapest, Hungary: ISES, Solar World Congress; 1993. p. 23−6.

[84] Li CH, Wang RZ, Dai YJ. Simulation and economic analysis of a solar-powered adsorption refrigerator using an evacuated tube for thermal insulation. Renewable Energy 2003;28(2):249−69.

[85] Sumathy K, Yeung KH, Yong L. Technology development in the solar adsorption refrigeration systems. Progress in Energy Combustion Science 2003;29:301−27.

[86] Luo HL, Wang RZ, Dai YJ, Wu JY, Shenand JM, Zhang BB. An efficient solar powered adsorption chiller and its application in low temperature grain storage. Solar Energy 2007;81(5):607−13.

[87] González MI, Rodríguez LR. Solar powered adsorption refrigerator with CPC collection system: collector design and experimental test. Energy Conversion and Management 2007;48(9):2587−94.

[88] Sapienza A, Glaznev IS, Santamaria S, Freni A, Aristov YI. Adsorption chilling driven by low temperature heat: new adsorbent and cycle optimization. Applied Thermal Engineering 2012;321:141−6.

[89] Lu ZS, Wang LW, Wang RZ. Experimental analysis of an adsorption refrigerator with mass and heat-pipe heat recovery process. Energy Conversion and Management 2012;53:291−7.

[90] Henning HM, Glaser H. Solar assisted adsorption system for a laboratory of the University Freiburg. 2003. http://www.bine.info/pdf/infoplus/uniklaircontec.pdf.

[91] El-Fadar A, Mimet A, Pérez-García M. Study of an adsorption refrigeration system powered by parabolic trough collector and coupled with a heat pipe. Renewable Energy 2009;34(10):2271−9.

[92] Hassan HZ, Mohamad AA, Bennacer R. Simulation of an adsorption solar cooling system. Energy 2011;36(1):530−7.

[93] Yeung MR, Yuen PK, Dunn A, Cornish LS. Performance of a solar-powered air conditioning system in Hong Kong. Solar Energy 1992;48:309−19.

[94] Balaras CA, Grossman G, Henning HM, Ferreira CA, Podesser E, Wang L, et al. Solar air conditioning in Europe − an overview. Renewable and Sustainable Energy Reviews 2007;11(2):299−314.

[95] Bujedo LA, Rodríguez J, Martínez PJ. Experimental results of different control strategies in a solar air-conditioning system at part load. Solar Energy 2011;85:1302−15.

[96] Agyenim F, Knight I, Rhodes M. Design and experimental testing of the performance of an outdoor LiBr/H$_2$O solar thermal absorption cooling system with a cold store. Solar Energy 2010;84:735−44.

[97] Lizarte R, Izquierdo M, Marcos JD, Palacios E. An innovative solar-driven directly air-cooled LiBr-H$_2$O absorption chiller prototype for residential use. Energy and Buildings 2012;47:1−11.

[98] Palacios E, Izquierdo M, Marcos JD, Lizarte R. Evaluation of mass absorption in LiBr flat-fan sheets. Applied Energy 2009;86:2574−82.

[99] Al-Alili A, Islam MD, Kubo I, Hwang Y, Radermacher R. Modeling of a solar powered absorption cycle for Abu Dhabi. Applied Energy 2012;93:160−7.

[100] Balghouthi M, Chahbani MH, Guizani A. Feasibility of solar absorption air conditioning in Tunisia. Building and Environment 2008;43:1459−70.

[101] Uppal AH, Norton B, Probert SD. A low-cost solar-energy stimulated absorption refrigerator for vaccine storage. Applied Energy 1986;25:167−74.

[102] Sierra FZ, Best R, Holland FA. Experiments on an absorption refrigeration system powered by a solar pond. Heat Recovery Systems CHP 1993;13:401−8.

[103] Pilatowski I, Rivera W, Romero JR. Performance evaluation of a monomethylamine/water solar absorption refrigeration system for milk cooling purposes. Applied Thermal Engineering 2004;24:1103−15.

[104] Lemmini F, Errougani A. Building and experimentation of a solar powered adsorption refrigerator. Renewable Energy 2005;30:1989−2003.

[105] Hammad M, Habali S. Design and performance study of a solar energy powered vaccine cabinet. Applied Thermal Engineering 2000;20:1785−98.

[106] Abu-Hamdeh NH, Alnefaie KA, Almitani KH. Design and performance characteristics of solar adsorption refrigeration system using parabolic trough collector: experimental and statistical optimization technique. Energy Conversion and Management 2013;74:162−70.

[107] Medini N, Marmottant B, El-Golli S, Grenier Ph. Etude d'une machine solaire autonome à fabriquer de la glace. International Journal of Refrigeration 1991:14.

[108] Sumathy K, Li ZF. Experiments with solar-powered adsorption ice-maker. Renewable Energy 1999;16:704—7.

[109] Zeo-Tech. Our developments and services. 2013. http://www.zeo-tech.de/index.php/en/developments.

[110] Ji X, Li M, Fan J, Zhang P, Luo B, Wang L. Structure optimization and performance experiments of a solar-powered finned-tube adsorption refrigeration system. Applied Energy 2014;30:440—51.

[111] Vasta S, Maggio G, Santori G, Freni A, Polonara F, Restriccia G. An adsorptive solar ice-maker dynamic simulation for north Mediterranean climate. Energy Conversion and Management 2008;49(11):3025—35.

[112] Badawy MT. Cycle analysis for solar ejector refrigeration and distillation system. In: Proceedings of World Renewable Energy Congress, Florence, Italy, vol. 4; 1998. p. 2076—9.

[113] Loehrke RI. A passive, vapour compression refrigerator for solar cooling. Transaction ASME, Journal of Solar Energy Engineering 1990;12:191—5.

[114] Besagni G, Mereu R, Inzoli F. Ejector refrigeration: a comprehensive review. Renewable and Sustainable Energy Reviews 2016;53:373—407.

[115] Smirnov HF, Kosoy BV. Refrigerating heat pipes. Applied Thermal Engineering 2001;21:631—41.

[116] Little AB, Garimella S. A review of ejector technology for refrigeration applications. International Journal of Air-Conditioning and Refrigeration 2011;19:1—15.

[117] Abdulateef JM, Sopian K, Alghoul MA, Sulaiman MY. Review on solar-driven ejector refrigeration technologies. Renewable and Sustainable Energy Reviews 2009;13:1338—49.

[118] Nehad AK. Experimental investigation of solar concentrating collectors in a refrigerant ejector refrigeration machine. International Journal of Energy Research 1997;21:1123—31.

[119] Nehad AK. An experimental study of an ejector cycle refrigeration machine operating on R113. International Journal of Refrigeration 1998;21(8):612—25.

[120] Eames IW, Aphornratana S, Haider H. A theoretical and experimental study of a small-scale steam jet refrigerator. International Journal of Refrigeration 1995;18(6):378—86.

[121] Vidal HS, Li M, Xu YX, Wu JY. Modelling and hourly simulation of a solar ejector cooling system. Applied Thermal Engineering 2006;26(7):663—72.

[122] Grazzini G, Rocchetti A. Numerical optimization of a two-stage ejector refrigeration plant. International Journal of Refrigeration 2002;25:621—33.

[123] Arbel A, Sokolov M. Revisiting solar-powered ejector air conditioner — the greener the better. Solar Energy 2004;77(1):57—66.

[124] Al-Alili A, Hwang Y, Radermacher R. Review of solar thermal air conditioning technologies. International Journal of Refrigeration 2014;39:4—22.

[125] Duffie JA, Beckman WA. Solar engineering of thermal processes. Hoboken, NJ: Wiley & Sons, Inc; 2013.

[126] Helm M, Keil C, Hiebler S, Mehling H, Schweigler C. Solar heating and cooling system with absorption chiller and low temperature latent heat storage: energetic performance and operational experience. International Journal of Refrigeration 2009;32:596—606.

[127] Schweigler C, Hiebler S, Keil C, Kobel H, Kren C, Mehling H. Low temperature heat storage for solar heating and cooling applications. Dallas, USA: ASHRAE Winter-Meeting; 2007.

[128] Mammoli A, Vorobieff P, Barsun H, Burnett R, Fisher D. Energetic, economic and environmental performance of a solar-thermal-assisted HVAC system. Energy and Buildings 2010;42:1524—35.

[129] Zhang XJ, Wang RZ. Design and performance simulation of new solar continuous solid adsorption refrigeration and heating hybrid system. Renewable Energy 2002;27(3):401−15.

[130] Zhang XJ, Wang RZ. A new combined adsorption-ejector refrigeration and heating hybrid system powered by solar energy. Applied Thermal Engineering 2002;22(11):1245−58.

[131] Suleiman R, Folayan C, Anafi F, Kulla D. Transient simulation of a flat plate solar collector powered adsorption refrigeration system. International Journal of Renewable Energy Research 2012;2(4):657−64.

[132] Chidambaram LA, Ramana AS, Kamaraj G, Velraj R. Review of solar cooling methods and thermal storage options. Renewable and Sustainable Energy Reviews 2011;15:3220−8.

[133] Sekret R, Turski M. Research on an adsorption cooling system supplied by solar energy. Energy and Buildings 2012;51:15−20.

Chapter 8

Solar Electric Cooling Systems

8.1 GENERALITIES

The search for cleaner, more sustainable energy sources is an ever-growing global concern because of escalating energy costs and global warming associated with fossil fuel sources [1−4]. Among the viable technologies for this purpose, solar electric cooling systems as solar photovoltaic (PV) systems are of increasing interest because these solid-state devices can convert light energy into electrical energy. To accommodate the huge demand for electricity, PV-based electricity generation has been rapidly increasing around the world alongside conventional power plants over the past two decades. Fig. 8.1 shows a comparative representation of the development of solar PV systems in different countries [5].

Besides vapor-compression cooling, some other types of electric cooling systems can be used in combination with PV panels. Conversely, solid-state thermoelectric (TE) devices can also change electrical energy supplied by solar PV panels into thermal energy for cooling using the Peltier effect. TE cooling possesses advantage that it can be powered by direct current (DC) electric sources, such as PV panels. Furthermore, because TE devices use no refrigerants or working fluids, they may be expected to produce negligible direct emissions of greenhouse gases (GHGs) over their lifetimes [6]. However, the main disadvantages of TE cooling are the high cost and low energy efficiency, which has restricted its application to cases where system cost and

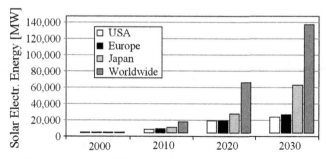

FIGURE 8.1 Global photovoltaic (PV)-based solar electricity production for four decades.

Solar Heating and Cooling Systems. http://dx.doi.org/10.1016/B978-0-12-811662-3.00008-6

energy efficiency are less important than energy availability, system reliability, and quiet operation environment.

TE cooling devices are widely used for electronic cooling such as personal computer (PC)—processors, portable food and beverage storages, temperature-control car seats, and even TE air conditioners (A/Cs).

There are good review papers on TE technologies and applications, including modeling and analysis of TE modules [7], solar-based TE technologies [1], cooling, heating, and generating power, and waste heat recovery [8,9].

Electrically driven thermoacoustic cooling systems are another option for solar cooling. These systems use pressure changes in acoustic waves to transfer heat between two reservoirs at different temperature levels. Efficiencies of thermoacoustic cooling systems are lower than those of vapor-compression systems [10].

This chapter covers solar electric cooling systems including the solar PV and TE systems. Thus, the utilization of solar PV panels coupled with a vapor-compression air-conditioning system is described, and a good amount of information regarding ecological refrigerant's trend is included. Additionally, this chapter presents the details referring to TE cooling parameters and formulations of the performance indicators, and focuses on the development of TE cooling systems in recent decade with particular attention on advances in materials and modeling approaches and applications.

8.2 SOLAR PHOTOVOLTAIC COOLING SYSTEMS

8.2.1 Description of the System

A PV panel is basically a solid-state semiconductor device that converts light energy into electrical energy. While the output of a PV panel is typically DC electricity, most domestic and industrial electrical appliances use alternating current (AC). Therefore, a complete PV cooling system typically consists of four basic components: PV modules, a battery, an inverter circuit, and a vapor-compression AC unit (Fig. 8.2) [11].

The *PV panel* consists of PV cells that enable photons to "knock" electrons out of a molecular lattice, leaving a freed electron and "hole" pair that diffuse in an electric field to separate contacts, generating DC electricity.

The *battery* is used for storing DC voltages at a charging mode when sunlight is available and supplying DC electrical energy in a discharging mode in the absence of daylight. A battery charge regulator can be used to protect the battery form overcharging.

The *inverter* is an electrical circuit that converts the DC electrical power into AC and then delivers the electrical energy to the AC loads.

The *vapor-compression AC unit* is actually a conventional cooling or refrigeration system that is run by the power received from the inverter.

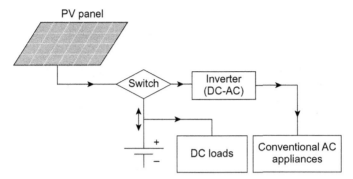

FIGURE 8.2 Schematic of a stand-alone PV system.

The PV system can perform as a stand-alone system (Fig. 8.2), a hybrid system (working with an oil/hydro/gas power plant) or as a grid or utility intertie systems. Though the efficiency of PV modules can be increased by using inverters, their coefficient of performance (COP) and efficiency are still not desirable.

8.2.2 Solar Photovoltaic-Driven Vapor-Compression Cooling Systems

8.2.2.1 Operation Principle and Energy Efficiency

A solar electric cooling system consists mainly of PV panels and an electrical cooling device. Most of the solar panels commercially available in the market are made from silicon as the ones shown in Fig. 8.2. Price of a solar panel varies widely in the market.

The biggest advantage of using PV panels for cooling is the simple construction and high overall efficiency when coupled with conventional vapor-compression system. A schematic diagram of such a system is given in Fig. 8.3.

Vapor-compression system consists of compressor, condenser expansion valve, and evaporator, connected with refrigerant pipelines. The basic vapor-compression cycle is considered to be one with isentropic compression, with no superheat of vapor and with no subcooling of liquid (Fig. 8.4).

Operational processes are the following:

1−2: isentropic compression in the compressor K, which leads to increased pressure and temperature from the values corresponding to evaporation p_e, t_e to those of the condensation p_c, $t_2 > t_c$;
2−2′: isobar cooling in the condenser C at pressure p_c from the temperature t_2 to $t_2' = t_c$;
2′−3: isotherm−isobar condensation in the condenser C at pressure p_c and temperature t_c;

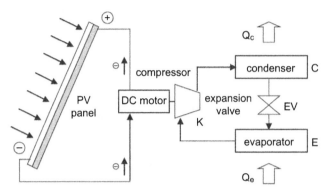

FIGURE 8.3   Schematic of a solar PV-driven vapor-compression air-conditioning system.

3—4: isenthalpic lamination in expansion valve EV, leading the refrigerant from State 3 of the liquid at p_c, t_c in State 4 of wet vapor at p_e, t_e; and 4—1: isotherm—isobar evaporation in the evaporator E at pressure p_e and temperature t_e.

In a theoretical vapor-compression cycle, the refrigerant enters the compressor at State 1 as saturated vapor and is compressed isentropically to the condensation pressure. The refrigerant temperature increases during this isentropic compression process to well above the temperature of the surrounding medium. The refrigerant then enters the condenser as superheated vapor at State 2 and leaves as saturated liquid at State 3 as a result of heat rejection to the surroundings. The refrigerant temperature at this state is still above the temperature of the surroundings. The saturated liquid refrigerant at State 3 is throttled to the evaporation pressure by passing it through an expansion valve. The refrigerant temperature drops below the temperature of the cold environment during this process. The refrigerant enters the evaporator at State 4 as a low-quality saturated mixture, and completely evaporates by

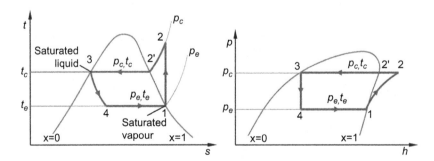

FIGURE 8.4   Single-stage vapor-compression process in t-s and p-h diagrams.

absorbing heat from the cold environment. The refrigerant then leaves the evaporator as saturated vapor and reenters the compressor, completing the cycle [12].

The specific compression work w, in kJ/kg, the specific cooling power q_e, in kJ/kg, the specific heat load at condensation q_c, in kJ/kg, volumetric refrigerating capacity q_{ev}, in kJ/m^3, and the COP are calculated for above presented processes as follows:

$$w = h_2 - h_1 \tag{8.1}$$

$$q_e = h_1 - h_4 = h_1 - h_3 \tag{8.2}$$

$$q_c = h_2 - h_3 \tag{8.3}$$

$$q_{ev} = \frac{q_e}{v_1} = q_e \rho_1. \tag{8.4}$$

The refrigerant mass flow rate m, in kg/s, is calculated from the required cooling capacity Q_e and the specific cooling power q_e:

$$m = \frac{Q_e}{q_e}. \tag{8.5}$$

The power necessary for the isentropic compression P_{is}, in kW, may be calculated as

$$P_{is} = mw. \tag{8.6}$$

The effective power P_{el} on the compressor shaft is larger and is defined as

$$P_{el} = \frac{P_{is}}{\eta_{is}}, \tag{8.7}$$

where η_{is} is the isentropic efficiency.

The COP is defined as follows:

$$COP = \frac{Q_e}{P_{el}} = \frac{q_e}{w} = \frac{h_1 - h_4}{h_2 - h_1}. \tag{8.8}$$

The overall COP of a solar electric cooling system is given by following equation:

$$COP_{sys} = \frac{Q_e}{Q_s}, \tag{8.9}$$

where Q_s is the solar power received by the solar collector surface.

8.2.2.2 Ecological Refrigerants

Environmental protection represents the fundamental condition of the society's sustainable development and is a high priority of national interest realized via institutional framework in which the legal norms regulate the development of activities with environmental impact and exert control over such activities.

One of the minor components of the atmosphere, the ozone layer, has a special importance in maintaining the ecological balance. Ozone is distributed primarily between the stratosphere (85—90%) and troposphere. Any perturbation of the atmospheric ozone concentration (which varies between 0 ppm and 10 ppm, depending on the regions) has direct and immediate effects upon life.

For most of the states, the problems of forming and maintaining the earth's ozone layer represents a major priority. In this context, during the past 30 years, the European Union has adopted a large number of laws and regulations concerning environmental protection to correct the pollution effects, frequently by indirect directives, through imposition of the levels of allowable concentrations by asking for government collaboration, programs, and projects for the regulation of industrial activities and productions. *The Alliance for Responsible Atmospheric Policy* is an industry coalition and leading voice for ozone protection and climate change policies, which maintains a brief summary of the regulations for some countries [13].

Refrigerants are the working fluids in A/C cooling and heat pump (HP) systems. They absorb heat from one area, such as an air-conditioned space, and reject it into another, such as outdoors, usually through evaporation and condensation. This section contains a good amount of information regarding ecological refrigerant trend.

Working fluids escaped through leakages from cooling equipment during normal operation (filling or emptying) or after accidents (damages) gather in significant quantities at high levels of the atmosphere (stratosphere). In the stratosphere, through catalytically decompounding, pollution from the working-fluid leakage depletes the ozone layer that normally filters the ultraviolet radiation from the sun, which is a threat to living creatures and plants on earth. Stratospheric ozone depletion has been linked to the presence of chlorine and bromine in the stratosphere. In addition, refrigerants contribute to global warming (also called global climate change) because they are gases that exhibit the greenhouse effect when in the atmosphere.

Vapor-compression—based systems are generally employed with halogenated refrigerants. The international protocols (Montreal and Kyoto) restrict the use of the halogenated refrigerants in the vapor-compression—based systems. As per Montreal Protocol 1987, the use of chlorofluorocarbons (CFCs) was completely stopped in most of the nations. However, hydrochlorofluorocarbons (HCFCs) refrigerants can be used until 2040 in the developing nations and the developed nations should phase out by 2030 [14]. To meet the global demand in A/C and HP sector, it is necessary to look for long-term alternatives to satisfy the objectives of international protocols. From the environmental, ecological, and health point of view, it is urgent to find some better substitutes for hydrofluorocarbon (HFC) refrigerants [15].

Hydrocarbon (HC) and HFC refrigerant mixtures with low environment impacts are considered as potential alternatives to phase out the existing halogenated refrigerants.

The design of the refrigeration equipment depends strongly on the properties of the selected refrigerant. Refrigerant selection involves compromises between conflicting desirable thermophysical properties. A refrigerant must satisfy many requirements, some of which do not directly relate to its ability to transfer heat. Chemical stability under conditions of use is an essential characteristic. Safety codes may require a nonflammable refrigerant of low toxicity for some applications. The environmental consequences of refrigerant leaks must also be considered. Cost, availability, efficiency, and compatibility with compressor lubricants and equipment materials are other concerns.

Safety properties of refrigerants considering flammability and toxicity are defined by ASHRAE Standard 34 [16]. Toxicity classification of refrigerants is assigned to classes A or B and by flammability refrigerants are divided into three classes (Table 8.1).

Global warming is a concern because of an increase in the greenhouse effect from increasing concentrations of GHGs attributed to human activities. Thus, the negative environmental impact of the working fluids, especially the effect of halogenated refrigerants on the environment, can be synthesized by two effects [17]:

- depletion of the ozone layer; and
- contribution to global warming at the planetary level via the greenhouse effect.

The measure of a material's ability to deplete stratospheric ozone is its *ozone depletion potential* (ODP), a relative value to that of R11, which has an ODP of 1.0.

The *global warming potential* (GWP) of a GHG is an index describing its relative ability to collect radiant energy compared to carbon dioxide (CO_2), which has a very long atmospheric lifetime. Therefore, refrigerants will be selected so that the ODP will be zero and with a reduced GWP.

TABLE 8.1 Safety Classification of Refrigerants

	Safety code	
Flammability	Lower Toxicity	Higher Toxicity
Higher flammability	A2	B2
Lower flammability	A2L	B2L
No flame propagation	A1	B1

During the last century, the halogenated refrigerants have dominated the vapor-compression—based systems due to its good thermodynamic and thermophysical properties. Thermodynamic properties of pure refrigerants are listed in Table 8.2 [18]. But, the halogenated refrigerants are having poor environmental properties with respect to ODP and GWP.

TABLE 8.2 Thermodynamic Properties of Pure Refrigerants

Refrigerant	Molecular Mass, M (g/mol)	Critical Temperature, t_{cr} (°C)	Critical Pressure, p_{cr} [MPa]	Boiling Point, t_{0n} [°C]
R11	137.37	198.0	4.41	23.7
R12	120.90	112.0	4.14	−29.8
R22	86.47	96.2	4.99	−41.4
R23	70.01	25.9	4.84	−82.1
R32	52.02	78.2	5.80	−51.7
R41	34.03	44.1	5.90	−78.1
R123	152.93	82.0	3.66	27.8
R124	136.48	122.3	3.62	−12.0
R125	120.02	66.2	3.63	−54.6
R134a	102.03	101.1	4.06	−26.1
R142b	100.49	137.2	4.12	−9.0
R143a	84.04	72.9	3.78	−47.2
R152a	66.05	113.3	4.52	−24.0
R161	48.06	102.2	4.70	−34.8
R170	30.07	90.0	4.87	−88.9
R218	188.02	71.9	2.68	−36.6
R290	44.10	96.7	4.25	−42.2
R600	58.12	152.0	3.80	−0.5
R600a	58.12	134.7	3.64	−11.7
R717	17.03	132.3	11.34	−33.3
R744	44.01	31.1	7.38	−78.4
R1270	42.08	92.4	4.67	−47.7

The traditional refrigerants (CFCs) were banned by the Montreal Protocol because of their contribution to the disruption of the stratospheric ozone layer. The Kyoto Protocol listed HCFCs as being with large GWPs.

With the phasing out of the use of CFCs, chemical substances such as the HCFCs and the HFCs were proposed and have been used as temporary alternatives.

The HFCs do not deplete the ozone layer and have many of the desirable properties of CFCs and HCFCs. They are being widely used as substitute refrigerants for CFCs and HCFCs. The HFC refrigerants have significant benefits regarding safety, stability, and low toxicity, being appropriate for large-scale applications.

Also, the HC and HFC refrigerant mixtures with low environment impacts are considered as potential alternatives to phase out the existing halogenated refrigerants. HC-based mixtures are environment friendly, which can be used as alternatives without modifications in the existing systems. However, HC refrigerant mixtures are highly flammable, which limits the usage in large capacity systems [19]. HFC mixtures are ozone-friendly, but have significant GWP.

The environmental impact of a heating, ventilation, air-conditioning, and refrigeration system is due to the release of refrigerant and the emission of GHGs for associated energy use. The *total equivalent warming impact* (TEWI) is used as an indicator for environmental impact of the system for its entire lifetime. TEWI is the sum of the direct refrigerant emissions, expressed in terms of CO_2 equivalents, and the indirect emissions of CO_2 from the system's energy use over its service life.

More environmentally friendly refrigeration systems have been investigated in recent years [20–23]. Two aspects are of particular concern, namely the use of ecological (environmentally friendly) refrigerants and the energy consumption issue.

The CFC refrigerants of R11 and R12 were substituted by simpler compound refrigerants R123 (HCFC) and R134a (HFC) with a reduced or even zero impact on the depletion of the ozone layer [24]. This alternative is attractive because the substitutes have similar properties (temperature, pressure) with the replaced ones and the changes that occur directly on the existing installations are realized with minimum of investments.

Additionally, the substitution of R123 or R11 refrigerants with R22 or R134a, having molecular masses lower by 50% (Fig. 8.5), leads to reduced dimensions of the cooling equipment by 25–30% [25].

For other refrigerants, no simple compound fluids, for example, for R502, could be replaced with a mixture of R115 (CFC) and R22 (HCFC) or in some cases only with R22, which is a fluid for temporary substitution. However, all these compounds are considered to be GHGs. As a response to these concerns, even more ecological refrigerants, mainly R1234yf [26] and natural refrigerants [27,28,29], particularly CO_2 and NH_3, have been proposed as substitutes.

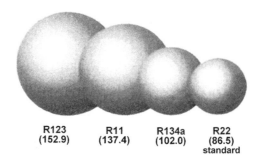

R123   R11   R134a   R22
(152.9)   (137.4)   (102.0)   (86.5)
                                 standard

FIGURE 8.5   Molecular mass of some halogenated refrigerants.

Very limited pure fluids are having suitable properties to provide alternatives to the existing halogenated refrigerants. The mixing of two or more refrigerants provides an opportunity to adjust the properties, which are most desirable. The three categories of mixtures used in air-conditioning and refrigeration applications are azeotropes, near-azeotropes (quasiazeotropes), and zeotropes [30].

Azeotropic mixture of the substances is the one that cannot be separated into its components by simple distillation. The azeotropic mixtures are having boiling points that are lower than either of their constituents. An azeotropic mixture maintains a constant boiling point and acts as a single substance in both liquid and vapor states.

The objective with near-azeotropic mixtures is to extend the range of refrigerant alternatives beyond single compounds. Near-azeotropes have most of the same attributes as azeotropes and provide a much wider selection possibilities. However, near-azeotropic mixtures may alter their composition and properties under leakage conditions.

Zeotropic refrigerant mixtures are blends of two or more refrigerants that deviate from perfect mixtures. A zeotropic mixture does not behave like a single substance when it changes state. Instead, it evaporates and condenses between two temperatures (temperature glide). The phase-change characteristics of the zeotropic refrigerant mixture (boiling and condensation) are nonisothermal. Zeotropic substances have greater potential for improvements in energy efficiency and capacity modulation. However, the major drawback of the zeotropic refrigerant mixture is the preferential leakage of more volatile components leading to change in mixture composition.

Fig. 8.6 illustrates a strategy concerning the refrigerants. Comparison of different refrigerants gives a good overview of achievable cycle performance for a basic referent thermodynamic cycle [31]. Table 8.3 gives comparison for refrigerants' reference cycle with evaporation temperature $t_e = -15\,°C$ and condensation temperature $t_c = +30°C$ [32].

Cycle data are available from different sources [17,18], or can be evaluated from suitable software such as REFPROP [33].

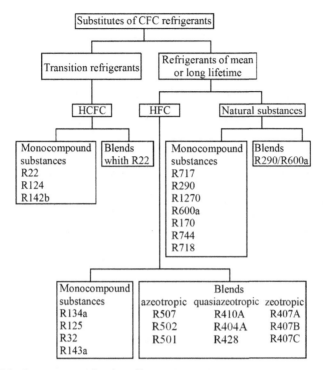

FIGURE 8.6 Strategy concerning the refrigerants.

The selection of refrigerants in Table 8.3 has been made to present the overview of cycle data for historically used natural inorganic refrigerants such as R717, R744, and R764 (which is not in use anymore), CFCs such as R11 or R12 and HCFCs such as R22, and mixture R502. Among the newly used refrigerants HFCs R32 and R134a are presented as well as zeotropic mixtures of HFCs R404 A, R407 C, R410 A, and azeotropic mixture of HFCs R507. Finally, natural HCs R600a and R290, together with propylene R1270 are listed.

As it can be seen from data presented in Table 8.3, pressures in the system are temperature-dependent and are different for each particular refrigerant. Evaporation and condensation temperatures are closely coupled with corresponding pressure for single-component refrigerants, whereas for zeotropic mixtures temperature glide appears during the phase change at constant pressure.

Pressures influence design and thus equipment costs, but also the power consumption for compression and thus operational costs. Refrigerant transport properties, such as liquid and vapor density, viscosity, and thermal conductivity, define heat transfer coefficients, and consequently, temperature differences in heat exchangers thus directly influence pressures in the system as well as the necessary heat transfer surface of heat exchangers. Molecular mass or volumetric refrigerating capacity of some refrigerants influences application of

TABLE 8.3 Parameters of −15/30 °C Cycle With Different Refrigerants

Refrigerant	p_e (bar)	p_c (bar)	p_c/p_e (−)	q_{ev} (kJ/m³)	COP (−)	t_2 (°C)	Safety Code
0	1	2	3	4	5	6	7
R717	2.362	11.672	4.942	2167.6	4.76	99.08	B2L
R744	22.90	72.10	3.149	7979.0	2.69	69.50	A1
R764	0.807	4.624	5.730	818.8	4.84	96.95	B1
R11	0.202	1.260	6.233	204.2	5.02	42.83	A1
R12	1.823	7.437	4.079	1273.4	4.70	37.81	A1
R22	2.962	11.919	4.024	2096.9	4.66	52.95	A1
R32	4.881	19.275	3.949	3420.0	4.52	68.54	A2L
R134a	1.639	7.702	4.698	1225.7	4.60	36.61	A1
R404 A	3.610	14.283	3.956	2099.1	4.16	36.01	A1
R407 C	2.632	13.591	5.164	1802.9	3.91	51.43	A1
R410 A	4.800	18.893	3.936	3093.0	4.38	51.23	A1
R502	3.437	13.047	3.796	2079.5	4.39	37.07	A1
R507	3.773	14.600	3.870	2163.2	4.18	35.25	A1
R600a	0.891	4.047	4.545	663.8	4.71	32.66	A3
R290	2.916	10.790	3.700	1814.5	4.55	36.60	A3
R1270	3.630	13.050	3.595	2231.1	4.55	41.85	A3

certain compressor types. For example, NH_3 systems are not suitable for application of centrifugal compressor due to the low molecular mass of NH_3. The higher the volumetric refrigeration capacity is, the smaller compressor displacement can be, which results in smaller compressors for refrigerants with high volumetric refrigeration capacities. A good example is R744, which has the highest volumetric capacity.

Achievable efficiency of the entire process is due in great part to the refrigerant used. Effective energy consumption or COP is not equal to the one of the theoretical cycle. Isentropic efficiency η_{is} in Eq. (8.7) is also dependent on refrigerant properties. Discharge temperature on the compressor outlet t_2 depends on refrigerant and system's pressures, and it must be limited to avoid deterioration of oil properties, or even the oil burnout. Behavior of some refrigerant during the compression can result is no or low superheating of the vapor at the end of the compression (e.g., R134a, which has low superheating, or R600a where final refrigerant state at the end of the compression can end in saturated area unless proper superheating at the compressor inlet is provided). Systems with such refrigerants are not suitable for utilization of superheated part of vapor heat content in refrigeration cycles with heat recovery for sanitary water heating during the cooling operation [31].

Pressure drop within heat exchangers and in pipelines connecting refrigeration machine components are essential for system efficiency and are also dependent of refrigerant properties.

Scientific research based on monocompound substances or mixtures will lead to the discovery of adequate substitutes for cooling applications that will not only be ecological (ODP $= 0$, reduced GWP), nonflammable, and nonpoisonous but also have favorable thermodynamic properties.

8.3 SOLAR THERMOELECTRIC COOLING SYSTEMS

TE cooling systems can be powered directly by a PV panel without the help of AC−DC inverter, which greatly reduces the costs. In solar electric cooling, power produced by the solar PV devices is supplied either to the Peltier cooling systems. It is possible to produce cool by TE processes, using the principle of producing electricity from solar energy through "Seebeck effect" and the principle of producing cool by "Peltier effect" [9]. It has produced such TE coolers, with the principal diagram in Fig. 8.7. TE generator consists of a small number of thermocouples that produce a low TE power but which can easily produce a high electric current. It has the advantage that can operate with a low level heat source and is therefore useful to convert solar energy into electricity. The TE refrigerator (cooler) is also composed of a small number of thermocouples made of two different semiconducting thermoelements through which run the current produced by the generator. The combination of the two parts is compatible with use as TE materials of the semiconductors based on bismuth telluride and antimony telluride alloys (Bi_2Te_3 and Sb_2Te_3) [34−37].

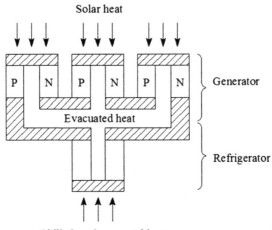

Solar heat

Evacuated heat

Chilled environmental heat

FIGURE 8.7   Schematic of solar thermoelectric (TE) cooling system.

Vella et al. [38] have shown that a TE generator, which draws its heat from solar energy, is a particularly suitable source of electrical power for the operation of a TE refrigerator. They developed the theory of the combined TE generator and refrigerator and determined the ratio of the numbers of thermocouples needed for the two devices. A four-couple TE generator has been used to power a single-couple refrigerator. Temperatures below 0°C have been achieved for a temperature difference across the generator of about 40 K.

The TE cooler is a unique cooling system, in which the electron gas serves as the working fluid. In recent years, concerns of environmental pollution due to the use of CFCs in conventional domestic refrigeration systems have encouraged increasing activities in research and development of domestic refrigerators using Peltier modules. Moreover, recent progress in TE and related fields have led to significant reductions in fabrication costs of Peltier modules and heat exchangers together with moderate improvements in the module performance. Although the COP of a Peltier module is lower than that of conventional compressor unit, efforts have been made to develop domestic TE cooling systems to exploit the advantages associated with this solid-state energy conversion technology [39]. Other applications of this technology are A/C and medical instruments. COP of this system is currently very low, ranging from 0.3 to 0.6.

8.3.1 Thermoelectric Cooler

A conventional TE cooler is composed of a number of N-type and P-type semiconductor junctions connected electrically in series by metallic interconnects (conducting strips, in general made of copper) and thermally in parallel, forming a single-stage cooler [40]. If a low-voltage DC power source is applied to a TE cooler, heat is transferred from one side of the TE cooler to the other side. Therefore, one face of the TE cooler is cooled and the opposite face is heated. Fig. 8.8 depicts a TE cooling module considered as a TE refrigerator, in which the electrical current flows from the N-type element to the P-type element [41]. The temperature T_c of the cold junction decreases and the heat is transferred from the environment to the cold junction at a lower temperature. This process happens when the transport electrons pass from a low-energy level inside the P-type element to a high-energy level inside the N-type element through the cold junction. At the same time, the transport electrons carry the absorbed heat Q_c to the hot junction, which is at temperature T_h. This heat is dissipated (Q_h) in the heat sink, while the electrons return at a lower energy level in the P-type semiconductor (the Peltier effect). If there is a temperature difference between the cold junction and hot junction of N-type and P-type thermoelements, a voltage (called Seebeck voltage) directly proportional to the temperature difference is generated [42,43].

The quality of a TE cooler depends on parameters such as the electric current I applied at the thermocouple of N-type and P-type thermoelements, the temperatures T_h and T_c of the hot and cold sides, the electrical contact resistance R between the cold side and the surface of the device, the thermal and electrical conductivities (λ, σ) of the thermoelement, and the thermal conductance K of the thermoelements in parallel [44].

The characteristics and performance of a TE refrigerator are described by parameters such as the figure of merit, the cooling capacity, and the COP [42].

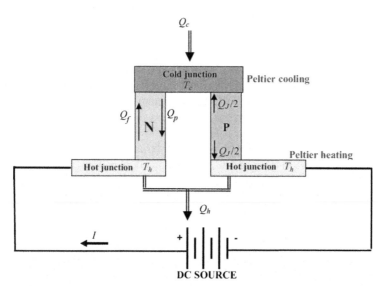

FIGURE 8.8 Schematic of a TE cooler.

8.3.2 Basic Definitions

8.3.2.1 Thermoelectric Figure of Merit

The TE figure of merit Z, in K^{-1}, indicates whether a material is a good TE cooler. It depends on three material parameters: electrical resistivity ρ, in $\Omega \cdot m$ (or electrical conductivity $\sigma = 1/\rho$, in $\Omega^{-1} m^{-1}$), Seebeck coefficient α, in V/K, and total thermal conductivity λ, in W/(m K), between the cold and hot sides.

$$Z = \frac{\alpha^2}{\rho \lambda} = \frac{\alpha^2 \sigma}{\lambda} = \frac{PF}{\lambda_\varphi + \lambda_\varepsilon} \qquad (8.10)$$

where the term $\lambda = \lambda_\varphi + \lambda_\varepsilon$ is the total thermal conductivity, composed of the phonon (or lattice) component λ_φ and the electronic component (electrons and holes) λ_ε; the product $PF = \alpha^2 \sigma$ is called the electrical power factor and depends on the Seebeck coefficient α and on the electrical conductivity σ [45].

Considering the absolute temperature T (which represents the mean temperature between the cold side and hot side of the TE module), a widely used parameter is the dimensionless product ZT as

$$ZT = \frac{\alpha^2 T}{\lambda} = \frac{\alpha^2 \sigma T}{\lambda}, \qquad (8.11)$$

An alternative expression of ZT takes into account the electrical resistance R of the thermoelements in series and the thermal conductance K of the thermoelements in parallel as [46]

$$ZT = \frac{\alpha^2 T}{RK}. \qquad (8.12)$$

The three transport parameters α, σ, and λ depend upon one another as a function of the band structure, carrier concentration, and many other factors. Fig. 8.9 illustrates the three main parameters, including the carrier concentration [46]. In particular, α and σ generally vary in a reciprocal manner making any improvement in the figure of merit Z difficult. In addition, the electrical conductivity and the Seebeck coefficient are inversely related, so it is not generally possible to increase the TE power factor above a particular optimal value for a bulk material [47].

Electrical resistivity ρ is an important material-dependent property that is usually a function of temperature. The value of ρ at room temperature is indicative of whether a material is an insulator (ρ is on the order of 10^6 Ωm or more) or a metal (ρ is on the order of 10^{-6} Ωm or less). In the latter case, if the lattice was perfect, the electron would travel infinitely through it, and the material would only exhibit finite conductivity because of the thermal motion of the lattice and the effect of impurities [48]. The resistivity of a semiconductor material falls between the metal and insulator regimes. It has been found that the optimum range of electrical resistivity for a TE material is from 10^{-3} to 10^{-2} Ωm [49].

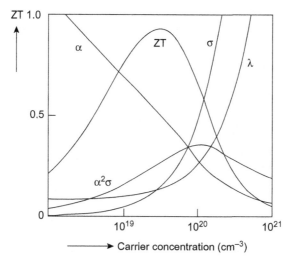

FIGURE 8.9   Influence of carrier concentration on TE figure of merit.

In practice, *ZT* represents the efficiency of the N-type and P-type materials that comprise a thermoelement. A TE material having a higher figure of merit *ZT* is more convenient, as it can carry out higher cooling power or temperature drop. The value of the figure of merit is approximately 1.0 in the TE cooling/heating modules and approximately 0.25 in the A/C applications with efficiency similar to the one of a classical system that uses the R134a refrigerant [8].

The three classes of materials (metals, semiconductors, and insulators) can be characterized by zero, small and large band gaps, respectively. To optimize the materials or compounds for TE applications, some key aspects are the maximization of the figure of merit and the optimization of some parameters of the material. In particular, *ZT* maximization can be achieved through both maximization of the power factor PF and minimization of the thermal conductivity λ as indicated in [50].

Semiconductor materials are promising for the construction of thermocouples because they have Seebeck coefficients in excess of 100 μV/K [51], and the only way to reduce λ without affecting α and σ in bulk materials, thereby increasing the *ZT*, is to use semiconductors of high atomic weight, such as Bi_2Te_3 and its alloys with Sb, Sn, and Pb.

The figure of merit of a semiconductor material limits the temperature difference between the hot and cold junctions, while the length-to-area ratio for an N-type and P-type semiconductor material defines the cooling capacity [52].

8.3.2.2 Cooling Capacity

The standard simplified energy equilibrium model for thermoelement is based on the balance of heat transfer and TE effects. Assumptions have been made that (1) TE material properties are temperature-independent, and (2) half of the Joule heat goes to the hot side while the other half goes to cold side.

The cooling capacity Q_c results from the energy balance at the cold side of the TE cooler as

$$Q_c = \underbrace{\alpha I T_c}_{Q_p} - \underbrace{K \Delta T}_{Q_f} - \underbrace{\frac{1}{2} R I^2}_{\frac{1}{2} Q_j},$$

(8.13)

where Q_p is the Peltier effect generated cooling at the thermoelement cold side, depending on the Seebeck coefficient α, the input current I, and the temperature of the cold junction T_c; Q_f is the heat flux conducted from the hot junction to the cold junction (according Fourier law), which depends on the thermal conductance K of the thermoelements in parallel and the temperature difference between hot and cold sides, $\Delta T = T_h - T_c$; and Q_j is the Joule heat, which depends on the electrical resistance R of the thermoelements in series and the input current I.

An improved simplified model that takes into consideration the Thomson effect is shown in Eq. (8.14) as

$$Q_c = \alpha I T_c - K \Delta T - \frac{1}{2} R I^2 + \frac{1}{2} \tau I \Delta T,$$

(8.14)

where τ is the Thomson coefficient.

In practice, starting from a relatively long thermoelement (e.g., over 1.5 mm) and reducing the thermoelement length, the cooling capacity at first grows until it reaches a maximum value, and then it starts decreasing. Furthermore, for long thermoelements, the contact resistances have a little effect on the cooling capacity, whereas for short thermoelements, the cooling capacity could change considerably by improving the contact resistances.

In the approach proposed in [53], analytical solutions are provided to express the cooling capacity in function of the junction temperature T_j and of the electric current I, by simplifying the thermal balance equations, for a TE device and for a TE module. In the latter case, the expression obtained is as follows:

$$Q_c = \frac{T_j \left(\alpha_m I^2 R_{h,a} - \alpha_m I - K_m \right) + K_m T_a + \left(R_m I^2 / 2 \right) \left(2 K_m R_{h,a} - \alpha_m I R_{h,a} + 1 \right)}{\left(\alpha_m I R_{h,a} - 1 \right) \left(\alpha_m I R_{j,c} - 1 \right) - K_m \left(R_{h,a} + R_{j,c} \right)},$$

(8.15)

where N is the number of thermoelements forming the module; S is the cross-sectional area; T_a is the ambient temperature; $R_{j,c}$ is the junction-to-cooler thermal resistance; $R_{h,a}$ is the hot side to ambient thermal resistance; $R_m = 2N\rho x/S$ is the electrical resistance of the module; $K_m = 2N\lambda S/x$ is the thermal conductance of the module; and $\alpha_m = 2N\alpha$ is the Seebeck coefficient of the module, in which x is the thermoelement length. The expressions obtained are used to solve the optimization of the TE cooler performance without requiring an iterative procedure.

8.3.2.3 Coefficient of Performance

The COP is the ratio between the cooling capacity Q_c and the electrical power consumption P_{el} as

$$COP = \frac{Q_c}{P_{el}}, \tag{8.16}$$

where the electrical power consumption of the TE cooler is given by

$$P_{el} = RI^2 + \alpha I \Delta T. \tag{8.17}$$

The electrical power consumption P_{el} in the thermoelement is the Joule resistance heating plus the work done in driving the current through the thermoelement against the Seebeck voltage caused by the temperature change [54].

The COP values mainly depend on the temperatures at the two sides of the TE element. This fact is well indicated starting from the definition of the (ideal) Carnot COP, here indicated as COP_C, that considers the temperatures of the hot source T_h and of the cold source T_c as

$$COP_C = \frac{1}{(T_h/T_c) - 1} = \frac{T_c}{T_h - T_c}. \tag{8.18}$$

The classical expression of the COP, corresponding to the maximum COP used for sizing the TE element [46,55], can be written by introducing a further relative term COP_r that takes into account the figure of merit of the module Z expressed as in Eq. (8.10) and the average temperature $T_m = (T_h + T_c)/2$, so that

$$COP = COP_C \, COP_r \tag{8.19}$$

with

$$COP_r = \frac{\sqrt{1 + ZT_m} - (T_h/T_c)}{\sqrt{1 + ZT_m} + 1}, \tag{8.20}$$

where ZT_m is the TE material figure of merit at average of hot and cold side temperature, T_m

Fig. 8.10 reports the Carnot COP, the COP, and the term COP_r for a TE device operating at fixed hot side temperature $T_h = 300$ K, in function of the cold side temperature T_c, using different values of Z. The classical COP expression can be simplified, as it neglects the following terms [36]:

- the Thomson effect;
- the dependence on temperature of the characteristics of the materials;
- the effects of the electrical contact resistances; and
- the effects of the thermal resistances.

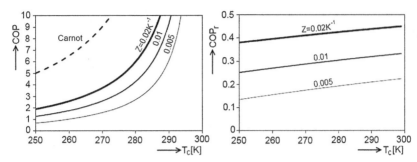

FIGURE 8.10 Variation of COP and of the relative component COP_r depending on T_c for different Z.

Impact of the Thomson effect. The Thomson effect is given by generation or absorption of a heat quantity in a homogeneous conductor, in which an electric current flows and where there is a temperature gradient. The heat absorption is achieved for one sense of electric current flowing through the conductor, and the heat generation is achieved for the reverse sense of electric current. Conventionally, the Thomson effect is considered to be

- positive, when the hot end has a high voltage and the cold end has a low voltage. Heat is generated when the current flows from the hotter end to the colder end, and heat is absorbed when the current flows from the colder end to the hotter end.
- negative, when the hot end has allowed voltage and the cold end has a high voltage. Heat is generated when the current flows from the colder end to the hotter end, and heat is absorbed when the current flows from the hotter end to the colder end. Some metals have negative Thomson coefficients (e.g., Co, Bi, Fe, Hg, etc.).

The representative parameter is the Thomson coefficient τ, according to which the generated heat flux Q is proportional to the thermal gradient ∇T and to the electric current I that flows through the conductor as

$$Q = -\tau I \nabla T. \tag{8.21}$$

For TE elements, the Thomson effect is taken into account in a simplified analytical study in [56], leading to the conclusion that its influence on the COP is approximately 2%.

Dependence on temperature of the characteristics of the materials. The analysis of the dependence on temperature of the characteristics of TE materials is considered in [57] to obtain a suitable design of TE devices able to guarantee a relevant performance.

Effects of the electrical contact resistances. The improved simplified model presented in [46] shows that the COP does not depend on the thermoelement length when the electrical and thermal contact resistances are neglected. By neglecting only the thermal contact resistances and taking into account the electrical resistance of the thermoelement, the figure of merit Z of a single module is written as

$$Z = \frac{zx}{x+n} \qquad (8.22)$$

in which the term n depends on the electrical contact resistivity between the thermoelements and the copper stripes and on the electrical resistivity of the thermoelement materials, while $z = \alpha^2/\rho\lambda$ is the figure of merit of the TE materials employed (which depends on the Seebeck coefficient of the TE materials, the electrical resistivity ρ, and the thermal conductivity λ of the thermoelement) and x is the thermoelement length.

By considering only the electrical contact resistance, the COP of a single TE module has the following expression [46]:

$$COP = COP_C \left(\frac{\sqrt{1 + (zx/(1+n))T_m} - (T_h/T_c)}{\sqrt{1 + (zx/(1+n))T_m} + 1} \right). \qquad (8.23)$$

The COP is a function of temperature difference. In particular, the COP increases when the temperature difference decreases. By considering a TE cooler used in household applications, generally the temperature difference has to reach $25-30$ K to obtain an appropriate cooling effect. In these conditions, the COP of the TE cooler can reach values approximately $0.5-0.7$, namely, approximately 50% of the COP of a vapor compressor cooler [40].

Rowe [40] and Chen et al. [58] show the COP dependence on the current by the following expression:

$$COP = \frac{\alpha_{P,N} I T_c - (RI^2/2) - \lambda\Delta T}{\alpha_{P,N} I \Delta T + RI^2}, \qquad (8.24)$$

where $\alpha_{P,N} = \alpha_P - \alpha_N$, in which α_P and α_N are the absolute Seebeck coefficients of N-type and P-type semiconductors, respectively; Eq. (8.24) emphasizes that the maximum COP does not depend on the number of TE cooler pairs in a module.

Goldsmid [59] proposed the following expression of the electric current corresponding to the maximum COP:

$$I_{COP_{max}} = \frac{(T_h - T_c)\alpha_{P,N}}{(\sqrt{1 + ZT_m} - 1)(R_P + R_N)}, \qquad (8.25)$$

where R_P and R_N are the electrical resistances of the P-type and N-type thermoelement legs, respectively.

8.3.3 Thermoelectric Cooling Modeling

8.3.3.1 One-Dimensional Thermoelement Modeling

Temperature-dependent parameters bring in two complications. One is the temperature-dependent Thomson effect. The other is that the Thomson effect,

Joule heat, and heat conduction have to be considered together since the temperature distribution is no longer linear through the thermoelement. More general one-dimensional transient model thus has been proposed and studied by several researchers [60,61], of which the governing equation is as follows:

$$\rho_e c_p \frac{\partial T(x,t)}{\partial t} = \frac{\partial}{\partial x}\left(\lambda(T)\frac{\partial T(x,t)}{\partial x}\right) + \frac{1}{\sigma(T)}j^2 - j\tau\frac{\partial T(x,t)}{\partial x} - \gamma S(T(x,t) - T_a)$$

(8.26)

in which

$$\gamma = 4\varepsilon\sigma_b T_a^3 + \alpha_c,$$

(8.27)

where ρ_e is the density of thermoelement, in kg/m^3; c_p is the specific heat at constant pressure, in J/(kg K); λ is the thermal conductivity, in W/(m K); σ is the electrical conductivity, in Ω^{-1}m^{-1}; $j = I/S$ is the electrical current density, in A/m^2; S is the cross-sectional area, in m^2; T_a is the ambient temperature, in °C; x is the thermoelement length, in m; t is the time, in s; γ is the combination heat transfer coefficient of radiation and convection; ε is the emissivity; $\sigma_b = 5.67 \times 10^{-8}$ W/(m^{2}K^4) is the Stefan–Boltzmann constant; and α_c is the convective heat transfer coefficient, in W/(m^{2}K).

In Eq. (8.26), terms on the right side represent, respectively, conduction heat transfer rate, Joule heat generation rate, the Thomson effect, and the combined radiation and convection heat transfer rate with ambient. In general, the last term may be negligible compared to the others.

The one-dimensional thermoelement model solves detailed temperature distribution along thermoelement length. During this modeling process, the electrical current density j is treated as uniform, and thus there is no need to solve the electric potential equation. However, in fact, temperature field and electric potential field are coupled in the thermoelement. If more precise numerical result is required, the temperature equation and electric potential equation should be coupled [62] as

$$\frac{\partial}{\partial x}\left(\sigma\left(\frac{\partial\phi}{\partial x} - \alpha\frac{\partial T}{\partial x}\right)\right) = 0$$

(8.28)

$$E = \frac{\partial\phi}{\partial x} + \alpha\frac{\partial T}{\partial x}$$

(8.29)

$$j = \sigma E,$$

(8.30)

where α is the Seebeck coefficient, in V/K, and ϕ is the electric scalar potential, in V.

Eq. (8.28) is the one-dimensional electric potential field equation. Once the electric potential ϕ is obtained, the electric field E and electric current density j can be calculated.

Fraisse et al. [63] presented a comparison study for four different modeling approaches applied for thermoelement, including standard simplified models, analytical models, model based on electrical analogy between heat transfer and electricity [64], and numerical models based on the finite element method. Selection between different models is highly depending on the modeling goal aimed as a trade-off exists between the modeling accuracy and the computational efforts.

8.3.3.2 Numerical Modeling of Thermoelectric Coolers

Since TE cooler consists of thermoelements, thus for TE cooler modeling, it is reasonable to numerically model every thermoelements in a TE cooler.

Instead of modeling each thermoelement individually, modeling the TE cooler as a single bulk is a much easier approach. This kind of the so-called compact models can handle the multiscale issue using fine mesh and coarse mesh at different regions respectively. Chen and Snyder [65] developed a compact modeling approach for TE coolers. It is demonstrated that with the results almost as accurate as the physical model (numerical study includes all coupled TE as well as components that provide losses and other parasitic effects), a significant amount of grid has been reduced and computational speed is roughly 100 times faster. The critical technique for compact modeling is to determine effective Seebeck coefficient and thermal and electrical conductivities for the compact TE cooler.

8.3.4 Thermoelectric Cooling Applications

TE cooling has various applications for cooling electronic devices and PV panels and for coolers and A/Cs in the households. A schematic view of a TE cooling application is shown in Fig. 8.11, in which the TE module is mounted on the top of the compartment to be cooled, and an additional heat exchanger with fins is placed above the hot side. The system can be integrated with additional forced air circulation through a fan located above the finned heat exchanger. Heat exchanger and fan can also be located inside the cooled compartment.

8.3.4.1 Cooling the Electronic Devices

In electronic cooling applications, a TE cooler serves as a device for transporting heat away from a surface that has a temperature higher than the ambient temperature.

Electronic devices such as PC processors generate a huge amount of heat during operation that poses great thermal management challenge because reliable operation temperature for electronic devices has to be maintained. In most cases, the maximum electronic device junction temperature needs to be held less than 85°C for reliable operation. The maximum heat flow from a high

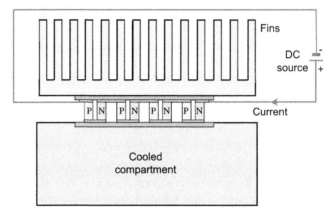

FIGURE 8.11 Schematic of a TE cooling application.

performance electronic package can be about 200 W and is still constantly increasing [66]. Conventional passive cooling technologies using air or water as working fluid, such as the microchannel sink, cannot fully meet the heat dissipation requirement and active cooling methods should be applied. Due to the limited installation space in electronic packages, conventional bulk cooling systems are too big. TE coolers combined with air cooling or liquid cooling approaches at the hot side show great potential because of their small size, high reliability, and no noise. Phelan et al. [67] reviewed current and future miniature refrigeration cooling technologies for high-power microelectronics and concluded that only TE coolers are now commercially available in small sizes.

Zhou and Yu [68] conducted detailed analyses of the optimal allocation of the finite thermal conductance between the cold side and hot side heat exchangers of a TE cooling system. The analysis results when the constraint of the total thermal conductance was considered to demonstrate that the maximum COP can exceed 1.5 when the finite total thermal conductance is optimally allocated. However, overall, the efficiency of the hot-side heat exchanger parameters is usually the predominant factor in determining the overall performance of a TE cooling system.

8.3.4.2 Thermoelectric Refrigeration

In refrigeration applications for which cost is not the main criterion, Peltier cooling appliances provide rapid cooling.

Generally, there are two types of TE refrigeration devices: domestic and portable refrigerators. The major difference between these two is the availability of electrical power. Both domestic [43,69] and portable [70,71] refrigerators have been extensively studied. Although the thermodynamic efficiency of TE cooler is only 1% compared to 14% of Stirling and reciprocating vapor-compression refrigeration systems [72], TE refrigerators offer advantages such as a more ecological system, more silent and robust, and more precise in temperature control [73]. TE refrigerators can be built into a limited space unit. Portable TE coolers have promising outdoor use, either using battery or powering by solar PV panels. The TE refrigerators reported in literature [37] are summarized in Table 8.4, where ΔT is the inside/outside temperature difference of the refrigerators.

One attractive application is for outdoor purposes, when operated in tandem with solar PV panels, as shown in Fig. 8.12. In the daytime, solar panels receive solar energy and turn it into electric power supplied to the TE refrigerator by means of the PV effect. If the amount of electric power produced is sufficient, the power surplus can be accumulated in a storage battery in addition to driving the refrigerator. If the solar panels cannot produce sufficient electric power, for example, on cloudy or rainy days, the storage battery may serve as a supplementary power source, and it can be used to power the refrigerator at night.

The COPs of present commercial TE refrigerators are typically between 0.3 and 0.7 for single-stage applications. Moreover, COPs greater than 1.0 can be achieved when the module is removing heat from an object that is warmer than the ambient temperature.

8.3.4.3 Thermoelectric Air Conditioners

TE A/Cs are environmentally friendly, simple, and reliable. They offer convenient installation and support complex water distribution pipes, and switching between the cooling and heating modes can be easily achieved by reversing the input current. However, these systems are still very expensive at present.

For A/C applications, Riffat and Qiu [74] report a comparative analysis of three domestic A/C devices (compression A/C, absorption A/C, and TE A/C). According to their study, the most efficient one is the vapor-compression A/C having the COP of 2.6–3.0, followed by absorption A/C with the COP of 0.6–0.7 (single-effect absorption) and TE with the COP of 0.38–0.45. However, the potential interest of using TE A/Cs depends on the fact that they work with DC input and as such can be directly connected to sources like PV cells, fuel cells, and in perspective batteries from plug-in electric vehicles. Further advantages of TE A/Cs are that their use can reduce the issues due to noise and adoption of ozone-depleting CFCs in vapor-compression A/Cs.

TE A/Cs are portable and low noise, but they have a relatively low COP, which is an additional factor that limits their application for domestic cooling. TE A/Cs, however, have a large potential market as A/Cs for small enclosures, such as cars and submarine cabins, where the power consumption would be low or safety and reliability would be important [74].

TABLE 8.4 Summary of Thermoelectric Refrigerators Reported in Literature

Dimension (m) or Volume (m³)	ΔT (°C)	COP (−)	Cooling Capacity (W)	Heat Sink Techniques	Refs.
0.23 × 0.18 × 0.32	22.0	0.16	15.3	Finned heat sink and fan at the hot side	[71]
0.115	20.0	0.30–0.50	50–100	Finned heat sink and fan in the cold side, liquid heat exchanger at hot side	[43]
—	20.0	0.23	12.0	Finned heat sink in the hot side	[70]
0.225	10.0	0.39	19.4	Apply phase change thermosiphon system in both hot and cold sides	[69]
0.055	23.9	0.56–0.64	30.0	Finned heat sink and fan at both hot and cold sides	[73]
0.056	19.8	0.20	12.5	Finned heat sink and fan at both hot and cold sides	[72]

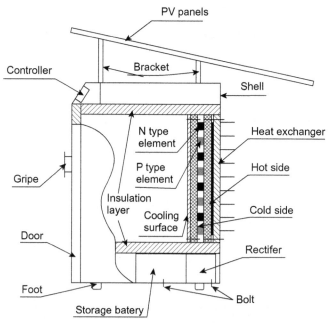

FIGURE 8.12   Schematic of solar PV-driven TE refrigerator.

A solar TE-cooled ceiling (STCC) combined with displacement ventilation system has been developed for space climate control, as presented in Fig. 8.13 [75]. The STCC adopts TE cooler instead of hydronic panels as radiant panels. The STCC is burdened with removal of a large fraction of sensible cooling load. The TE modules are connected in series and sandwiched between the aluminum radiant panel and heat pipe sinks in STCC.

The heat sinks are used for dissipating the heat of TE modules. The fan can provide forced air convection to help the TE modules to release heat more

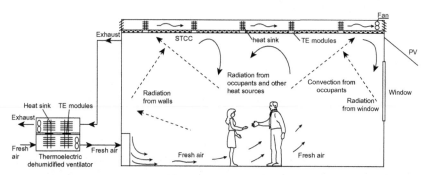

FIGURE 8.13   Schematic of solar thermoelectric cooled ceiling combined with displacement ventilation system.

efficiently into the atmosphere. By controlling the direction of the current, the functions of cooling and heating can be both achieved.

The performance of the TE cooled ceiling was investigated under both the cooling and heating modes. Results indicated that increasing the operating voltage increased the total heat flux. Decreasing the temperature difference between ambient temperature and indoor temperature significantly increased the total heat flux and slightly increased the COP_{sys} in both the cooling and heating modes. The total heat flux of the STCC system in cooling mode was higher than 60 W/m^2 and the COP_{sys} could reach 0.9 under operating voltage of 5 V. In the heating mode, the total heat flux of the STCC system under operating voltage of 4 V was over 110 W/m^2 and the COP of the system could reach 1.9 [75].

TE A/C with heat storage system has been developed as shown in Fig. 8.14. The TE cooling system primarily consists of a TE cooling unit, a shell-and-tube phase change material (PCM) heat storage unit, an air—water heat exchanger, and a piping system. Heat absorbed from the indoor environment through the TE cooling unit can be released through the air—water heat exchanger with water as the heat transfer fluid. The system can realize two operating modes, which are dissipating generated heat to outdoor air through the air—water heat exchanger (Mode 1) and releasing heat to the shell-and-tube PCM heat storage unit (Mode 2). The two modes can be easily switched over through manually controlling valves [76].

The working principle is as follows: if outdoor air temperature is relatively low, such as in the early morning or late afternoon, the working Mode 1 will be in operation and heat generated by space cooling will be dissipated to the outdoor environment. When outdoor air temperature is high, the PCM heat storage unit will be activated and the system will convert to Mode 2. At night, the PCM heat storage unit will be discharged by using relatively cool outdoor air. Therefore, PCM with appropriate melting temperature suitable for local weather conditions would be preferred for its advantage of using "free cooling" at night to "regenerate" the PCM.

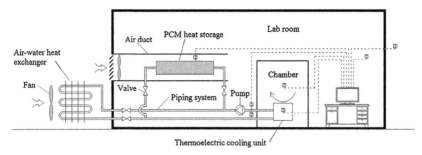

FIGURE 8.14   Schematic of the prototype TE cooling system.

The TE module used in this study was RC12-8. PCM was RT22, which had a melting temperature of 19−23°C (with main peak at 22°C and heat storage capacity of 200 kJ/kg).

The experiment results showed that the average COP of the TE A/Cs was 0.8, and the maximum COP value was 1.22. The maximum cooling capacity achieved 210 W. Comparison experimental study showed that 35.3% electrical energy could be saved in the prototype TE cooling system by using PCM heat storage on the condition that outdoor air temperature was in the range of 30−33°C and temperature of the conditioned space was set at 24°C [76].

8.3.5 Conclusion

The performance of solar TE cooling system can be improved by selecting TE and PV systems with higher performances. The figure of merit ZT value of TE module is only approximately 0.7 at present. If the ZT factor of the TE material could be further improved and increased to, for example, 2.0 in the next decade, the COP value and energy saving of this TE cooling system would be further increased. In the near future, solar TE cooling system will make a significant contribution, especially in zero-energy buildings, in reducing fossil fuel consumption and carbon emissions.

REFERENCES

[1] Xi H, Luo L, Fraisse G. Development and applications of solar-based thermoelectric technologies. Renewable and Sustainable Energy Reviews 2007;11:923−36.

[2] Thirugnanasambandam M, Iniyan S, Goic R. A review of solar thermal technologies. Renewable and Sustainable Energy Reviews 2010;14:312−22.

[3] Afshar O, Saidur R, Hasanuzzaman M, Jameel M. A review of thermodynamics and heat transfer in solar refrigeration system. Renewable and Sustainable Energy Reviews 2012;16:5639−48.

[4] Kalkan N, Young EA, Celiktas A. Solar thermal air conditioning technology reducing the footprint of solar thermal air conditioning. Renewable and Sustainable Energy Reviews 2012;16:6352−83.

[5] Solangi KH, Islam MR, Saidur R, Rahim NA, Fayaz H. A review on global solar energy policy. Renewable and Sustainable Energy Reviews 2011;15:2149−63.

[6] Tsubota T, Ohno T, Shiraishi N, Miyazaki Y. Thermoelectric properties of $Sn_{1-x-y}Ti_ySb_xO_2$ ceramics. Journal of Alloys and Compounds 2008;463:288−93.

[7] Lineykin S, Ben-Yaakov S. Modeling and analysis of thermoelectric modules. IEEE Transactions on Industry Applications 2007;43(2):505−12.

[8] Bell LE. Cooling, heating, generating power, and recovering waste heat with thermoelectric systems. Science 2008;321:1457−61.

[9] Riffat SB, Ma X. Thermoelectrics: a review of present and potential applications. Applied Thermal Engineering 2003;23:913−35.

[10] Poese ME, Smith RW, Garett SL, van Gerwen R, Gosselin P. Thermoacoustic refrigeration for ice cream sales. In: Proceedings of 6th Gustav Lorentzen natural working fluids conference, Glasgow, Scotland; 2004.

[11]   Sarbu I, Sebarchievici C. Review of solar refrigeration and cooling systems. Energy and Buildings 2013;67(12):286−97.

[12]   Cengel Y, Boles M. Thermodynamics: an engineering approach. New York: McGraw Hill; 2014.

[13]   ARAP Web Site. Alliance for responsible atmospheric policy, <http://www.arap.org>.

[14]   Richard LP. CFC phase out; have we met the challenge. Journal of Fluorine Chemistry 2002;114:237−50.

[15]   Johnson E. Global warming from HFC. Environmental Impact Assessment Review 1998;18:485−92.

[16]   ASHRAE Standard 34. Designation and safety classification of refrigerants. Atlanta, GA: American Society of Heating, Refrigerating and Air-Conditioning; 2007.

[17]   ASHRAE handbook. Fundamentals. Atlanta, GA: American Society of Heating, Refrigerating and Air-Conditioning; 2013.

[18]   Calm JM, Hourahan GC. Refrigerant data summary. Engineering Systems 2001;18:74−88.

[19]   Palm B. Hydrocarbons as refrigerants in small heat pump and refrigeration systems − a review. International Journal of Refrigeration 2008;31:552−63.

[20]   Minor B, Spatz M. HFO-1234yf low GWP refrigerant update. In: International Refrigeration and Air-Conditioning Conference at Purdue, West Lafayette, IN, USA, paper no. 1349; 2008.

[21]   Aprea C, Greco A, Maiorino A. An experimental evaluation of the greenhouse effect in the substitution of R134a with CO_2. Energy 2012;45:753−61.

[22]   Zhao Y, Liang Y, Sun Y, Chen J. Development of a mini-channel evaporator model using R1234yf as working fluid. International Journal of Refrigeration 2012;35:2166−78.

[23]   Subiantoro A, Ooi KT. Economic analysis of the application of expanders in medium scale air-conditioners with conventional refrigerants, R1234yf and CO_2. International Journal of Refrigeration 2013;36:1472−82.

[24]   Agrawal AB, Shrivastava V. Retrofitting of vapour compression refrigeration trainer by an eco-friendly refrigerant. Indian Journal of Science and Technology 2010;3(4):6837−46.

[25]   Wright B. Environment forum. New York, NY: Carrier Air Conditioning Company; 1992.

[26]   Honeywell <http://www.1234facts.com>.

[27]   Hwang Y, Ohadi M, Rademacher R. Natural refrigerants. Mechanical Engineering 1998;120:96−9.

[28]   Lorentzen G. Ammonia: an excellent alternative. International Journal of Refrigeration 1988;11:248−52.

[29]   Lorentzen G. Revival of carbon dioxide as a refrigerant. International Journal of Refrigeration 1994;17(5):292−301.

[30]   Didion DA, Bivens DB. Role of refrigerant mixtures as alternatives to CFCs. International Journal of Refrigeration 1990;13:163−75.

[31]   Pavkovic B. Refrigerant − Properties and air-conditioning applications. Rehva Journal 2013;50(5):7−11.

[32]   Sarbu I. A review on substitution strategy of non-ecological refrigerants from vapour compression-based refrigeration, air-conditioning and heat pump systems. International Journal of Refrigeration 2014;46(10):123−41.

[33]   Lemmon EW, Huber ML, McLinden MO. REFPROP Reference fluid thermodynamic and transport properties. NIST Standard Reference Database 23 Version 9.1. US Secretary of Commerce; 2013.

[34]   Sarbu I, Valea E, Sebarchievici C. Solar refrigerating systems. In: Material researches and energy engineering. Advanced material research, 772; 2013. p. 581−6.

[35] Elsheikh MH, Shnawah DA, Sabri MFM, Said SBM, Hassan MH, Ali Bashir MB, et al. A review on thermoelectric renewable energy: principle parameters that affect their performance. Renewable and Sustainable Energy Reviews 2014;30:337−55.

[36] Enescu D, Virjoghe EO. A review on thermoelectric cooling parameters and performance. Renewable and Sustainable Energy Reviews 2014;38:903−16.

[37] Zhao D, Tan G. A review of thermoelectric cooling: materials, modeling and applications. Applied Thermal Engineering 2014;66:15−24.

[38] Vella GJ, Harris LB, Goldsmid HJ. A solar thermoelectric refrigerator. Solar Energy 1976;18(4):355−9.

[39] Saidur R, Masjuki H, Mahlia M, Tan C, Ooi J, Hasanuzzaman M, et al. Performance investigation of a solar powered thermoelectric refrigerator. International Journal of Mechanical and Materials Engineering 2008;3:7−16.

[40] Rowe DM. Thermoelectrics handbook − macro to nano. Boca Raton (FL): CRC Press Taylor&Francis; 2006.

[41] Riffat SB, Ma XL. Improving the coefficient of performance of thermoelectric cooling systems: a review. International Journal of Energy Research 2004;28(9):753−68.

[42] Min G, Rowe DM. Peltier device as a generator. In: CRC handbook of thermoelectric. New York: CRC Press; 1995.

[43] Min G, Rowe DM. Experimental evaluation of prototype thermoelectric domestic-refrigerators. Applied Energy 2006;83:133−52.

[44] Chein R, Huang G. Thermoelectric cooler application in electronic cooling. Applied Thermal Engineering 2004;24(14−15):2207−17.

[45] Snyder GJ, Toberer ES. Complex thermoelectric materials. Nature Materials 2008;7:105−14.

[46] Min G, Rowe DM. Improved model for calculating the coefficient of performance of a Peltier module. Energy Conversion and Management 2000;2(41):163−71.

[47] Zide J, Vashaee D, Bian Z, Zeng G, Bowers J, Shakouri A, et al. Demonstration of electron filtering to increase the Seebeck coefficient in $In_{0.53}Ga_{0.47}As/In_{0.53}Ga_{0.28}Al_{0.19}As$ superlattices. Physical Review B 2006:74.

[48] Bulusu A, Walker DG. Review of electronic transport models for thermoelectric materials. Superlattices and Microstructures 2008;44:1−36.

[49] Cadoff IB, Miller E. Thermoelectric materials and devices. New York: Reinhold Publishing Corporation; 1960.

[50] Zlatic V, Hewson AC. Properties and applications of thermoelectric materials. The search for new materials for thermoelectric devices. In: Proceedings of the NATO advanced research workshop on properties and application of thermoelectric materials B: physics and biophysics. Berlin: Springer; 2008.

[51] Dughaish ZH. Lead telluride as a thermoelectric material for thermoelectric power generation. Physica B 2005;322:205−23.

[52] Rowe DM. Handbook of thermoelectric − Introduction. Boca Raton (FL): CRC Press; 1995.

[53] Zhang HY. A general approach in evaluating and optimizing thermoelectric coolers. International Journal of Refrigeration 2010;33(6):1187−96.

[54] Parrot JE, Penn AW. The design theory of thermoelectric cooling elements and units. Solid-State Electronics 1961;3:91−9.

[55] Goldsmid JH. Introduction to thermoelectricity. In: Series in material science. Berlin Heidelberg: Springer; 2010.

[56] Chen J, Yan Z, Wu L. Non equilibrium thermodynamic analysis of a thermoelectric device. Energy 1997;22(10):979−85.

[57] Yamashita O. Effect of linear and non-linear components in the temperature dependences of thermoelectric properties on the cooling performance. Applied Energy 2009;86:1746−56.

[58] Chen WH, Liao CY, Hung CI. A numerical study on the performance of miniature thermoelectric cooler affected by Thomson effect. Applied Energy 2012;89:464—73.

[59] Goldsmid JH. Electronic refrigeration. London: Pion; 1986.

[60] Huang M-J, Yen R-H, Wang A-B. The influence of the Thomson effect on the performance of a thermoelectric cooler. International Journal of Heat and Mass Transfer 2005;48:413—8.

[61] Reddy BVK, Barry M, Li J, Chyu MK. Mathematical modeling and numerical characterization of composite thermoelectric devices. International Journal of Thermal Sciences 2013;67:53—63.

[62] Perez-Aparicio JL, Palma R, Taylor RL. Finite element analysis and material sensitivity of Peltier thermoelectric cells coolers. International Journal of Heat and Mass Transfer 2012;55:1363—74.

[63] Fraisse G, Ramousse J, Sgorlon D, Goupil C. Comparison of different modeling approaches for thermoelectric elements. Energy Conversion and Management 2013;65:351—6.

[64] Fraisse G, Lazard M, Goupil C, Serrat JY. Study of a thermoelement's behavior through a modeling based on electrical analogy. International Journal of Heat and Mass Transfer 2010;53:3503—12.

[65] Chen GM, Snyder J. Analytical and numerical parameter extraction for compact modeling of thermoelectric coolers. International Journal of Heat and Mass Transfer 2013;60:689—99.

[66] Zhang HY, Mui YC, Tarin M. Analysis of thermoelectric cooler performance for high power electronic packages. Applied Thermal Engineering 2010;30:561—8.

[67] Phelan PE, Chiriac VA, Lee T-YT. Current and future miniature refrigeration cooling technologies for high power microelectronics. IEEE Transactions on Components and Packaging Technologies 2002;25(3):356—65.

[68] Zhou Y, Yu J. Design optimization of thermoelectric cooling systems for applications in electronic devices. International Journal of Refrigeration 2012;35:1139—44.

[69] Vian JG, Astrain D. Development of a thermoelectric refrigerator with two phase thermosiphons and capillary lift. Applied Thermal Engineering 2009;29:1935—40.

[70] Dai YJ, Wang RZ, Ni L. Experimental investigation and analysis on a thermoelectric refrigerator driven by solar cells. Solar Energy Materials and Solar Cells 2003;77:377—91.

[71] Abdul-Wahab SA, Elkamel A, Al-Damkhi AM, Al-Habsi IA, Al-Rubai'ey HS, Al-Battashi AK, et al. Design and experimental investigation of portable solar thermoelectric refrigerator. Renewable Energy 2009;34:30—4.

[72] Hermes CJL, Barbosa JR. Thermodynamic comparison of Peltier, Stirling, and vapor compression portable coolers. Applied Energy 2012;91:51—8.

[73] Astrain D, Vian JG, Albizua J. Computational model for refrigerators based on Peltier effect application. Applied Thermal Engineering 2005;25:3149—62.

[74] Riffat SB, Qiu G. Comparative investigation of thermoelectric air-conditioners versus vapour compression and absorption air-conditioners. Applied Thermal Engineering 2004;24(14—15):1979—93.

[75] Liu ZB, Zhang L, Gong GC. Experimental evaluation of a solar thermoelectric cooled ceiling combined with displacement ventilation system. Energy Conversion and Management 2014;87:559—65.

[76] Zhao D, Tan G. Experimental evaluation of a prototype thermoelectric system integrated with PCM (phase change material) for space cooling. Energy 2014;68:658—66.

Chapter 9

Solar-Assisted Heat Pumps

9.1 GENERALITIES

Concerning the use of high-efficiency heating/cooling systems and the integration of renewable energy sources (RES), the heat pump (HP) is one of the most advantageous systems to be considered in a heating, ventilating, and air-conditioning plant.

The amount of ambient energy E_{res} captured by HPs to be considered as renewable energy shall be calculated in accordance with the following equation [1]:

$$E_{res} = E_U \left(1 - \frac{1}{SPF} \right), \qquad (9.1)$$

where E_U is the estimated total usable thermal energy delivered by HPs and SPF is the estimated seasonal performance factor (SPF) for these HPs.

Only HPs for which SPF $> 1.15/\eta$ shall be taken into account, where η is the ratio between the total gross production of electricity and the primary energy consumption for electricity production. For European Union countries, the average η is 0.4, meaning that the minimum value of the SPF should be 2.875.

As it is well known, designing an HP system needs particular care concerning the selection of both the heating system (in order to lower heat supply temperature) and the heat source (external ambient air is the most diffused but the worst from thermodynamic point of view as the buildings' heating loads generally increase as air temperature decreases) [2]. In particular, this second aspect should be carefully evaluated when designing an HP system as the potential advantages of alternative heat sources could be significant. This is why there is an increasing interest in dual source systems during the last decades. The idea of utilizing different RES for an HP at a single-family house is presented in Fig. 9.1.

The main idea in dual source systems is that the HP absorbs heat by two heat sources. Two arrangements widely studied in literature are air-source HP (ASHP)/solar collectors and ground-source HP (GSHP)/solar collectors. The three typical configurations for the operation of such systems are "in series" (the two sources are aligned in series so that the former raises the temperature

Solar Heating and Cooling Systems. http://dx.doi.org/10.1016/B978-0-12-811662-3.00009-8

347

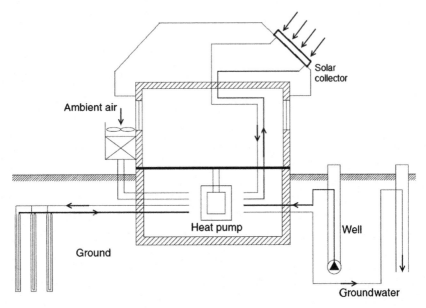

FIGURE 9.1   Different RES for a heat pump.

before that heat is taken from the latter), "in parallel" (solar energy is used to supply the heating load directly) or "dual source" (the HP takes heat choosing time by time the most favorable source from the thermodynamic point of view) [3].

Solar energy has the characteristics of intermittence and low density, which largely restrict the application of solar heating. The solar-assisted heat pump (SAHP) heating system, which combines HP technology with solar heating technology, can solve the intermittent problem of solar energy.

HP systems can extract low-grade thermal energy from the environment and waste heat for use in water/space heating applications. On the other hand, photovoltaic/thermal (PV/T) solar collectors that simultaneously produce electricity and heat are currently considered the most efficient devices to harness the available solar energy. Waste heat recovery from solar cells for the evaporator of SAHP for water/space heating would mutually improve the PV/T collector efficiency and the COP of SAHPs.

Thygesen and Karlsson [4] simulated and analyzed three different SAHP systems in Swedish near-zero energy single-family houses. The analyzed systems were: a PV-system and an HP, an HP and a solar thermal system and an HP, a PV-system and a solar thermal system. The conclusion was that a PV system in combination with an HP was a superior alternative to a solar thermal system in combination with an HP.

This chapter presents the operation principle of a HP, discusses the vapor-compression-based HP systems, and describes the thermodynamic cycle and they calculation, as well as operation regimes of a vapor-compression HP with electro-compressor. The calculation of greenhouse gas (GHG) emissions of HPs and energy and economic performance criteria that allow for implementing an HP in a heating/cooling system is considered. A detailed description of the HP types and GSHP development is presented and important information on the selection of the heat source and HP systems are discussed. Additionally, other approach is to integrate the solar thermal system on the source side of the HP so that the solar thermal energy is either the sole heat source for the HP or provides supplementary heat. Additionally, the operation principle and calculation of the thermodynamic cycle for a solar-assisted absorption HP are also briefly analyzed. Finally, analytical and experimental studies are performed on a direct-expansion solar-assisted heat pump (DX-SAHP) water heating system. The effect of various parameters, including solar radiation, ambient air temperature, collector area, storage volume and speed of compressor, have been investigated on the thermal performance of the DX-SAHP system. A novel heating, ventilating, and air-conditioning system consisting of a solar-assisted absorption ground-coupled HP is also described and some of the influence parameters on its energy efficiency are analyzed. A model of the experimental installation is developed using the TRNSYS software and validated with experimental results obtained in the installation for its cooling-mode operation.

9.2 OPERATION PRINCIPLE OF A HEAT PUMP

An HP is a thermal installation that is based on a reverse Carnot thermodynamic cycle, which consumes drive energy and produces a thermal effect.

Any HP moves (pumps) heat E_S from a source with low temperature t_s to a source with a high temperature t_u, consuming the drive energy E_D.

- A heat source can be
  - a gas or air (outdoor air, warm air from ventilation, or hot gases from industrial processes);
  - a liquid called "generic water": surface water (river, lake, or sea), groundwater, or discharged hot water (domestic, technologic, or recirculated in cooling towers); or
  - ground, with the advantage of accessibility.
- *Heat consumer.* The heat pump yields thermal energy at a higher temperature, depending on the application of the heat consumer. This energy can be used for
  - space heating, which is related to low-temperature heating systems: radiant panels (floor, wall, ceiling, or floor-ceiling), warm air, or convective systems; or
  - water heating (pools, domestic, or technologic hot water).

The heat consumer is recommended to be associated with a cold consumer. This can be performed with either a reversible (heating−cooling) or a double-effect system. In cooling mode, an HP operates exactly like a central air-conditioner (A/C).

- *Drive energy.* HPs can be used to drive different energy forms as
  - electrical energy (electro-compressor);
  - mechanical energy (mechanical compression with expansion turbines);
  - thermomechanical energy (steam ejector system);
  - thermal energy (absorption cycle); or
  - thermo-electrical energy (Peltier effect).

The most used HP systems are electrically driven (vapor compression-based) and absorption HPs.

9.3 VAPOR COMPRESSION-BASED HEAT PUMP SYSTEMS

The air- and GSHPs are those with electro-compressors. The process of elevating low-temperature heat to over 38°C and transferring it indoors involves a cycle of evaporation, compression, condensation, and expansion (Fig. 9.2A). A non-chlorofluorocarbon refrigerant is used as the heat-transfer medium, which circulates within the HP.

9.3.1 Thermodynamic Cycle

The basic vapor-compression cycle is considered to be the one with isentropic compression and subcooling of liquid, and with no superheat of vapor (Fig. 9.2B).

Operational processes are outlined next:

$1-2$: isentropic compression in the compressor K, which leads to increased pressure and temperature from the values corresponding for evaporation p_e, t_e to those of the condensation p_c, $t_2 > t_c$;

$2-2'$: isobar cooling in the condenser C at pressure p_c from the temperature t_2 to $t_{2'} = t_c$;

$2'-3$: isothermisobar condensation in the condenser C at pressure p_c and temperature t_c;

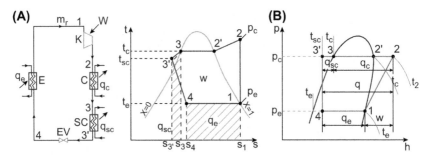

FIGURE 9.2   Schematic of an HP (A) and vapor-compression processes in t-s and p-h diagrams (B).

3–3': isobar subcooling in the subcooler SC at pressure p_c from the temperature t_c to $t_{sc} < t_c$;

3'–4: isenthalpic lamination in expansion valve EV, leading the refrigerant from the three states of the liquid at p_c, t_c in the four states of the wet vapor at p_e, t_e;

4–1: isothermisobar evaporation in the evaporator E at pressure p_e and temperature t_e.

The specific compression work w, in kJ/kg, the specific cooling power q_e, in kJ/kg, the specific heat load at condensation q_c, in kJ/kg, the specific subcooling power q_{sc}, in kJ/kg, volumetric refrigerating capacity q_{ev}, in kJ/m^3, the coefficient of performance COP are calculated for above presented processes as follows:

$$w = h_2 - h_1 \tag{9.2}$$

$$q_e = h_1 - h_4 = h_1 - h_{3'} \tag{9.3}$$

$$q_c = h_2 - h_3 \tag{9.4}$$

$$q_{sc} = h_3 - h_{3'} \tag{9.5}$$

$$q = q_c + q_{sc} = h_2 - h_{3'} \tag{9.6}$$

$$q_{ev} = \frac{q_e}{v_1} = q_e \rho_1 \tag{9.7}$$

$$COP = \frac{q}{w} = \frac{h_2 - h_{3'}}{h_2 - h_1} \tag{9.8}$$

Thermal power (capacity) of heat pump Q_{hp}, in kW, is expressed as

$$Q_{hp} = m_r(q_c + q_{sc}), \tag{9.9}$$

where m_r is the mass flow rate of refrigerant, in kg/s.

The power necessary for the isentropic compression P_{is}, in kW, may be calculated using the equation

$$P_{is} = m_r w. \tag{9.10}$$

The effective electrical power P_{el} on the compressor shaft is larger and is defined as

$$P_{el} = \frac{P_{is}}{\eta_{is}}, \tag{9.11}$$

where η_{is} is the isentropic efficiency.

The operational scheme of a solar PV-powered HP system is similarly to the operational scheme of a solar electric compression air-conditioner illustrated in Fig. 8.3.

9.3.2 Operation Regimes of a Heat Pump

The operation regime of an HP is adapted to the existing heating system of buildings. If the supply temperature is higher than the maximum supply temperature of the HP (55°C), then the heat pump will only operate in addition to traditional sources of heat. In new buildings, a distribution system should be selected with a maximum supply temperature of 35°C.

The following operating regimes are described next:

- *Monovalent regime*: For the univalent regime, the HP system meets the entire heat demand of the building at all times. The distribution system should be designed for a supply temperature below the maximum supply temperature of the HP. This operation regime is well suited for applications with supply temperatures of up to 65°C. Systems with groundwater or ground heat-source collectors are operated as monovalent systems.
- *Bivalent regime*: A bivalent heating plant (Fig. 9.3) has two sources of heat. An HP with electrical action is combined with at least one heat source for solid, liquid or gaseous fuels, or solar source. This regime can be bivalent-parallel (HP operates simultaneously with another heat source) or bivalent-alternate (usage of either the HP system or the other heat source).
  - *Bivalent-parallel*. The HP heats independently to a certain set point, at which an auxiliary heating system (electric element or boiler) is turned on and the two systems operate in parallel to meet the heating demand for a maximum supply temperature of up to 65°C. This operation is used mainly with new air source systems or in renovations of old buildings.

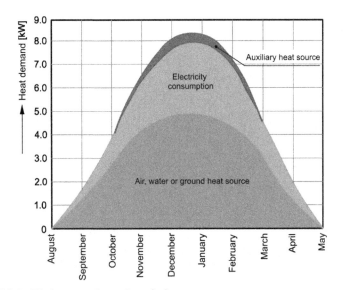

FIGURE 9.3 Bivalent operation regime of a heat pump.

- *Bivalent-alternate.* The HP heats independently to a certain set point. Once this point is reached, a boiler meets the full heating demand. This operation is suitable for supply temperatures of up to 90°C and is typically installed in renovated buildings.
- *Monoenergetic regime* is a bivalent operation regime in which the second heat source (auxiliary source) functions with the same type of energy (electricity) as the HP.

To make an economic operation of a heating system with HPs possible, in some countries, the electricity supplier provides special electricity tariffs for HPs. These prices usually assume that the electricity supply for HPs can be interrupted when the network is overloaded. For example, electricity supply for HP systems with the univalent operation regime may be discontinued three times in 24 h for more than 2 h. The operating time between two interruptions should not be less than the previous interruption. In the case of HP systems with bivalent operation, the electricity supply may be interrupted during the heating period for up to 960 h.

For existing buildings, the bivalent operation regime is recommended because a heat source exists, which can usually be used to further cover the peak loads of cold winter days with required supply temperatures of over 55°C.

For new buildings, the univalent operation regime has proven useful because it may be interrupted. The HP can cover the annual heat demand, and the periods of interruption do not lead to disturbances in operation because, for example, floor-heating interruption may not cause changes in the comfort temperature.

9.3.3 Performance and CO_2 Emission of HP

9.3.3.1 Coefficient of Performance

The operation of an HP is characterized by the coefficient of performance (COP) defined as the ratio between useful thermal energy E_t and electrical energy consumption E_{el} as

$$COP = \frac{E_t}{E_{el}}. \tag{9.12}$$

The overall COP of a solar-assisted HP (SAHP) system is given by the following equation:

$$COP_{sys} = \frac{E_t}{E_s}, \tag{9.13}$$

where E_s is the solar thermal energy received by the solar collector surface.

Seasonal coefficient of performance ($COP_{seasonal}$) or average COP over a heating (cooling) season, often indicated as the SPF or annual efficiency, is obtained if in Eq. (9.12) is used in summation of both usable energy and consumed energy during a season (year).

In the heating operate mode, the COP of HP is defined by the following equation as

$$COP_{hp} = \frac{Q_{hp}}{P_{el}},$$ (9.14)

where Q_{hp} is the thermal power of heat pump, in W and P_{el} is the electric power consumed by the compressor of HP, in W.

In the cooling mode, an HP operates exactly like a central A/C. The energy efficiency ratio (EER) is analogous to the COP but describes the cooling performance. The EER_{hp}, in Btu/(Wh), is defined as

$$EER_{hp} = \frac{Q_e}{P_{el}},$$ (9.15)

where Q_e is the cooling power of HP, in British Thermal Unit per hour (Btu/h) and P_{el} is the compressor power, in W.

The COP of an HP in cooling mode is obtained by the following equation:

$$COP_{hp} = \frac{EER_{hp}}{3.412},$$ (9.16)

where value 3.412 is the transformation factor from Watt in Btu/h.

The GSHP systems intended for ground-water or oven-system applications have heating COP ratings ranging from 3.0 to 4.0 and cooling EER ratings between 11.0 and 17.0. Those systems intended for closed-loop applications have COP ratings between 2.5 and 4.0 and EER ratings ranging from 10.5 to 20.0 [5]. The characteristic values of the SPF of modern GSHPs are commonly assumed to be approximately 4 each, meaning that four units of heat are gained per unit of consumed electricity.

The sizing factor (SF) of the HP is defined as the ratio of the HP capacity Q_{hp} to the maximum heating demand Q_{max} as

$$SF = \frac{Q_{hp}}{Q_{max}}.$$ (9.17)

The SF can be optimized in terms of energy and economics, depending on the source temperature and the used adjustment schedule.

9.3.3.2 Profitability and Capabilities of Heat Pump

The factors that can affect the life-cycle efficiency of an HP are (1) the local method of electricity generation; (2) the local climate; (3) the type of HP (ground or air source); (4) the refrigerant used; (5) the size of the HP; (6) the thermostat controls; and (7) the quality of work during installation.

Considering that the HP has over-unit efficiency, evaluation of the consumed primary energy uses a synthetic indicator as [6]

$$\eta_s = \eta_g COP_{hp} \qquad (9.18)$$

in which

$$\eta_g = \eta_p \eta_t \eta_{em}, \qquad (9.19)$$

where η_g is the global efficiency and η_p, η_t, and η_{em} are the electricity production, the transportation, and the electromotor efficiency, respectively.

For justifying the use of an HP, the synthetic indicator has to satisfy the condition $\eta_s > 1$. Additionally, the use of an HP can only be considered if the $COP_{hp} > 2.78$.

The COP of an HP is restricted by the second law of thermodynamics:

- in heating mode:

$$COP \leq \frac{t_u}{t_u - t_s} = \varepsilon_C, \qquad (9.20)$$

- in cooling mode:

$$COP \leq \frac{t_s}{t_u - t_s}, \qquad (9.21)$$

where t_u and t_s are the absolute temperatures of the hot environment (condensation temperature) and the cold source (evaporation temperature), respectively, in K.

The maximum value ε_C of the efficiency can be obtained in the reverse Carnot cycle.

In many cases, HP systems can be successfully combined with solar thermal systems so that the solar thermal energy can be used to meet a large proportion of the hot water requirements in summer and part of heating load during transitional periods. Alternatively, the COP of HPs increases significantly when the temperature of the heat source is increased with solar thermal energy.

9.3.3.3 Economic Indicators

In the economic analysis of an HP system, different methods could be used to evaluate the systems. Some of them are presented in Chapter 6.

9.3.3.4 Calculation of Greenhouse Gas Emissions

Due to the diversity in each country with respect to heating practices, direct energy use by HPs, and primary energy sources for electricity, country-specific calculations are provided.

The annual heating energy provided by HPs is defined as E_t. The annual primary energy consumption from HP electricity use is then

$$E_{el} = \frac{E_t}{SPF}. \qquad (9.22)$$

Because HP electricity consumption is considered the most important source for GHG emissions [7], other potential contributors (e.g., HP life cycle, HP refrigerant, and borehole construction) are neglected. Applying an emission factor g_p, in kg CO_2/kWh, the annual GHG emissions G_{HP}, in kg CO_2, from an HP operation can be obtained as

$$C_{HP} = g_p E_{el}. \qquad (9.23)$$

The emission factor typically varies among different countries and characterizes the GHG intensity of electricity production. Note that although carbon dioxide (CO_2) represents the most important GHG, there are several other compounds that contribute similarly to climate change. Their combined impact is commonly normalized to the specific effect of CO_2, and all emissions are expressed in CO_2 equivalents. For the sake of readability, however, the emissions are expressed only in kg CO_2.

Thus, the CO_2 emissions C_{CO_2} of the HP during its operation can be evaluated with the following equation:

$$C_{CO_2} = g_{el} E_{el}, \qquad (9.24)$$

where g_{el} is the specific CO_2 emission factor for electricity. The average European CO_2 emission factor for electricity production is 0.486 kg CO_2/kWh and for Romania is 0.547 kg CO_2/kWh [8].

Theoretical emissions C_{sub}, in kg CO_2, from the substituted energy by HP are determined by E_t and the emission factor g_f representative for the substituted heat mix as

$$C_{sub} = g_f E_t. \qquad (9.25)$$

The substituted heat is a mix from different energy carriers, i. The emission factor thus depends on the portion $e_{sub,i}$ of each energy carrier in the substituted heat mix as

$$g_{sub} = \sum_i g_{f,i} e_{sub,i}. \qquad (9.26)$$

The portions $e_{sub,i}$ ($\Sigma_i e_{sub,i} = 1$) are also termed substitution factors [7]. The annually saved emissions ΔC_{GHG} are obtained by

$$\Delta C_{GHG} = C_{sub} - C_{HP}. \qquad (9.27)$$

According to Eq. (9.27), $C_{GHG} = 0$ indicates no savings and negative values denote increased GHG emissions from HPs in comparison to conventional heatings.

9.3.4 Types of Heat Pumps

HPs are classified by (1) the heat source and sink; (2) the heating and cooling distribution fluids; and (3) the thermodynamic cycle. The following classifications can be made according to [9]:

- function: heating, cooling, domestic hot water (DHW) heating, ventilation, drying, heat recovery, etc.
- heat source: ground, ground-water, water, air, exhaust air, etc.
- heat source (intermediate fluid)-heat distribution: air-to-air, air-to-water, water-to-water, antifreeze (brine)-to-water, direct expansion-to-water, etc.

9.3.4.1 Air-to-Air Heat Pumps

Air-to-air HPs are the most common and are particularly suitable for factory-built unitary HPs. These HPs are also found in controlled dwelling ventilation applications to enable an increase in the heat recovery from the exhaust air, and can even allow for the cooling of selected rooms. For these applications, various units are used. Air-to-air HPs have a full-hermetic compressor, finned heat exchangers for the evaporator and condenser, and an expansion valve as well as the necessary safety mechanisms. As the outdoor air temperature decreases, the heat demand increases and the HP capacity substantially decreases due to the efficiency reduction.

9.3.4.2 Water-to-Air Heat Pumps

Description of the System. Water-to-air HPs rely on water as the heat source and sink, and use air to transmit heat to or from the conditioned space. They include:

- surface-water heat pumps (SWHPs), which use surface water from a lake, pond, or stream as a heat source or sink;
- ground-water heat pumps (GWHPs), which use ground-water from wells as a heat source and/or sink;
- SAHPs, which rely on low-temperature solar energy as the heat source.

Solar-assisted HPs. An advanced HP system was developed, where solar energy is coupled to the evaporator side of a water-source HP incorporated into an air distribution system common in North America. A schematic of this system is shown in Fig. 9.4. Solar collectors mounted on the south facing roof inject heat into a storage tank (ST) located on the evaporator side of the HP. The solar collector circulation pump is controlled to operate when heat gain from the solar collectors is available and there is a storage capacity for heat. The ST on the evaporator side of the HP is maintained up to 45°C, with any additional solar heating used to offset the DHW load. In the event the evaporator side ST falls below 5°C and there is a demand for heat, the HP is stopped and an electric duct heater is activated to maintain comfort conditions.

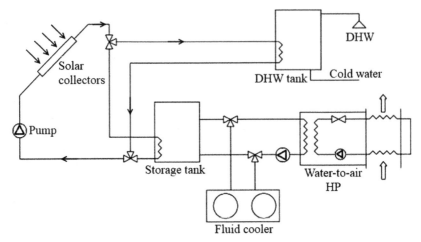

FIGURE 9.4   Schematic of a solar-assisted water-to-air heat pump.

In cooling mode, heat is rejected through a fluid cooler and any solar heating is used to meet the DHW load. The DHW tank has an electric heater to ensure adequate DHW temperatures are attained.

Typically, these systems have used a water tank for storage, connected in series between the solar loop and the HP evaporator. In this configuration, heat may also be stored in the tank when the HP operates in cooling mode.

9.3.4.3 Air-to-Water Heat Pumps

Air-to-water HPs [10] use outdoor air as their heat source and are mostly operated in bivalent heating systems, as well as for cooling, heat recovery, and DHW production. The indoor unit contains the substantial components and is fitted indoors, protected from weather and freezing temperatures. The outdoor unit is connected to the indoor unit via refrigeration lines. Through the elimination of air ducts, extremely quiet, energy-efficient fans are made possible.

9.3.4.4 Water-To-Water Heat Pumps

Water-to-water HPs operate as GWHPs or SWHPs. Water-to-water HPs use water (e.g., aquifer-fed boreholes, lakes, or water bodies) as the heat source and sink for heating and cooling. Antifreeze-to-water HPs are used in closed-loop ground-coupled installations. Water-to-water and antifreeze-to-water HPs are used for monovalent heating operation, as well as cooling, heat recovery, and DHW production.

Heating/cooling changeover can be performed in the refrigerant circuit, but it is often more convenient to perform the switch in the water circuits. Several water-to-water HPs can be grouped together to create a central cooling and heating plant to serve several air-handling units. This application has advantages of better control, centralized maintenance, redundancy, and flexibility.

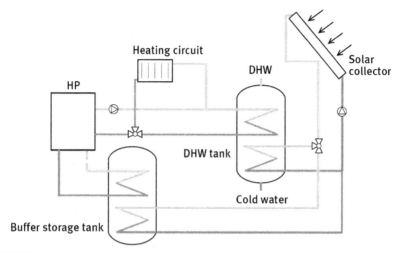

FIGURE 9.5   Schematic of a solar-assisted water-to-water HP system.

Fig. 9.5 shows a system that exclusively uses the solar thermal system as the heat source for the HP. The flat-plate collector (FPC) gives precedence to directly heating the DHW ST. If the solar thermal energy generated surpasses the requirements or the generated temperatures are too low, the solar heat is fed into a buffer ST that is used as the heat source for the HP. The buffer ST ensures that the solar thermal energy can be used at a later stage following its generation.

9.3.4.5 Ground-Coupled Heat Pumps

Ground-coupled HPs use the ground as a heat source and sink. An HP may have a refrigerant-to-water heat exchanger or may be direct-expansion (DX). In systems with refrigerant-to-water heat exchangers, a water or antifreeze solution is pumped through horizontal, vertical, or coiled pipes embedded in the ground. DX ground-coupled HPs use refrigerant in DX, flooded, or recirculation evaporator circuits for the ground pipe coils. A loop of suitable pipe containing the refrigerant and lubricant is put in direct contact with the ground or water body. The compressor operation circulates the refrigerant directly around this loop, thus eliminating the heat transfer losses associated with the intermediate water-DX heat exchanger found in conventional water source HPs [10]. There is also no need for a source-side circulation pump as the compressor fulfills this role. However, care must be taken to ensure that the DX loops are totally sealed and corrosion resistant and that the lubricant is adequately circulated to meet the needs of the compressor.

A hybrid ground-coupled HP is a variation that uses a cooling tower or air-cooled condenser to reduce the total annual heat rejection to the ground coupling.

9.3.5 Selection of Heat Source and Heat Pump System

An HP heating system consists of a heat source, an HP unit, and a heat delivery system. For system planning, all of the components must be designed to interact optimally to ensure the highest level of performance and reliable operation [10].

As a rule, the heat source with the highest temperature levels should be selected to ensure the highest possible COP and thereby the lowest operation costs.

If ground-water is available at a reasonable depth, temperature, acceptable quality, and in sufficient quantity, the highest COP can be achieved. The best ground-water heat source system is an open system, which may require approval.

If the use of ground-water is not available, the ground can function as an efficient, effective thermal storage medium with a relatively high temperature. If sufficient surface area is available, horizontal collectors offer the most cost-effective solution. If space is limited, vertical loops using geothermal energy can be effective. These heat sources are closed systems, meaning that the antifreeze solution (brine) stays within the buried tube system.

In direct expansion systems, the heat stored in the ground is absorbed directly by the refrigerant. Horizontal collectors are mainly used with this system.

If ground-water or ground-source systems cannot be used, air as a heat source is available practically anywhere. These systems are particularly suitable for retrofits or combined with another heat source (i.e., bivalent operation).

The selection of the heat source determines to a great extent the required HP type and operation. The exact size selection of the HPs is important because oversized systems operate with lower efficiencies, leading to excessive costs. The required thermal power of an HP must cover the total heat demand as

$$Q_{\text{req}} = Q_{\text{h}} + Q_{\text{v}}, \tag{9.28}$$

where Q_{h} is the heating demand and Q_{v} is the ventilation demand.

The determination of the heat demand can be performed according to European standard EN 12,831 or national standards, for example DIN 4701 and EnEV 2002 (Germany); ÖNORM 7500 (Austria); SIA 380-1 and SIA 384.2 (Switzerland); and SR 1907 (Romania). The values of the specific heat demand provided in Table 9.1 are based upon experience in European buildings. The specific heat demand is multiplied by the area of the heated space to give the total heat demand.

With active cooling, the refrigeration cycle of the HP is reversed using a four-way valve. The condenser is transformed into an evaporator and actively removes the unwanted heat (cooling) from the floor/wall areas and sends it into the ground.

TABLE 9.1 Specific Heat Demand for the Building Heating

No.	Type of Building	Specific Heat Demand (W/m²)
1	Passive house	15
2	Low-energy building	40
3	New building with good insulation	50
4	Old building with standard insulation	80–90
5	Old building without insulation	120

The cooling capacity during cooling operation is often higher than the heating requirement during heating operation. The determination of the cooling demand can be performed according to European standard EN 15,243 or national standards, for example VDI 2078 (Germany) and SR 6648-1,2 (Romania).

Utilities sometimes offer a reduced price for electricity for an HP. In return, they maintain the right to interrupt the supply at certain periods during the day. The power supply can be interrupted three times for 2 h within a 24-h period. Therefore, the HP must meet the required daily heat demand in the time in which electricity is delivered. The required heat capacity of the HP Q_{hp}, in W, can be calculated as

$$Q_{hp} = \frac{24 Q_{req}}{18 + 2}. \tag{9.29}$$

To achieve maximum system efficiency, a separate, independent HP should be planned for DHW heating. This should be optimally sized and installed, and can provide additional functions (ventilation, cooling, and dehumidification).

The heating HP is oversized for DHW heating during the summer and must operate at a higher temperature difference in the winter due to the water heating priority. The average DHW demand can be assumed to be 30 l/person/day at 45°C. The maximum operating temperature in the distribution network is 60°C. If the heating system is to be used for DHW production, notice should be taken of the following [10]:

- A heating capacity of 0.25 kW per person for hot water heating should be accounted for when sizing the HP (single-family house).
- A hot water tank for a three-to-five-person household is approximately 300 l.
- Plan the heat exchanger for a 5-K temperature difference with a supply temperature of 55°C or 65°C. This results in an approximately 50 or 60°C hot water temperature produced by the HP.

- For internal heat exchangers, a surface area of $0.4-0.7$ m^2/kW heating capacity is recommended. If this cannot be achieved, then plate heat exchangers will be required.

An exact calculation of the heat demand for DHW can be made using international or national norms.

Numerous compact units that combine heating, hot water production, and even controlled ventilation are on the market. In some cases, combination STs are used in these applications. In these tanks, the domestic water and heating loop water are stored separately. The heat sources range from outdoor air to exhaust air to ground or ground-water. In many cases, various heat sources are combined.

With the use of a conventional combination ST, it should be noted that it is less efficient (and more costly) to heat the DHW via the heating water loop. This is because the HP is forced to operate with a higher temperature difference.

Once the total required heating capacity of the HP is determined, an appropriate HP model can be selected, based upon the technical characteristics data (see Chapter 5) provided by HP manufacturers.

9.3.6 Air-Source Heat Pump Systems

9.3.6.1 Description of the Systems

The main components of an ASHP system are listed in the following. Generally a fully hermetic compressor (piston or scroll) with built-in, internal overload protection is used in ASHPs. Stainless-steel flat-plate heat exchangers are used for the condenser. Expansion valve. Weather-dependent defrost mechanism is preferably hot gas. Copper-tube aluminum-finned evaporator. For quiet operation, an axial fan with low speed should be used. Electrical components and the controller are either integrated or externally mounted, depending on manufacturer and model. Control of the heating system is commonly integrated.

The heat demand of a building depends of the climate zone in which it is located. In temperate climate conditions such as in Romania, the heat demand Q_{req} evolves from the minimum values in the provisional seasons (spring and autumn) to the maximum values in the cold season (Fig. 9.6). The annual number of hours with the minimum outdoor air temperature represents approximately $10-15\%$ of the total time for heating, which is why the selection of an HP to cover the integral peak load is not recommended.

To reduce costs, the HP is selected to cover only $70-75\%$ of the maximum heat demand of the building. The rest of the heat demand is produced by an auxiliary traditional source (i.e., electric heater or oil/gas boiler). In this case, the HP operates in bivalent mode (Fig. 9.6), distinguishing three situations:

1. if the outdoor air temperature t_a is lower than the limit heating temperature t_{lim}, the HP provides the full heat demand up to the balance point temperature t_{ech};

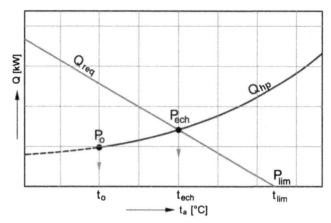

FIGURE 9.6 Heat demand provided in function of source temperature P_o-stop-heating point; P_{ech}-balance point; P_{lim}-limit heating point.

2. if the outdoor air temperature t_a is lower than t_{ech}, the HP provides part of the heat demand and the rest is provided by the peak traditional source;
3. when the outdoor air temperature t_a reaches the stop-heating temperature t_o, the HP is switched off and the traditional source meets the full heat demand.

Usually, the balance point corresponds to an outdoor temperature of 0 to +5°C. For temperate climate zones, the HP covers approximately two-thirds of the annual heat demand.

Fig. 9.7 illustrates a system with an ASHP and an FPC. The solar thermal energy principally heats a DHW ST. If the generated temperature is not sufficient, the solar thermal energy is directly used as an additional heat source for the HP during the HP operation. This increases the source temperature of the HP, which for ASHPs is only very low on cold days during the heating season.

An advanced HP system was developed, where solar energy is coupled to the evaporator side of an ASHP incorporated into an air distribution system. A schematic of this system is shown in Fig. 9.8.

The objective of this concept is to reduce the temperature difference across the evaporator and condenser of an air-to-air HP through the use of solar energy, thereby improving the system performance and heating capacity at lower ambient temperatures. Solar collectors mounted on the south facing roof heat the heat transfer fluid (HTF), preheating the ambient air through the outdoor unit during heating mode. The circulating pump is controlled to operate when heat gain from the solar collector is available. In the event the HP is not in operation, the HTF is diverted to the DHW ST used to preheat the DHW for the household. An electric heater ensures DHW is maintained at adequate temperatures. When the HP is unable to meet the heat demand of the house, the electric baseboard heaters are activated to maintain comfort conditions.

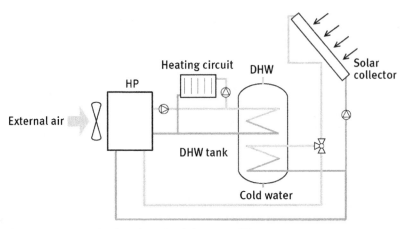

FIGURE 9.7 Schematic of a solar-assisted air-to-water HP system.

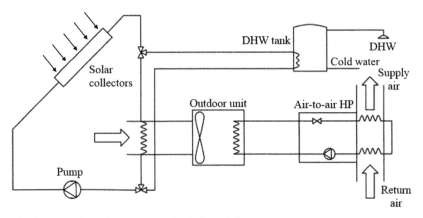

FIGURE 9.8 Schematic of a solar-assisted air-to-air heat pump.

9.3.6.2 Modeling the Solar-Assisted ASHP System for the Building Heating

Because of the flexibility, convenience and low investment of ASHP, thermal solar energy was combined with ASHP to form a new solar-assisted ASHP system for the building heating [11]. The whole heating system consisted of two parts: ASHP unit and solar collection unit. The schematic of the solar-assisted ASHP system is shown in Fig. 9.9. The output water temperature of the system (t_{wo}) is 45°C and the return water temperature of the system (t_{wr}) is 40°C. In the loop of hot water, the condenser of ASHP is in series with solar collector, and the ASHP is in front of the solar collector. The system works by three modes: (1) single ASHP mode, which is operated when the sun is not available or solar radiant intensity is low, then the available heat of radiation is

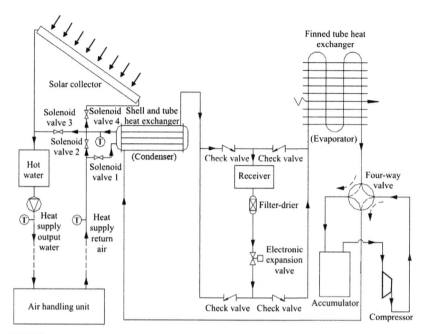

FIGURE 9.9   Schematic of the solar-assisted air-source HP heating system.

smaller than or equal to the heat dissipation of solar collector with the water temperature of 45°C (the available heat of radiation refers to the heat that can be absorbed by water in the solar collector); (2) combination mode, which is operated when the available heat of radiation is more than the heat dissipation of solar collector with the water temperature of 45°C, and the available heat of radiation is not adequate to increase the water temperature by 5°C; (3) single solar collector mode, which is operated when the solar collector can provide adequate heat to increase the water temperature by 5°C. The specific process of each mode is as follows:

1. *Single ASHP mode*: the solenoid valves 2 and 4 are closed, and the solenoid valves 1 and 3 are opened; the return water passes through the solenoid valve 1 and enters the condenser (shell and tube heat exchanger), then the water cools down refrigerant and absorbs the heat, whose temperature increases to 45°C; after it goes out of the condenser, it gets through the solenoid valve 3 to reach the hot water tank, and flows out of system after being pressurized by the pump to the air handling unit (AHU).

2. *Combination mode*: the solenoid valves 2 and 3 are closed, and the solenoid valves 1 and 4 are opened; the return water passes through the solenoid valve 1 and enters the condenser, then the water cools down refrigerant and absorbs the heat, whose temperature increases; after it goes out of the shell and tube condenser, it gets through the solenoid valve 4 to

reach the solar collector and absorbs the heat; when the temperature increases to 45°C, it enters the hot water tank and flows out the system after being pressurized by the pump to the AHU.

3. *Solar collector direct-heat supply mode*: the solenoid valves 1 and 3 are closed, and the solenoid valves 2 and 4 are opened; the return water passes through the solenoid valves 2 and 4, then the water directly enters the solar collector to absorb the heat; when the temperature rises to 45°C, it enters the hot water tank and flows out the system after being pressurized by the pump to the AHU.

The compressor in the ASHP is a variable-capacity compressor. Under the combination mode, the compressor runs at variable capacity to maintain the t_{wo} at 45°C. The higher the solar radiant intensity is, the lower the capacity the compressor runs with, and vice versa. The testing results by solar radiant intensity instrument are used as the reference for the switching between the modes.

Under the combination mode, the temperature of output water from the condenser of the ASHP t_{wo} declines with the increase of solar radiant intensity. In this way, the condensing pressure and condensing temperature of the HP system will be decreased, which make the improvement of HP performance.

9.3.6.2.1 Model of the ASHP Unit

To obtain the operational characteristic and the performance rules of the system at steady state, a steady-state model of the heating system was built up [11]. The solar-assisted ASHPS was designed without heat storage, and its rated heating capacity was 10 kW. So the heating capacity of the ASHP was 10 kW under the rated heating condition. The compressor of the HP was a frequency-conversion compressor, and the refrigerant of HP was R22. The evaporator was a finned-tube heat exchanger, the condenser was a shell and tube heat exchanger, and the throttle was an electronic expansion valve. In the ASHP model, the following assumptions about heat exchangers were made: (1) it is thermal isolation between the shell and tube heat exchanger and the environment; (2) the pressure inside heat exchanger and the temperature of refrigerant inside two-phase zone are constant; (3) only pure refrigerant is considered; (4) axial heat transfer is neglected; and (5) the refrigerant from each channel of the heat exchanger is fully mixed.

9.3.6.2.2 Model of the Condenser

1. Heat transfer coefficients of refrigerant in single-phase zone: The heat transfer coefficient of superheated zone α_{sh}, in $W/(m^2 \cdot K)$, and the heat transfer coefficient of subcooled zone α_{sc}, in $W/(m^2 \cdot K)$, can both be calculated by the standard Dittus−Boelter formula as

$$\alpha_{sh} = 0.023 \frac{\lambda_v}{d_i} \mathrm{Re}_v^{0.8} \mathrm{Pr}_v^{0.3} \tag{9.30}$$

$$\alpha_{sc} = 0.023 \frac{\lambda_l}{d_i} \mathrm{Re}_l^{0.8} \mathrm{Pr}_l^{0.3}, \tag{9.31}$$

where λ_v and λ_l are the thermal conductivities, in W/(m·K), of vapor and liquid, respectively; d_i is the internal diameter of pipe, in m; Re_v, Re_l are the Reynolds numbers for vapor and liquid; and Pr_v, Pr_l are the Prandtl numbers for vapor and liquid.

2. Heat transfer coefficient of refrigerant in the two-phase zone: Local heat transfer convection coefficient in the two-phase zone $\alpha_{tp}(x)$, in W/(m²·K), can be evaluated by Dobson formula as

$$\alpha_{tp}(x) = 0.023 \frac{\lambda_l}{d_i} \mathrm{Re}_l^{0.8} \mathrm{Pr}_l^{0.4} \left(1 + \frac{2.22}{X_t^{0.889}} \right) \tag{9.32}$$

in which

$$X_t = 0.551 \frac{p}{p_{cr}} \left(\frac{1-x}{x} \right)^{0.9}, \tag{9.33}$$

where X_t is Martinelli value; p is the pressure, in Pa; p_{cr} is the critical pressure of refrigerant, in Pa; and x is the dryness ratio of refrigerant.

3. Heat transfer coefficient of external water: The shell side of the shell and tube heat exchanger was equipped with baffle plates. Heat exchange happened by the water flowing vertically and horizontally outside the pipes. The heat transfer coefficient of external water α_w, in W/(m²·K), can be calculated using following equation:

$$\alpha_w = 0.25 \frac{\lambda_w}{d_e} \mathrm{Re}_w^{0.6} \mathrm{Pr}_w^{0.33}, \tag{9.34}$$

where d_e is the external diameter of pipe, in m; λ_w is the thermal conductivity of water, in W/(m·K); Re_w and Pr_w are the Reynolds and Prandtl numbers, respectively for water.

9.3.6.2.3 Model of the Evaporator

1. Heat transfer coefficient of refrigerant in the superheated zone: In the superheated zone, the refrigerant was in the state of superheated vapor and moved in the form of turbulent flow inside the pipes. Heat transfer coefficient of refrigerant α_{sh}, in W/(m²·K), can be evaluated by Petukhov–Popov equation as

$$\alpha_{\text{sh}} = \frac{\lambda_v}{d_i} \cdot \frac{(f/8)\,\text{Re}_v\,\text{Pr}_v}{1.07 + 12.7(f/8)^{0.5}\left(\text{Pr}_v^{2/3} - 1\right)} \tag{9.35}$$

in which

$$f = (1.82\lg\text{Re}_v - 1.64)^{-2}, \tag{9.36}$$

where f is the friction factor for pipes.

2. Average heat transfer coefficient of refrigerant in two-phase zone: The local heat transfer coefficient in the two-phase zone $\alpha_{\text{tp}}(x)$, which was related to the refrigerant dryness ratio x, may be evaluated by

$$\alpha_{\text{tp}}(x) = \frac{3.0}{X_t^{2/3}}\alpha_l \tag{9.37}$$

in which

$$X_t = \left(\frac{\mu_l}{\mu_v}\right)^{0.1}\left(\frac{1-x}{x}\right)^{0.9}\left(\frac{\rho_v}{\rho_l}\right)^{0.5} \tag{9.38}$$

$$\alpha_l = 0.023\frac{\lambda_l}{d_i}\text{Re}_l^{0.8}\,\text{Pr}_l^{0.3}, \tag{9.39}$$

where α_l is the heat transfer coefficient of refrigerant in pure liquid phase, in W/(m$^2\cdot$K); μ_l and μ_v are the dynamic viscosities, in Pa$\cdot$s, of liquid refrigerant and vapor refrigerant, respectively; and ρ_l and ρ_v are the densities, in kg/m^3, of liquid refrigerant and vapor refrigerant, respectively.

The average heat transfer coefficient of the two-phase zone α_{tp} may be evaluated by the following formula as

$$\alpha_{\text{tp}} = \frac{\int_{x_i}^{x_0}\,\mathrm{d}x}{\int_{x_i}^{x_0}\left(1/\alpha_{\text{tp}}(x)\right)\mathrm{d}x}. \tag{9.40}$$

3. Heat transfer coefficient of external air: The fin type of the finned tube exchanger is a straight one-piece fin. The heat transfer coefficient of external air α_L in W/(m^{2}K) can be obtained as [11]

$$\alpha_L = \frac{\lambda_L}{d_b}\text{Nu} \tag{9.41}$$

in which

$$\text{Nu} = 0.982\text{Re}_L^{0.424}\left(\frac{s}{d_b}\right)^{-0.0887}\left(\frac{N\cdot s_1}{d_b}\right)^{-0.1590}, \tag{9.42}$$

where λ_L is the thermal conductivity of air, in W/(m$\cdot$K); d_b is the diameter of root of fins, in m; Nu is the Nusselt number; Re$_L$ is the Reynolds number for air; s is the fin interval, in m; s_1 is the interval between tubes along the air direction, in m; and N is the row number of finned tube.

9.3.6.2.4 Model of the Compressor

The suction pressure and discharge pressure of compressor were assumed as the evaporation pressure and condensation pressure of the system respectively. The mass flow rate of refrigerant in compressor m_k, in kg/m^3, depended on the working capacity of compressor V_k, in m^3, the capacity efficiency of compressor η_{ck}, and the specific volume of refrigerant at the entrance of compressor v_k as

$$m_k = \frac{V_k n \eta_{ck}}{60 v_k} \tag{9.43}$$

in which n is the rotational speed of compressor, in rev/min, and η_{ck} is provided by the compressor manufacturer as a constant.

The input power of compressor P_k, in kW, can be calculated by the following equation:

$$P_k = \frac{m_k(h_{dis} - h_{suc})}{\eta_k}, \tag{9.44}$$

where h_{dis} is the enthalpy at the compressor discharge, in kJ/kg; h_{suc} is the enthalpy at the compressor suction, in kJ/kg; and and η_k is the shaft efficiency of compressor.

9.3.6.2.5 Model of the Electronic Expansion Valve

Under steady working condition, the refrigerant mass flow rate of the electronic expansion valve m_{ev}, in kg/s can be calculated using equation

$$m_{ev} = C_{ev}A_{ev}\sqrt{\rho_i(p_i - p_o)}, \tag{9.45}$$

where C_{ev} is the flow coefficient of electronic expansion valve; A_{ev} is the circulation area of expansion valve, in m^2; ρ_i is the refrigerant density, in kg/m^3 at the expansion valve inlet; and p_i and p_o are the pressures, in Pa, at the expansion valve inlet and the expansion valve outlet, respectively.

9.3.6.2.6 Model of the Solar Collection Unit

For an evacuated tube collector of the solar collection unit (type BTZ22), the instantaneous efficiency of collector η_c can be obtained as [11]

$$\eta_c = 0.682 - 2.32\frac{t_i - t_a}{I_T}, \tag{9.46}$$

where t_i is the hot water temperature inside collector tube, in K; t_a is the ambient temperature, in K; and I_T is the solar radiation intensity, in W/m^2.

9.3.6.2.7 Validation of the Mathematical Model

To validate the mathematical model of the system, an ASHP unit was developed. The compressor of the unit was a frequency conversion rotary compressor, type THS20MC6-Y, and the compressor frequency can be adjusted from 15 to 110 Hz by a general frequency converter, the main parameters of the compressor are shown in Table 9.2.

The condenser of the unit was RER-20 type, whose main parameters are shown in Table 9.3 in detail. The shape of the evaporator of the unit was U-shaped with flat fin, and the row number was 2. The main parameters of the evaporator are shown in Table 9.4.

The layout of measuring spots is shown in Fig. 9.9. The temperature was measured by the platinum-resistance temperature sensors with an accuracy of 0.1°C and the measuring range from −50 to 150°C. The measuring range of the pressure sensors was from 0 to 2.5 MPa with an accuracy of 0.25%. The water flow rate was measured by the turbine flow meter, type LWY-15C, with an accuracy of 1% and the measuring range 0−6 m³/h. The power of the compressor was measured by the power meter type WT230, with an accuracy of 0.1%. All the data for every spot were collected in real-time by a data collector Agilent model 34,970a.

To get a steady experimental environment for the unit, the unit was placed in an environmental chamber as shown in Fig. 9.10, which was used to obtain constant temperature and relative humidity of the air. The air temperature could be manually adjusted from −5 to 43°C with an accuracy of 0.2°C and the air relative humidity could be adjusted from 5% to 95% with an accuracy of 2%.

The simulations and tests of the unit have been done under the ambient temperature of 7°C, the return water temperature t_{wr} of the ASHP of 40°C, and a variety of the output water temperatures t_{wo} of the ASHP. Fig. 9.11 shows the effect of t_{wo} on the relative COP of the ASHP, which took the COP of ASHP under the rated condition as a benchmark. In this figure, it can be seen that the variation trend of the simulation results is in good agreement with the experimental data, and the maximum deviation is less than 8%.

9.3.7 Ground-Source Heat Pump Systems

Recently, the ground-source heat pump (GSHP) system has attracted more and more attention due to its superiority of high energy efficiency and environmental friendliness [9, 12−14]. A GSHP system includes three principle components: (1) a ground connection subsystem, (2) HP subsystem, and (3) heat distribution subsystem.

The GSHPs comprise a wide variety of systems that may use ground-water, ground, or surface water as heat sources or sinks. These systems have been basically grouped into three categories by ASHRAE [15]: (1) ground-water heat pump (GWHP) systems, (2) surface-water heat pump (SWHP) systems,

TABLE 9.2 The Main Parameters of Compressor

Type	Refrigerant	Output Power (W)	Displacement (ml/rev)	Range of Frequency (Hz)	Cooling Capacity (W)	Input Power (W)	COP
THS20MC6-Y	R22	2300	30.4	15–110	8150/9900	2580/3190	3.16/3.1

TABLE 9.3 The Parameters of Shell and Tube Heat Exchanger

Type	Shell Diameter (mm)	Pipe Diameter (mm)	Effective Length of Single Pipe (mm)	Pipe Number	Arrangement of Pipes	Tube Pitch (mm)	Number of Passes
THS20MC6-Y	168 × 6	12 × 0.7	700	68	Triangular distribution	16	4

TABLE 9.4 The Parameters of Finned-Tube Heat Exchanger

Pipe Diameter	Effective Length of Single Pipe (m)	Pipe Number	Circuit Number	Pipe Spacing (mm)	Front Face Area (m²)	Thickness of Fin (mm)	Fin Spacing (mm)
11 × 0.5	1.8	28 × 2	8	25	1.8 × 0.72	0.2	1.90

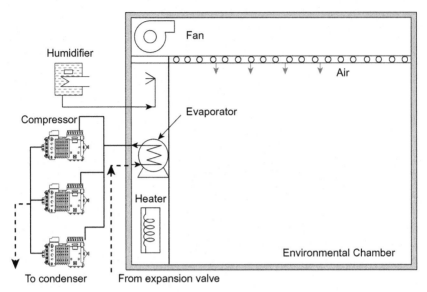

FIGURE 9.10 Schematic of the environmental chamber.

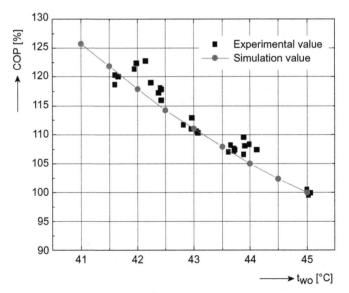

FIGURE 9.11 Effect of t_{wo} on the COP of air-source heat pump.

and (3) ground-coupled heat pump (GCHP) systems. The schematics of these different systems are shown in Fig. 9.12. Many parallel terms exist: geothermal heat pump (GHP), earth energy system (EES), and ground-source system (GSS).

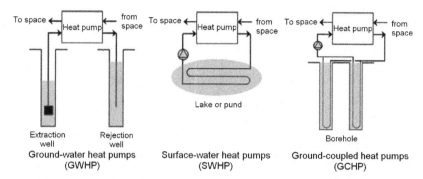

FIGURE 9.12 Schematics of different ground-source heat pumps.

The GWHP system, which utilizes ground-water as heat source or sink, has some marked advantages including low initial cost and minimal requirement for ground surface area over other GSHP systems [16]. However, a number of factors seriously restrict the wide application of the GWHP systems, such as the limited availability of ground-water and the high maintenance cost due to fouling corrosion in pipes and equipment. In an SWHP system, heat rejection/extraction is accomplished by the circulating working fluid through high-density polyethylene (HDPE) pipes positioned at an adequate depth within a lake, pond, reservoir, or the suitable open channels. The major disadvantage of the system is that the surface water temperature is more affected by weather condition, especially winter.

Among the various GSHP systems, the vertical GCHP system has attracted the greatest interest in research field and practical engineering. Several literature reviews on the GCHP technology have been reported [17,18].

In a GCHP system, heat is extracted from or rejected to the ground via a closed-loop, that is, ground heat exchanger (GHE), through which pure water or antifreeze fluid circulates. The GHEs commonly used in the GCHP systems typically consist of HDPE pipes that are installed in either vertical boreholes (called vertical GHE) or horizontal trenches (horizontal GHE). In direct expansion systems, the heat stored in the ground is absorbed directly by the working fluid (refrigerant). This result in an increased COP. Horizontal GHE is mainly used with this system.

The GSHPs work best with heating systems, which are optimized to operate at lower water temperature than radiator and radiant panel systems (floor, wall, and ceiling). GSHPs have the potential to reduce cooling energy by 30–50% and reduce heating energy by 20–40% [19]. The GSHPs tend to be more cost-effective than conventional systems in the following applications:

- in new construction where the technology is relatively easy to incorporate, or to replace an existing system at the end of its useful life;

- in climates characterized by high daily temperature swings, or where winters are cold or summers hot, and where electricity cost is higher than average; and
- in areas where natural gas is unavailable or where the cost is higher than electricity.

9.3.7.1 Description of SWHP Systems

Surface water bodies can be very good heat source and sinks, if properly used. The maximum density of water occurs at 4.0°C, not at the freezing point of 0°C. This phenomenon, in combination with the normal modes of heat transfer to and from takes, produces temperature profile advantageous to efficient HP operation. In some cases, lakes can be the very best water supply bodies for cooling. Various water circulation systems are possible and several of the more common are presented next.

In a closed-loop system, a water-to-air HP is linked to a submerged coil. Heat is exchanged to or from the lake by the refrigerant circulating inside the coil. The HP transfers heat to or from the air in the building.

In an open-loop system, water is pumped from the lake through a heat exchanger and returned to the lake some distance from the point at which it was removed. The pump can be located either slightly above or submerged below the lake water level. For HP operation in the heating mode, this type is restricted to warmer climates. Entering lake water temperature must remain above 5.5°C to prevent freezing.

Thermal stratification of water often keeps large quantities of cold water undisturbed near the bottom of deep lakes. This water is cold enough to adequately cool buildings by simply being circulated through heat exchangers. An HP is not needed for cooling, and energy use is substantially reduced. Closed-loop coils may also be used in colder lakes. Heating can be provided by a separate source or with HPs in the heating mode. Precooling or supplemental total cooling are also permitted when water temperature is between 10 and 15°C.

Advantages of closed-loop SWHPs are (1) relatively low cost because of reduced excavation costs, (2) low pumping energy requirements, and (3) low operating cost. Disadvantages are (1) the possibility of coil damage in public lakes and (2) wide variation in water temperature with outdoor conditions.

9.3.7.2 Description of GWHP Systems

A GWHP system removes ground-water from a well and delivers it to an HP (or an intermediate heat exchanger) to serve as a heat source or sink [5]. Both unitary and central plant designs are used. In the unitary type, a large number of small water-to-air HPs are distributed throughout the building. The central plant uses one or a small number of large-capacity chillers supplying hot and chilled water to a two- or four-pipe distribution system. The unitary approach is more common and tends to be more energy-efficient.

Direct systems (in which ground-water is pumped directly to the HP without an intermediate heat exchanger) are not recommended except on the very smallest installations. Although some installations of this system have been successful, others have had serious difficulty even with ground-water of apparently benign chemistry. The specific components for handling ground-water are similar. The primary items include (1) wells (supply and, if required, injection), (2) a well pump (usually submerged), and (3) a ground-water heat exchanger. The use of a submerged pump avoids the possibility of introducing air or oxygen into the system. A back-washable filter should also be installed. The injection well should be located from 10 to 15 m in the downstream direction of the ground-water flow.

In an open-loop system, the intermediate heat exchanger between the refrigerant and the ground-water is subject to fouling, corrosion, and blockage. The required flow rate through the intermediate heat exchanger is typically between 0.027 and 0.054 l/s. The ground-water must either be reinjected into the ground by separate wells or discharged to a surface system such as a river or lake. To avoid damage due to corrosion, the conductivity of the water should not exceed 450 μS per cm.

Commissioning must be performed by the manufacturer's customer service department or qualified representatives. The heat collection, heat distribution, and electrical connections must all be in working order before commissioning.

The drill diameter should be at least 220 mm (larger for sandy conditions to prevent sand entry). Fig. 9.13 shows the construction of a drilled well.

The ground-water flow rate G, in m^3/s, must be capable of delivering the full capacity required from the heat source. This depends on the evaporator cooling power Q_e, in W, and the water cooling degree and is given by the following equation:

$$G = \frac{Q_e}{\rho_w c_w (t_{wi} - t_{we})} \qquad (9.47)$$

in which ρ_w is the water density, in kg/m^3; c_w is the specific heat of water, in J/(kg·K); and t_{wi} and t_{we} are the water temperatures, in K, at the HP inlet and the HP outlet, respectively.

The values of the flow rate for each HP model are usually provided in the manufacturer's data sheets.

Table 9.5 summarizes the calculated COP values of GWHP and SWHP systems, operating as water-to-water heat pumps.

The installation of a GWHP that uses a safety refrigerant is possible in any space that is both dry and protected from freezing temperatures. The system should be installed on an even, flat surface and the construction of a free-standing base is recommended. The placement of the unit should be such that servicing and maintenance are possible. Generally, only flexible connections to the HP should be implemented.

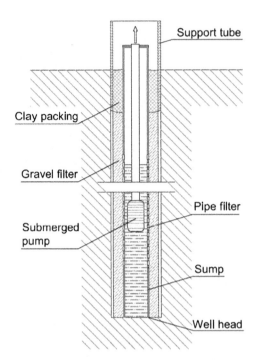

FIGURE 9.13   Drilled-Well Construction.

TABLE 9.5 The COP of Water-To-Water GWHP and SWHP Systems

Water Temperature at Evaporator Inlet t_s (°C)	Water Temp. at Condenser Outlet, t_u (°C)				
	30	35	40	45	50
5	4.55	4.10	3.70	3.40	3.15
10	5.30	4.65	4.15	3.75	3.45
15	6.25	5.35	4.70	4.20	3.85
20	7.70	6.35	5.45	4.80	4.30
25	9.95	7.80	6.45	5.55	4.85
30	14.10	10.10	7.95	6.55	5.60

9.3.7.3 Description of GCHP Systems

The ground serves as an ideal heat source for monovalent HP systems. The GCHP is a subset of the GSHP and is often called a closed-loop HP. A GCHP system consists of a reversible vapor-compression cycle that is linked to a

GHE buried in the soil (Fig. 9.12). The heat-transfer medium, an antifreeze solution (brine), is circulated through the GHE (collector or loop) and the HP by an antifreeze solution pump. The GHE size needs to take into account the total annual heating demand, which, for domestic heating operation, is typically between 1700 and 2300 h in central Europe.

9.3.7.3.1 Types of Horizontal GHEs

Horizontal GHEs can be divided into at least three subgroups: single-pipe, multiple-pipe, and spiral. Single-pipe horizontal GHEs consist of a series of parallel pipe arrangements laid out in trenches. Typical installation depths in Europe vary from 0.8 to 1.5 m. Consideration should be given to the local frost depth and the extent of snow cover in winter. Horizontal GHEs are usually the most cost-effective when adequate yard space is available and the trenches are easy to dig. Antifreeze fluid runs through the pipes in a closed system. The values of the specific extraction/rejection power q_E for ground [20,21] are given in Table 9.6. For a specific power of extraction/rejection q_E, the required ground area A can be obtained as follows [22]:

$$A = \frac{Q_e}{q_E}, \qquad (9.48)$$

where $Q_e = Q_{hp} - P_{el}$ is the cooling power HP.

The values of the cooling power for each HP model are usually provided in the manufacturer's data sheets.

To save required ground area, some special GHEs have been developed [23]. Multiple pipes (two, four, or six), placed in a single trench, can reduce the amount of required ground area. The trench collector is vilely used in North America, and less in Europe.

The spiral loop (Fig. 9.14) is reported to further reduce the required ground area. This consists of pipe unrolled in circular loops in trenches with a horizontal configuration. For the horizontal spiral loop layout, the trenches are generally a depth of 0.9−1.8 m. The distance between coil tubes is of 0.6−1.2 m. The length of collector pipe is of 125 m per loop (up to 200 m).

TABLE 9.6 Specific Extraction Power for Ground

No.	Type of Ground	q_E (W/m^2)
1	Dry sandy	10−15
2	Moist sandy	15−20
3	Dry clay	20−25
4	Moist clay	25−30
5	Ground with ground-water	30−35

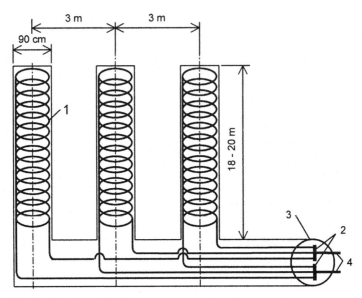

FIGURE 9.14   Spiral ground coil.

The ends of parallel coils 1 are arranged by a manifold-collector 2 in a heart 3, and then the antifreeze fluid is transported by main pipes of 4 at HP. For the trench collector, a number of pipes with small diameters are attached to the steeply inclined walls of a trench several meters deep.

Horizontal ground loops are the easiest to install while a building is under construction. However, new types of digging equipment allow horizontal boring, thus making it possible to retrofit such systems into existing houses with minimal disturbance of the topsoil and even allowing loops to be installed under existing buildings or driveways.

For all horizontal GHEs in the heating-only mode, the main thermal recharge is provided by the solar radiation falling on the earth's surface. Therefore, it is important not to cover the surface above the ground heat collector.

Disadvantages of the horizontal systems are: (1) these systems are more affected by ambient air temperature fluctuations because of their proximity to the ground surface, and (2) the installation of the horizontal systems needs much more ground area than vertical system.

9.3.7.3.2 Types of Vertical GHEs

There are two basic types of vertical GHEs or borehole heat exchangers (BHE): U-tube and concentric- (coaxial-) tube system configurations (Fig. 9.15). BHEs are widely used when there is a need to install sufficient heat exchanger capacity under a confined surface area, such as when the earth is rocky close to the surface, or where minimum disruption of the landscape is desired.

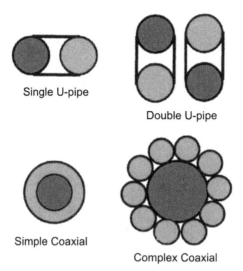

FIGURE 9.15 Comsmon vertical GHE designs.

The U-tube vertical GHE may include one, tens, or even hundreds of boreholes, each containing single or double U-tubes through which heat exchange fluid are circulated. Typical U-tubes have nominal diameters in the range of 20–40 mm and each borehole is normally 20–200 m deep with a diameter ranging from 100 to 200 mm. Concentric pipes, either in a very simple method with two straight pipes of different diameters or in complex configurations, are commonly used in Europe. The borehole annulus is generally backfilled with some special material (grout) that can prevent contamination of ground-water.

Geological analysis should be completed before drilling to give an indication of the underground layers and an exact collection capacity. The drilling and the insertion of the loop pipe should be completed by a specialized and licensed drilling company. The hole will be refilled and the tubing secured according to industry standard and should include proper sealing in the case of ground-water.

A typical borehole with a single U-tube is illustrated in Fig. 9.16. The required borehole length L can be calculated by the following steady-state heat transfer equation [15]:

$$L = \frac{qR_g}{t_g - t_f},\tag{9.49}$$

where: q is the heat transfer rate, in kW; t_g is the ground temperature, in K; t_f is the heat carrier fluid (i.e., antifreeze, refrigerant) temperature, in K; and R_g is the effective thermal resistance of ground per unit length, in $(m \cdot K)/kW$.

The GHE usually are designed for the worst conditions by considering that the need to handle three consecutive thermal pulses of various magnitude and

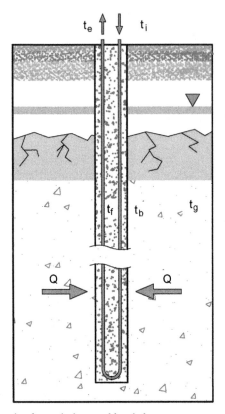

FIGURE 9.16   Schematic of a vertical grouted borehole.

duration: yearly average ground load q_a for 20 years, the highest monthly ground load q_m for 1 month, and the peak hourly load q_h for 6 h. The required borehole length to exchange heat at these conditions is given by [24]

$$L = \frac{q_h R_b + q_a R_{20a} + q_m R_{1m} + q_h R_{6h}}{t_g - (t_f + \Delta t_g)}, \tag{9.50}$$

where R_b is the effective borehole thermal resistance; R_{20a}, R_{1m}, R_{6h} are the effective ground thermal resistances for 20-year, 1-month, and 6-h thermal pulses, respectively; and Δt_g is the increase of temperature because of the long-term interference effect between the borehole and the adjacent boreholes.

The effective ground thermal resistance depends mainly on the ground thermal conductivity and, to a lesser extent, on the borehole diameter and the ground thermal diffusivity. Alternative methods of computing the thermal borehole resistance are presented by Bernier [24] and Hellström [25].

The antifreeze solution (brine) circulation loop normally consists of the following components: collector pipes, manifold, vent, circulation pump, expansion vessel, safety valve, insulation or condensate drainage, and flexible connections to the HP unit (closed system).

All of the components should be corrosion-free materials and should be insulated with closed-cell insulation in the heating space to prevent condensation.

The piping should be sized such that the fluid velocity does not exceed 0.8 m/s.

The antifreeze mass flow rate must be capable of transporting the full thermal capacity required from the heat source. The mass flow rate m_b, in kg/s, is given by

$$m_b = \frac{3600Q_e}{c_b\Delta t}, \qquad (9.51)$$

where Q_e is the cooling power of HP, in kW; c_b is the specific heat of antifreeze, in kJ/(kg·K); and Δt is the temperature difference, in K (e.g., 3 K).

The ground loop circulation pump must be sized to achieve the minimum flow rate through the HP.

The nominal volume V_N of the diaphragm expansion vessel, in liters, for the antifreeze solution circulation loop can be calculated as

$$V_N = \frac{p_{si} + 0.05}{p_{si} - (p_{st} + 0.05)}(\beta + 0.005)V_T, \qquad (9.52)$$

where p_{si} is the safety valve purge pressure, equal to 0.3 MPa; p_{st} is the nitrogen preliminary pressure (0.05 MPa); β is the thermal expansion coefficient ($\beta = 0.01$ for Tyfocor); and V_T is the total volume of the system (heat exchanger, inlet duct, HP), in l.

The installation of a GCHP that uses a safety refrigerant is possible in any space that is both dry and protected from freezing temperatures. The system should be installed on an even, horizontal surface, and the construction of a free-standing base is recommended. The placement of the unit should be such that servicing and maintenance are possible. Generally, only flexible connections to the HP should be implemented.

Advantages of the vertical GCHP are that it (1) requires relatively small ground area, (2) is in contact with soil that varies very little in temperature and thermal properties, (3) requires the smallest amount of pipe and pumping energy, and (4) can yield the most efficient GCHP system performance. Disadvantage is a higher cost because of the expensive equipment needed to drill the borehole.

9.3.7.3.3 Simulation Models of GHEs

The main objective of the GHE thermal analysis is to determine the temperature of the heat-carried fluid, which is circulated in the U-tube and the HP, under certain operating conditions. Actually, the heat transfer process in a GHE involves a number of uncertain factors, such as the ground thermal properties, the ground-water flow rate, and building loads over a long lifespan of several or even tens of years. In this case, the heat transfer process is rather complicated and must be treated, on the whole, as a transient one. In view of the complication of this problem and its long time scale, the heat transfer process may usually be analyzed in two separated regions. One is the solid soil/rock outside the borehole, where the heat conduction musty be treated as a transient process. Another sector often segregated for analysis is the region inside the borehole, including the grout, the U-tube pipes and the circulating fluid inside the pipes. This region is sometime analyzed as being steady-state and sometime analyzed as being transient. The analyses on the two spatial regions are interlinked on the borehole wall. The heat transfer models for the two separate regions and computer programs for GCHP design/simulation are discussed in [6,9].

9.3.7.4 Solar Thermal Energy as a Heat Source for GSHPs

Fig. 9.17 shows a system configuration that combines a solar thermal system with a GCHP. The solar thermal system charges a DHW ST that, whenever required, is also heated by the HP. In this example the space heating is directly and exclusively provided by the HP, that is, without a heat ST.

It is important that the HP control system prioritizes the solar heat generation. The high SPF of the solar thermal system means that the electrical energy requirement lowers and the system efficiency increases. The improvement in efficiency and the operational cost savings depend on many parameters such as the solar irradiance, the type of main components, and their sizing. With systems that use GSHPs, the solar thermal system reduces the heat absorbed from the ground in summer. The extent to which this reduces the drop in ground temperature during the heating operation as a result of this, or

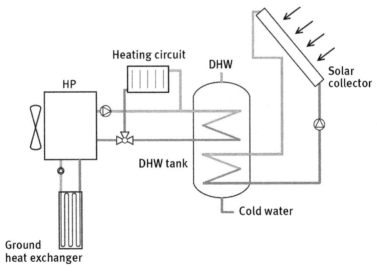

FIGURE 9.17   Schematic of a GCHP with an FPC connected to the DHW storage tank.

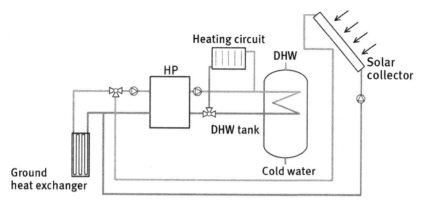

FIGURE 9.18   Schematic of a GCHP with an unglazed solar collector that feeds solar heat to the ground or directly to HP as a second heat source.

the extent to which the BHE or horizontal GHE can be made smaller, depends among other things on the relative reduction in the heat removal and the existence of ground-water flows.

Fig. 9.18 shows a system with a GCHP that incorporates unglazed collectors to provide heat cost-effective at a low-temperature level. The solar heat is exclusively injected on the source side. The solar thermal energy stored in summer can be used for regenerating the ground. This stabilizes the heat source against any possible, unforeseeable increases in the heat removed and (slightly) increases the heat source temperature of the HP.

HP with schematic diagram in Fig. 9.19 operates in monovalent regime and extracts the heat absorbed by HTF (antifreeze solution) in the solar collector ensuring the heating and DHW production for a building.

In winter the internal temperature of latent heat storage can be decreased to values in the range $-4.8 - +7°C$ with the ice forming because the absence of solar radiation and the heat extraction by HP, and the ground temperature can have the values of $+1$ to $+9°C$. In this situation a heat rush appears from ground to storage producing the ice melting.

In summer, in latent heat storage can be reached higher temperatures in the range $+25 - +40°C$ due the solar heat gain. In this case the ground is heated and thermal flux is directed from the storage to the surrounding ground layers more cool ($+18 - +22°C$). The fluid used for the phase change in storage is the water.

9.4 ENERGY PERFORMANCES OF A DX-SAHP WATER HEATING SYSTEM

This study focuses on developing a simplified model, which can estimate the annual performances of a DX-SAHP water heating system and can also be used to study the effect of various parameters that influence system performances [26].

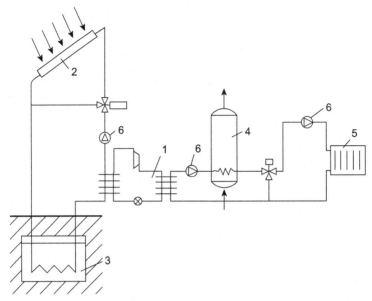

FIGURE 9.19   Schematic of a heating and DHW system with HP and latent heat storage buried in ground (1) HP; (2) solar collector; (3) latent heat storage; (4) DHW tank; (5) heating circuit; (6) circulating pump.

9.4.1 System Description

Fig. 9.20 shows the schematic diagram of a simple DX-SAHP water heating system. The system consists of an unglazed solar collector as an evaporator, a hot water ST with an immersed heat exchanger as condenser, a thermostatic expansion valve, and a small hermetic refrigeration compressor.

To begin with, the refrigerant from the condenser passes through an expansion valve directly into the solar collector where it gets evaporated by incident solar energy. The ambient air acts as an additional heat source or sink, depending on whether the refrigerant temperature is higher or lower than that of ambient. The evaporated refrigerant passes through the compressor, and finally the high-temperature vapor is pumped into the condenser where it gets condensed. The energy rejected by the condenser contributes to load requirements (hot water applications) through a refrigerant-to-water heat exchanger immersed in the hot water ST. The HP cycle on a $p-h$ diagram is illustrated in Fig. 9.2.

9.4.2 Experimental Plant

A prototype DX-SAHP water heating system was designed under climate conditions of Shanghai, China [26]. A 2 m^2 solar collector without any glazing or back insulation was used as a heat source as well as an evaporator for the

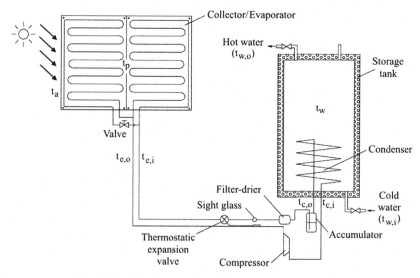

FIGURE 9.20   Schematic of the DX-SAHP water heater.

refrigerant R22. It consists of two aluminum absorber plates in parallel, which were made by a special process, in which the piping network design is laid between two sheets of aluminum and retained after the sheets are bonded by rolling them together, and then the tubes are formed by overpressurizing the network so that the serpentine fluid circuit is within the fin. As a result, the collector/evaporator is light in weight and very thin, so that it can be mounted easily anywhere. The collector/evaporator was fixed vertically on a facing south wall. A small R22 reciprocating-type hermetic compressor with 1/3 HP (248.6 W) rated input power was used in the system. The condenser was made up of a copper tube coil, which was immersed in the ST (of 150 l capacity and insulated by 50 mm thick polyurethane), and a receiver as well as a filter-drier were installed downstream the condenser. The HP system was controlled by thermostat located inside the ST, while the thermostatic expansion valve regulates refrigerant flow rate to the solar collector/evaporator.

9.4.3 Measuring Apparatus

The ambient temperature, the collector surface temperature, the refrigerant temperature at various locations of the system, and the water temperatures in ST were measured by copper—constantan thermocouples. The low and the high pressures across the compressor were measured with pressure gages. To protect the compressor from overload, low-pressure- and high-pressure cut-off switches were connected to the compressor. A solar pyranometer was mounted near the collector to measure the instantaneous solar radiation and the power consumption of the system was measured by a wattmeter. The above-measuring processes were monitored and controlled by a personal computer (PC)-based system. The data were recorded at every 5-min interval in a data logger, which was later used for analysis.

9.4.4 Simulation Model

A simple one-dimensional mathematical model has been developed [26] to predict the thermal performance of the system, based on the following assumptions:
- The system is at quasi-steady state within the chosen time interval;
- Pressure drop is negligible in evaporator, condenser as well as in piping;
- Refrigerant is considered at saturation both at the exits of the condenser and the evaporator;
- Compression of refrigerant vapor is assumed to follow a polytropic process;
- Expansion of refrigerant liquid is considered to be isenthalpic; and
- Hot water ST is assumed to be no stratified.

Based on the above assumptions, the governing equations describing the thermal performance of various components of the proposed system have been formulated and solved by an iterative numerical method that takes into account the interactions between the various system components.

9.4.4.1 Unglazed Solar Collector/Evaporator

The useful energy Q_u collected by an unglazed collector operating at steady-state conditions can be evaluated using Eq. (3.19).

The collector heat loss coefficient U_L is mainly due to the convection and radiation heat transfer from the top surface of the collector to the surroundings, shown as follows:

$$U_L = \alpha_c + \alpha_r, \tag{9.53}$$

where α_c and α_r are the convective and the radiative heat transfer coefficients between collector surface and environment, in $W/(m^2 \cdot K)$.

The convective coefficient α_c, due to wind, is determined using the following experimental correlation [26]:

$$\alpha_c = 2.8 + 3.0 u_w, \tag{9.54}$$

where u_w is the wind velocity, in m/s.

The radiative heat transfer coefficient α_r is expressed as [26]

$$\alpha_r = \varepsilon \sigma \left(t_p^2 + t_a^2 \right) (t_p + t_a), \tag{9.55}$$

where ε is the hemispherical emittance and $\sigma = 5.67 \times 10^{-8}$ $W/(m^2 \cdot K^4)$ is the Stefan Boltzmann constant.

The collector efficiency factor F can be estimated using Eq. (3.76). The internal heat transfer coefficient U_i of two-phase flows in horizontal tubes is evaluated using the following correlation [27]:

$$U_i = \frac{0.0082\lambda_l}{D_i}\left(\text{Re}^2 J \Delta x L_f / L\right)^{0.40}, \tag{9.56}$$

where λ_l is the thermal conductivity of liquid refrigerant, in W/(m·K); D_i is the internal diameter of the collector tube, in m; Re is the Reynolds number; J is a dimensionless constant with the value 778; Δx is the change in quality of the refrigerant from inlet to outlet; L_f is the latent heat, in kJ/kg; and L is the tube length, in m.

Using Eq. (3.19), the mean fluid temperature in the collector tube t_{cm}, which is assumed to be same as evaporation temperature of the refrigerant, can be expressed as t_e:

$$t_e = \frac{1}{F}\left[t_p - (1-F)\left(\frac{\alpha I_T}{U_L} + t_a\right)\right]. \tag{9.57}$$

The efficiency of the collector/evaporator is defined by first equality of Eq. (3.13). The energy collected by solar collector/evaporator can also be expressed in terms of the enthalpy (h) change of the refrigerant from inlet to outlet of the evaporator and be expressed as Q_e

$$Q_e = m_r(h_1 - h_4), \tag{9.58}$$

where m_r is the mass flow rate of refrigerant, in kg/s.

9.4.4.2 Compressor

For a constant compressor speed operation, the mass flow rate m_r, in kg/s, of refrigerant pumped and circulated by the compressor is given by the following equation:

$$m_r = \frac{D_0 \lambda_k}{v_1}, \tag{9.59}$$

where D_0 is the theoretical displacement of compressor, in m^3/s; λ_k is the volumetric efficiency of compressor; and v_1 is the specific volume of absorbed vapor, in m^3/kg.

The displacement D_0, in m^3/s, for a reciprocating type compressor is given by [6]

$$D_0 = \frac{\pi d_k^2}{4}\frac{1}{60} S N n_k, \tag{9.60}$$

where d_k is the cylinders' diameter, in m; N is the number of cylinders; S is the piston path, in m; and n_k is the compressor speed, in rev/min.

The volumetric efficiency λ_k can be estimated using the equation [26]

$$\lambda_k = 1 + C - C\left(\frac{p_d}{p_s}\right)^{1/n}, \tag{9.61}$$

where C is the clearance volumetric ratio; p_d is the discharge pressure; p_s is the suction pressure; and n is the polytropic index.

Since the compression of refrigerant vapor is assumed to be a polytropic process, the specific compression work w can be calculated as follows [28]:

$$w = \frac{n}{n-1}\frac{p_s v_1}{\lambda_k}\left[\left(\frac{p_d}{p_s}\right)^{(n-1)/n} - 1\right]. \tag{9.62}$$

However, it can also be expressed in terms of the enthalpy change of the refrigerant from inlet to outlet of the compressor according Eq. (9.2). Electrical power consumed by the compressor P_{el}, in kW, is given by the following equation:

$$P_{el} = m_r w. \tag{9.63}$$

9.4.4.3 Hot-Water Storage Tank/Condenser

An energy balance on the nonstratified ST with the immersed condenser yields as

$$M_w c_{pw}\frac{dt_w}{d\tau} = Q_c - U_t A_t (t_w - t_r), \tag{9.64}$$

where M_w is the water mass, in kg; c_{pw} is the specific heat of water, in J/(kg·K); t_w is the water temperature, in K; t_r is the refrigerant temperature, in K; τ is the time, in s; Q_c is thermal power of condenser, in W; U_t is the tank heat loss coefficient, in W/(m^2·K); and A_t is the tank area, in m^2.

The total heat rejection from the condenser can be estimated as

$$Q_c = A_c U_c (t_c - t_w), \tag{9.65}$$

where t_c is the condensation temperature, in K; A_{cc} is the area of the cooling coil, in m^2; and U_{cc} is the overall heat transfer from the cooling coil, in W/(m^2 K) as given below:

$$U_{cc} = \cfrac{1}{\cfrac{A_{cc}}{\alpha_{cc}A_{ci}} + \cfrac{\delta_m A_{cc}}{\lambda_m A_{cm}} + \cfrac{1}{\alpha_w}}. \tag{9.66}$$

Neglecting the thermal resistance of the metal tubes (thickness of the tubes <3 mm), Eq. (9.66) can be simplified as

$$U_{cc} = \frac{1}{(A/\alpha_{cc})/(1/\alpha_w)}. \tag{9.67}$$

where $A = A_{cc}/A_{ci}$ and A_{ci} is the internal area of cooling coil.

The convective heat transfer coefficient α_{cc} of refrigerant inside the horizontal tubes, during condensation process, for two-phase annular flow can be estimated using following equation [29]

$$\alpha_{cc} = 0.0265 \frac{\lambda_l}{d_i} \left(\frac{G_r d_i}{\mu_l} \right)^{0.8} \left(\frac{c_{pr} \mu_l}{\lambda_l} \right)^{0.3}, \tag{9.68}$$

where λ_l is the thermal conductivity of liquid refrigerant, in W/(m·K); d_i is the internal tube diameter of condenser, in m; G_r is the mass velocity for two-phase flow, in kg/(s·m^2); μ_l is the dynamic viscosity of liquid, in Ns/m^2; and c_{pr} is the specific heat of refrigerant, in J/(kg·K).

The water-side convective heat transfer coefficient α_w is given by

$$\alpha_w = 0.5 \frac{\lambda_w}{d_o} \left(\frac{g \beta_t \Delta t d_o^3 \rho_w^2 c_{pw}}{\mu_w \lambda_w} \right)^{0.25}, \tag{9.69}$$

where λ_w is the thermal conductivity of water, in W/(m·K); d_o is the external tube diameter of condenser, in m; g is the gravitational acceleration, in m/s^2; β_t is the coefficient of thermal expansion, in K^{-1}; ρ_w is the water density, in kg/m^3; μ_w is the dynamic viscosity of water, in Ns/m^2; and c_{pw} is the specific heat of water, in J/(kg·K).

Heat dissipated to the cooling medium at the condenser Q_c includes heat gain at the evaporator Q_e and the power consumed by the compressor P_k, as shown by the following equation:

$$Q_c = Q_e + P_K. \tag{9.70}$$

Finally, the COP of the HP system is defined as

$$\mathrm{COP} = \frac{Q_c}{P_{el}}. \tag{9.71}$$

Based on the above-detailed analysis of each component of the DX-SAHP system, a FORTRAN program was developed [26] to estimate thermal performance of the said system.

9.4.5 Results and Discussions

9.4.5.1 Long-Term Thermal Performance of the System

The long-term system performance is usually determined by carrying out the transient system analysis that determines the thermal performance of the system at some chosen time interval, representing a reasonable compromise between computational effort and accuracy. As a part of the simulation strategy, the simulated system is assumed to operate only when the radiation is above 250 W/m^2, because at low radiation levels, the collector temperature is significantly below the ambient temperature and the COP of the system deteriorates rapidly [26]. The various input parameters used in the simulation program are listed in Table 9.7.

Fig. 9.21 shows the predicted variation in the monthly mean COP and collector efficiency η_c for a year-round performance. It is seen that during winter months (December—February) the COP of the system is higher (approximately 6.0) compared to the summer months (June—August) (approximately 4.0), which seems somewhat contradictory. This is because, though the solar radiation intensity as well as the initial water temperature in the ST is higher during summer, the rate of increase in the evaporation temperature is not that appreciable during summer, compared to winter period. The compressor speed n_k was 3000 rev/min. To improve the COP of the system during summer, it is

TABLE 9.7 Parameters Used in the Long-Term Performance Prediction

Parameter	Value
1. Latitude	31.17 degrees N
2. Collector area (A_c)	2 m^2
3. Absorptance (α)	0.96
4. Emittance (ε)	0.10
5. Ground reflectance (ρ_g)	0.25
6. Slope of collector (β)	90 degrees
7. Collector azimuth (γ)	0°
8. Storage volume (V_w)	150 l
9. Thermal conductivity of insulation (λ_i)	0.0346 W/(m K)
10. Thickness of insulation (δ_i)	0.05 m
11. Load temperature ($t_{w,o}$)	50°C
12. Polytropic index R22 (n)	1.18
13. Type of compressor	Reciprocating-type hermetic compressor
14. Displacement volume of compressor	0.0000743 m^3
15. Wind speed (u_w)	3.0 m/s (summer)
	3.2 m/s (winter)
16. Initial water temperature in the tank ($t_{w,i}$)	20°C (summer)
	10°C (winter)
17. Time-step size	300 s

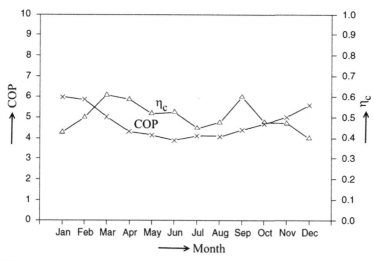

FIGURE 9.21 Variation in the monthly mean COP and collector efficiency in a year.

recommended to modulate the compressor capacity by lowering the compressor speed during the said period.

The monthly mean collector efficiency η_c ranges from 40% to 60% (Fig. 9.21), and the difference between the summer and the winter values is small. This is due to the fact that the temperature difference between the collector and the ambient temperatures does not vary much during any period. The results also indicate that an unglazed collector/evaporator has relatively higher collector efficiency compared to a conventional solar collector. This is mainly due to the direct circulation of the refrigerant in the collector/evaporator tubes, where the phase change also occurs in the same section. This produces a quenching effect on the collector and causes its temperature to drift to a lower value, thereby improves the collector efficiency significantly.

9.4.5.2 Effect of Various Parameters on the Performances of the System

Having established a set of typical results on a year-round performance of the system, the effect of various operating parameters, such as insolation, ambient temperature, collector area, storage volume and compressor speed, is studied subsequently.

- *Effect of solar radiation and ambient temperature.* Fig. 9.22 shows the effect of solar radiation intensity I_T and ambient temperature t_a on the system performances. In this case the collector area (A_c), wind speed (u_w), and initial water temperature ($t_{w,i}$) in the tank are chosen to be 2 m^2, 3 m/s, and 20°C, respectively. It is clearly seen that, with the increase in solar radiation, both the COP and collector efficiency η_c increase. This is mainly

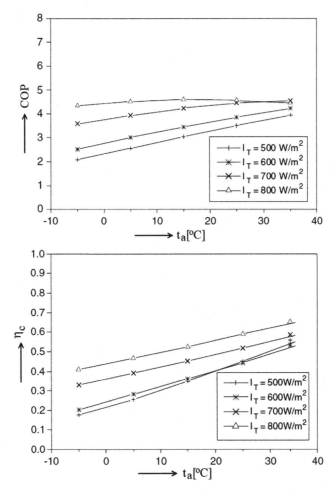

FIGURE 9.22   Effect of solar radiation and ambient temperature on the system performances.

because of two reasons: (1) an increase in intensity enables to attain a higher evaporating temperature of the refrigerant, consequently resulting in a higher system COP; and (2) the higher radiation intensity enables the collector to obtain much more useful energy collected and relatively small increase in heat loss from the collector, and hence, increasing the η_c. As indicated in Fig. 9.22, with the increase in I_T from 500 to 800 W/m^2, for a given t_a of 5°C, the system COP and collector efficiency η_c increases from 2.6 to 4.5 and from 26% to 47%, respectively.

Similarly, the system performances improve with increase in ambient temperature. It can be contributed to the fact that the rising t_a lowers the heat loss from the collector and increases the fluid temperature in the collector, which results in higher COP as well as η_c. From Fig. 9.22 it should also be noted that when I_T is higher, the increase in t_a does not greatly influence the system performances. This is because the useful energy gained by the collector is much higher compared to the reduction in heat loss from the collector to the environment.

- *Effect of collector area and storage volume.* Fig. 9.23 shows the combined effect of collector area A_c and storage volume on the system performances. In this case the radiation intensity (I_T), wind velocity (u_w), initial water temperature ($t_{w,i}$) in the tank, and ambient temperature (t_a) are chosen to be 600 W/m^2, 3 m/s, 20°C and 20°C, respectively. From Fig. 9.23 is seen that, initially, both the COP and η_c increase rapidly with the increase in storage volume, and then the increase is slow with further increase in the storage

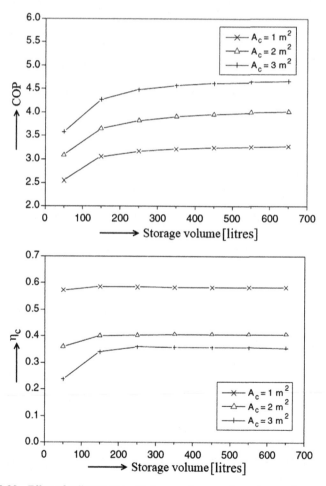

FIGURE 9.23 Effect of collector area and storage volume on the system performances.

volume. Nevertheless, for a given ST, with the increase in collector area, the COP increases and the η_c decreases significantly. This is mainly because of two reasons: (1) for a given A_c, if the storage volume increases, the condensation temperature decreases, which leads to a slight decrease in evaporation temperature causing an increased energy collected by the collector. A decrease in condensation temperature, but a relatively smaller decrease in evaporation temperature causes a reduction in compression work leading to a higher system COP. Meanwhile, the lower evaporation temperature of the refrigerant in the solar collector/evaporator reflects lower heat loss, thereby resulting in higher η_c; (2) for a given ST, larger A_c increases the fluid temperature in the collector, and hence, increasing the system COP but lowers the collector efficiency η_c.

It is interesting to note that, if the storage volume increases beyond a certain value, both the COP and the collector efficiency have very little improvement. This indicates that, while designing a DX-SAHP water heating system, an optimum size of ST should be chosen for a given collector area.

- *Effect of compressor rotational speed.* The diurnal and seasonal variations in solar radiation and ambient temperature lead to a large fluctuation in the load on the collector/evaporator panel, which results in degraded thermal performance of the system, due to the mismatch between the variable load on the collector and the constant capacity of the compressor. Fig. 9.24 shows the effect of compressor speed n_k on the system performances. In this case the radiation intensity (I_T), collector area (A_c), wind velocity (u_w), initial water temperature ($t_{w,i}$) in the tank, and ambient temperature (t_a) are chosen to be 600 W/m^2, 2 m^2, 3 m/s, 20°C, and 20°C, respectively. It is seen that, with the increase in n_k, the system COP decreases while the η_c increases rapidly. This is because, when the rotational speed of compressor

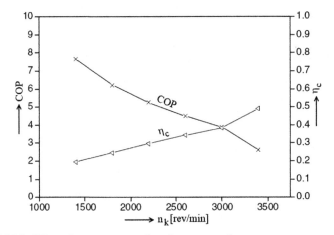

FIGURE 9.24   Effect of compressor speed on the system performances.

is higher, the mass flow rate of refrigerant increases, which leads to a higher compressor work, thereby resulting in lower values of the COP. On the other hand, a higher mass flow rate through the collector produces a strongly quenching effect on the collector, which in turn drops the collector temperature, thereby resulting in higher collector efficiencies η_c. This mismatch between the COP and η_c implies that, for a given A_c and ambient conditions, there exists an optimum n_k to enable the system to obtain a reasonable compromise between the COP and η_c.

9.4.6 Conclusions

Long-term thermal performance of the system has been predicted by a simulation program, which is based on a simple mathematical model of the DX-SAHP system.

Theoretically, it is shown that the thermal performances of the system are affected significantly by the variation of solar radiation, collector area, and compressor speed.

To minimize the mismatch between the variable load on the collector and the constant capacity of the compressor, a variable-speed compressor or an electronic expansion valve with a controller is recommended to be included in the system, which would positively improve the annual system thermal performance of the DX-SAHP system.

9.5 SOLAR-ASSISTED ABSORPTION HEAT PUMPS

9.5.1 Operation Principle

Absorption is the process by which a substance changes from one state into a different state. It requires very low or no electric input. Absorption HPs are thermally driven. The schematic diagram of a solar-assisted absorption HP system is shown in Fig. 9.25. The basic components of an absorption HP system are the generator G, absorber Ab, condenser C, evaporator E, expansion valve EV, and a solution pump P. The generator (desorber) G receives heat Q_g from the solar collector SC to regenerate the absorbent that has absorbed the refrigerant in the absorber. To ensure continuous operation and reliability of the system, a hot water ST is used. The extra thermal energy separates the refrigerant vapor from the rich solution with concentration ξ_r. The refrigerant vapor, with higher concentration ξ'' ($\xi'' \approx 1$), generated in this process are condensed in the condenser C, rejecting the condensation heat Q_c to a carried fluid (e.g., water, air, etc.), and then laminated in expansion valve EV_1. The regenerated absorbent from the generator (weak solution with concentration ξ_w) is sent back to the absorber Ab, where the absorbent-rich solution absorbs the refrigerant vapor from the evaporator E, rejecting the sorption heat Q_a to carried fluid. During this process, the absorber is cooled to keep its pressure at a low level. Then, the solution pump increases the pressure of the refrigerant/

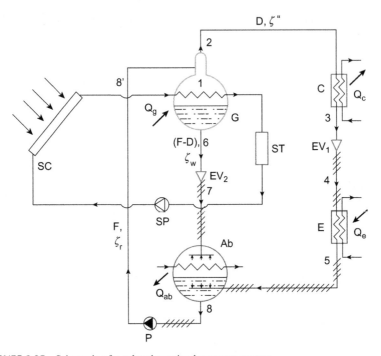

FIGURE 9.25   Schematic of a solar absorption heat pump system.

absorbent mixture (rich solution) to the high-pressure level. The solution pumps' electrical power requirement is much less than that of the compressors in the vapor-compression systems. In the evaporator, the liquefied refrigerant from the condenser evaporates, removing the heat Q_e from the cooling load. Most common working pairs are NH_3/H_2O and $H_2O/LiBr$.

Along with the basic components, certain heat recovery components are added to the absorption HP system to increase its COP. These heat recovery components are a solution heat exchanger (SHX) and a refrigerant precooler (PC). Also, additional refrigerant rectification equipment such as a rectifier and a dephlegmator is added in the design to rectify the refrigerant vapor in the case of a volatile absorbent. Rectification equipment is added to restrict the volatile absorbent (water) within the generator and absorber, thus preventing it from entering into the evaporator.

9.5.2 Thermodynamic Cycle

In a theoretical absorption cycle, the pressure losses are neglected and it is assumed that the generator pressure p_G is equal to the condenser temperature p_c (i.e., $p_G = p_c$) and also the absorber pressure p_{ab} is equal to the evaporator pressure p_e (i.e., $p_{ab} = p_e$). Therefore, the HP system has two pressures: higher pressure p_c (full line) and lower pressure p_e (locked line) (Fig. 9.25). To simplify the thermodynamic analysis of the system processes, a common temperature for providing the heat by HP condenser (t_c) and absorber (t_{ab}) is considered.

The concentration ξ of the refrigerant into binary solution is defined as a ratio between the refrigerant mass and the mixture mass as

$$\xi = \frac{M_{rf}}{M_{rf} + M_{ab}} \tag{9.72}$$

where M_{rf} is the refrigerant mass and M_{ab} is the absorbent mass.

The operational processes are illustrated in Fig. 9.26 for an absorption HP system working with NH_3/H_2O pairs.

The cooling power Q_e, in kW, the heat load at condensation Q_c, in kW, the absorber thermal power Q_{ab}, in kW, the generator thermal power Q_g, in kW, and the circulation factor f are calculated for previously presented processes using following thermal balance equations applied for E, C, Ab, and G [28]:

$$Q_e = D(h_5 - h_4) \tag{9.73}$$

$$Q_c = D(h_2 - h_3) \tag{9.74}$$

$$Q_{ab} = Dh_5 + (F - D)h_8 - Fh_8 \tag{9.75}$$

$$Q_g = Dh_2 + (F - D)h_8 - Fh_8 \tag{9.76}$$

and the mass balance equation applied for G:

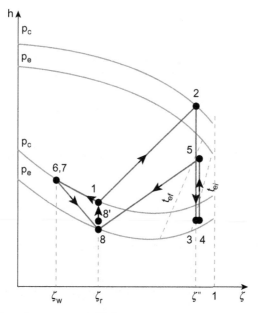

FIGURE 9.26 Absorption processes in hξ diagram.

$$F\xi_r = D\xi'' + (F - D)\xi_w \qquad (9.77)$$

$$f = \frac{F}{D} = \frac{\xi'' - \xi_w}{\xi_r - \xi_w}, \qquad (9.78)$$

where D is the mass flow rate of NH_3 vapor, in kg/s; F is the mass flow rate of rich solution (NH_3/H_2O), in kg/s; h is the specific enthalpy, in kJ/kg; ξ'' is the NH_3 vapor concentration; ξ_r is the concentration of rich solution; and ξ_w is the concentration of weak solution.

Thermal power (capacity) of HP Q_{hp}, in kW, is expressed as

$$Q_{hp} = Q_c + Q_{ab}. \qquad (9.79)$$

9.5.3 Energy Performance

Because the electrical power consumed by solution pump is negligible, in the heating operate mode the COP of HP is defined as

$$COP_{hp} = \frac{Q_{hp}}{Q_g}, \qquad (9.80)$$

where Q_{hp} is the thermal power of HP, in kW, and Q_g is the thermal power of generator, in kW.

The overall efficiency of the solar absorption HP system (COP_{sys}) or the instantaneous solar COP (COP_{sol}) is obtained from Eq. (9.81):

$$COP_{sys} = COP_{sol} = \frac{Q_{hp}}{Q_s}, \qquad (9.81)$$

where Q_s is the solar power received by the solar collector surface.

In the cooling mode, the EER is analogous to the COP but describes the cooling performance. The EER_{hp}, in Btu/(W·h) is defined as

$$EER_{hp} = \frac{Q_e}{Q_g}, \qquad (9.82)$$

where Q_e is the cooling power of HP, in British Thermal Unit per hour (Btu/h), and Q_g is the thermal power of generator, in W. The COP of an absorption HP in cooling mode is obtained by Eq. (9.16).

9.6 PERFORMANCES OF AN EXPERIMENTAL SOLAR-ASSISTED ABSORPTION GROUND-COUPLED HEAT PUMP IN COOLING OPERATION

The studies of solar absorption cooling systems integrated with geothermal energy through ground-coupled heat pump (GCHP) are really scarce. The purpose of this study is to analyze some of the influence parameters on the energy efficiency of a solar-assisted absorption GCHP in its cooling-mode operation [30].

9.6.1 Experimental Plant

The experimental plant is located in Valladolid, Spain (41 degrees 31′09″ N, 4 degrees 43′03″ W), in a tertiary-use building that has an area of 1200 m², a location with a continental Mediterranean climate. The building is occupied only on weekdays (Monday—Friday) from 7:00 to 15:00. In summer it is cooled by a radiant floor system [30].

Fig. 9.27 shows the experimental installation setup. It consists of absorption H_2O/LiBr HP (Thermax LT1) thermally driven by a solar collector array and by a natural gas-fired condensing boiler as support in times of low solar radiation. The HP is coupled to the ground through a closed-loop geothermal system. For the cooling operation in summer time, the evaporator is connected to the radiant floor system of the building and the condenser exchanges heat with the geothermal system. The main characteristics of the experimental plant are summarized in Table 9.8.

The absorption HP generator is powered by an 84 m² solar array with heat pipe technology (Vitosol 300) and a high-efficiency natural gas-fired condensing boiler (Viessman, Vitocrossal 300). Generator arrangement is complemented with a system of 8 m³ of thermal accumulation by water, distributed in four tanks of 2 m³ each. The storage was designed through a set of valves and pipe connections, which allows different configurations (one, two,

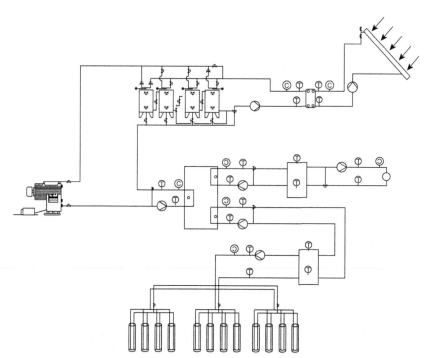

FIGURE 9.27 Schematic of the experimental setup.

TABLE 9.8 Main Characteristics of the Experimental Plant

Variable	Value	Unit
1. Building		
Area	1200	m^2
2. Absorption Chiller (H_2O/LiBr Thermax LT1 Single-Effect)		
Nominal cooling capacity	35	kW
COP nominal	0.70	
3. Solar Array		
Heat pipe collector (Vitosol 300)	84	m^2
Flow rate of the primary circuit	5.5	m^3/h
Flow rate of the secondary circuit	4.0	m^3/h
Storage volume	8.0	m^3
4. Geothermal Heat Exchanger (Mouvitech Collector PEN 32 × 3.0)		
Number of boreholes	12	
Depth borehole	100	m
Center-to-center half distance	0.0254	m
5. Radiant Floor		
Area	1000	m^2
Pipe spacing (center to center)	0.25	m
Pipe outside diameter	0.016	m

three or four in series, two by two in parallel, etc.). Because of the use of glycol in the solar circuit, there is a heat exchanger to separate this primary circuit from the secondary circuit, which goes up to the generator. Pumps, for both primary and secondary circuits, have variable-speed drives in order to implement different control strategies for the reduction of electrical energy demand.

In summer operation, the nominal 35 kW evaporator is connected to the radiant floor cooling system of the building. The nominal 80 kW condenser exchanges heat with the closed-loop geothermal system, which consists of 12 boreholes (100 m each) grouped in three blocks of four tubes. The first group has two simple and two double probes. The second group has four tubes with different fillings. The third group has two double and two high-turbulence probes. To avoid sudden changes in the operation of the system, the condenser and evaporator of the absorption HP are connected to two STs of 2 m^3 capacity each.

The installation has a complete instrumentation that provides instantaneous information on its actual behavior. The monitoring system consists mainly of temperature probes and flow meters located in the main five water circuits of the installation: the three circuits that enter the absorption GCHP (generator, condenser, and evaporator), the radiant floor circuit, and the geothermal circuit. The temperature sensors are TAC 100-100 with an accuracy of 1.3°C. Six temperature probes are located at the inlet and outlet of the three water circuits that exchange energy with the GCHP: generator ($t_{g,i}$, $t_{g,o}$), condenser ($t_{c,i}$, $t_{c,o}$), and evaporator($t_{e,i}$, $t_{e,o}$). Two temperature probes measure the inlet and outlet temperatures of the radiant floor circuit ($t_{f,i}$, $t_{f,o}$) and two more temperature probes measure the inlet and outlet temperatures of the ground heat exchanger (GHE) ($t_{GHE,i}$, $t_{GHE,o}$). Furthermore, there is an electromagnetic flow meter (ABB FXE4000, model DE43 F) in each one of these circuits, which allows to measure the water flow rate that goes through the circuits connected to the generator (m_g), to the evaporator (m_e), to the condenser (m_c), to the radiant floor (m_f), and to the GHE (m_{GHE}), with an accuracy of 0.5% of the measured value.

9.6.2 Development and Validation of the Simulation Model

A model of the experimental plant described above has been implemented with TRNSYS [31] to evaluate the plant its performances for different operating parameters.

9.6.2.1 TRNSYS Model

TRNSYS is a transient system simulation program with a modular structure that was designed to solve complex energy systems problems by breaking the problem down into a series of smaller components known as *Types*. Fig. 9.28 depicts the TRNSYS model scheme developed to simulate the experimental plant.

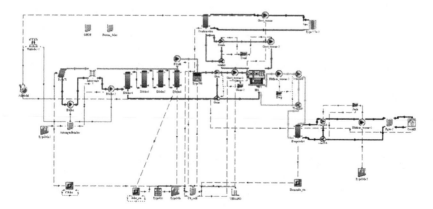

FIGURE 9.28   Schematic of the TRNSYS model of the experimental plant.

To simulate the performance of the Thermax LT1 absorption chiller, Type 107 and Type 71 were used to simulate the solar collectors of the solar array. Since standard efficiency curves are calculated for a single solar collector in clear days, at normal sun incidence and nominal flow rate of the water, a few correction factors are introduced in the model, in order to account for series connection, clouds, etc.

The natural gas-fired boiler used as auxiliary heater is represented by Type 700 in TRNSYS library. The boiler is activated only when its upstream water temperature is lower than a fixed set point.

A vertical GHE model must analyze the thermal interaction between the duct system and the ground, including the local thermal process around a pipe and the global thermal process through the storage and the surrounding ground. GHE has been modeled using the known as "duct ground heat storage model." This model assumes that the boreholes are placed uniformly within a cylindrical storage volume of ground. There is convective heat transfer within the pipes, and conductive heat transfer to the storage volume. Also, U-tube pipe parameters correspond to the properties of polyethylene pipes DN 32 mm PE100.

Type 4 is used to simulate the thermal energy STs. Simulation is based on the assumption that the tanks can be divided into N fully mixed equal sub-volumes. The tanks are also equipped with a pressure relief valve, in order to account for boiling effects. The model takes into account the energy released by the fluid flowing through the valve, whereas the corresponding loss of mass is neglected. The temperatures of the N nodes are calculated on the basis of unsteady energy and mass balances.

For the radiant floor cooling system in the building, TRNSYS Type 56 component was used in the model. Building was simulated using TRNBuild interface taking into account internal gains by occupancy and working hours.

9.6.2.2 Control Strategy

The control strategy that has been developed for the summer-operation simulation may be separated in four principal components, each one with its own conditions.

- *Solar array.* The solar field consists of three elements: solar collectors, primary pump, and heat exchanger. The primary pump will switch on between 9:00 and 20:00 h every day, and the secondary pump has two conditions, which it has to carry out before the switch on: first condition is that primary pump is switched on, and second condition is that the outlet temperature of the solar collectors is above the solar array STs' temperature.
- *Absorption HP.* The control for the switch on of the absorption HP has to achieve three simultaneous conditions:

    Schedule: As it was said before, the building is occupied between Monday and Friday from 7:00 to 15:00, so the HP only operates in these days.

Solar conditions: In this simulation the installation operates only with solar energy, without the boiler back up. The solar condition is that temperature of ST is above the set point generator temperature.

Demand condition: HP is in only operation when evaporator ST temperature is between 8 and 15°C.

- *GHE.* The control for the switch-on of the pump of the GHE has to fulfill two conditions: that absorption HP is in operation and that the temperature of the condenser ST is above the condenser set-point temperature.
- *Radiant floor.* The control for the switch on of the pump of the radiant floor circuit has to obey two conditions: the building occupancy schedule and the temperature of the evaporator ST being between 10 and 15°C.

9.6.2.3 Model Validation

A comparison between measured energy flows of the experimental plant and its predictions from the TRNSYS model described above is used to validate the model.

Experimental data were acquired during 13 days in the summer of 2011. Selected days were nonconsecutive: five days were at the end of June, five in middle July, and three more in different weeks of August. Experimental measurements were performed under real working conditions, during working hours in the office building, and for a continuous operation time of 4—5 h each day. Temperature and mass flow rate data were acquired in 5-min intervals.

The inlet and outlet temperatures (t_i, t_o) measured for the water circuits connected with generator, condenser, and evaporator of the HP, for the whole operation time $\Delta\tau$, and the mass flow rate m for each one of these circuits allow to calculate, using Eq. (9.83), the thermal energy Q, in kWh, to or from generator, condenser, and evaporator. The same stands for the GHE circuit and the radiant floor as

$$Q = mc_p(t_i - t_o)\Delta\tau, \tag{9.83}$$

where c_p is the specific heat at constant pressure, in kJ/(kg K).

Energy flows for each element (generator, condenser, evaporator, GHE, and radiant floor) were summed for the whole operation time ($\Delta\tau$) in the 13 days selected for model validation. Results were compared with those obtained with the TRNSYS model. A summary of the comparison is presented in Table 9.9. Deviation of the results provided by the model from the experimental data is within an acceptable limit, equivalent to other simulation results presented in literature for similar equipment [32].

9.6.3 Results

Simulation was run for the summer operation period of the installation (from 1st June to 30th September).

TABLE 9.9 Comparison Between Experimental and Simulated Data of Energy Transferred in the Installation

Specification	Q_g (kWh)	Q_c (kWh)	Q_e (kWh)	Q_{GHE} (kWh)	Q_f (kWh)	COP_{hp}
Measured data	1965.5	2601.7	662.5	2558.5	647.5	0.34
Simulation	2195.9	3020.3	742.2	2944.4	666.8	0.34
Deviation (%)	10.5	13.9	10.7	13.1	2.9	0.30

Performance of the installation will be evaluated in terms of COP of the HP (COP_{hp}), defined by Eq. (9.84), and the overall efficiency of the solar cooling system COP_{sys} defined by Eq. (9.85) as

$$COP_{hp} = \frac{Q_e}{Q_g} \tag{9.84}$$

$$COP_{sys} = \frac{Q_e}{Q_s} \tag{9.85}$$

in which

$$Q_s = \Sigma m_p c_p \left(t_{c,o} - t_{c,i} \right) \Delta \tau, \tag{9.86}$$

where Q_e and Q_g are the energies, in kWh, transferred to the evaporator and generator of the HP calculated by integrating numerically the respective heat flows; Q_s is the solar energy, in kWh, captured by the solar array during the considered period of time; m_p is the fluid flow rate through the solar array primary circuit in kg/s; c_p is the antifreeze fluid (water–glycol) specific heat, kJ/(kg·K); $t_{c,i}$ is the inlet temperature to the solar array, in °C; $t_{c,o}$ is the outlet temperature to the solar array, in °C; and $\Delta \tau$ is the time interval, in s.

9.6.3.1 Influence of the Generator Temperature on the System Performance

The generator temperature working range for the HP, according to the manufacturer specifications, is from 75 to 90°C. Simulation with TRNSYS model of the system were run for t_g 76, 78, 80, 82, 84, 85, and 90°C. The temperature at the inlet of condenser was fixed to 28°C, which is the nominal value according to manufacturer. The temperature at the inlet of the evaporator was fixed to 15°C as a design parameter. Fig. 9.29 depicts graphically the influence of the generator temperature $t_{g,i}$ on the COP_{hp} and on the COP_{sys}.

The COP_{hp} value has a very small increase from 0.23 to 0.24 when the generator temperature increases from 76 to 80°C. For higher t_g, the COP_{hp} increases more strongly with temperature, reaching a maximum value of 0.43 for t_g of 90°C. On the other hand, the COP_{sys} remains near-constant with a mean value of 0.135 for t_g from 76 to 84°C. For higher generator temperature, the COP_{sys} starts to decrease when t_g increases, reaching a minimum value of 0.063 when t_g is 90°C.

Results stands that the optimum working t_g for the HP in terms of its best COP_{hp} is not the best working temperature in terms of COP_{sys} for the overall system. This deals with the fact that trying to reach higher temperatures for the t_g in the solar field reduces the number of hours that the system is working. So the low COP_{hp} for low generator temperatures is compensated with the increasing number of hours that the system can work exclusively with solar energy for these low temperatures.

9.6.3.2 Influence of the Condenser Temperature on the System Performance

The condenser temperature working range for the HP, according to the manufacturer specifications, is from 25 to 29°C. Simulations with TRNSYS model of the installation were run for $t_{c,i}$ 25, 26, 27, 28, and 29°C. The temperature at the inlet of generator was fixed to 85°C, which is the HP nominal value according to manufacturer. The temperature at the inlet of the evaporator was fixed to 15°C as a design parameter. Fig. 9.30 illustrates the influence of the condensation temperature t_c on the COP$_{hp}$ and on the COP$_{sys}$.

As expected, Fig. 9.30 shows that the lower the condensation temperature is, the higher are the COP$_{hp}$ and the COP$_{sys}$. If the temperature at the inlet of the condenser decreases from the nominal 28 to 25°C, the COP$_{sys}$ increases

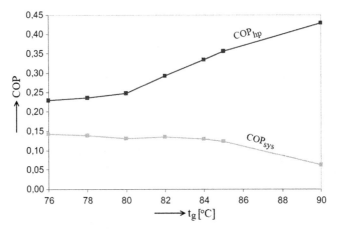

FIGURE 9.29   COP$_{hp}$ and COP$_{sys}$ depending on the generator temperature.

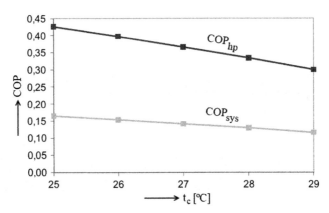

FIGURE 9.30   COP$_{hp}$ and COP$_{sys}$ depending on the condenser temperature.

from 0.13 to 0.17, and the COP_{hp} increases from 0.33 to 0.43, that is a 30% increase in both magnitudes.

9.6.4 Conclusions

The COP_{sys} for high temperatures at the condenser ($t_{c,i}$ 28 and 29°C) is near-independent of the generator temperature when the t_g is between 76 and 85°C. Nevertheless, for low temperatures at the condenser ($t_{c,i}$ 25 and 26°C), the influence of the generator temperature is significant. For a condenser temperature of 25°C the COP_{sys} increases nearly by 40% when the t_g decreases from 84 to 76°C.

REFERENCES

[1] Seppänen O. European parliament adopted the directive on the use of renewable energy sources. Rehva Journal 2009;46(1):12−4.

[2] Lazzarin RM. Dual source heat pump systems: operation and performance. Energy and Buildings 2012;52:77−85.

[3] Kaygusuz K. Performance of solar-assisted heat pump systems. Applied Energy 1995;51:93−109.

[4] Thygesen R, Karlsson B. Economic and energy analysis of three solar assisted heat pump systems in near zero energy buildings. Energy and Buildings 2013;66:77−87.

[5] Heinonen EW, Tapscott RE, Wildin MW, Beall AN. NMERI 96/15/32580 Assessment of anti-freeze solutions for ground-source heat pump systems. New Mexico Engineering Research Institute; 1996.

[6] Sarbu I, Sebarchievici C. Ground-source heat pumps: fundamentals, experiments and applications. Oxford: Elsevier; 2015.

[7] Bayer P, Saner D, Bolay S, Rybach I, Blum P. Greenhouse gas emission savings of ground source heat pump systems in Europe. Renewable and Sustainable Energy Reviews 2012;16:1256−67.

[8] IEE, Intelligent Energy Europe. http://ec.europa.eu/energy/environment; 2013.

[9] Sarbu I, Sebarchievici C. General review of ground-source heat pump system for heating and cooling of buildings. Energy and Buildings 2014;70(2):441−54.

[10] Ochsner K. Geothermal heat pumps: a guide to planning & installing. London-Sterling: Earhscan; 2007.

[11] Liang C, Zhang X, Li X, Zhu X. Study on the performance of a solar assisted air source heat pump system for building heating. Energy and Buildings 2011;43:2188−96.

[12] Pahud D, Mattthey B. Comparison of the thermal performance of double U-pipe borehole heat exchanger measured in situ. Energy and Buildings 2001;33(5):503−7.

[13] Bose JE, Smith MD, Spitler JD. Advances in ground source heat pump systems—an international overview. In: Proceedings of the 7th International conference on energy agency heat pump, Beijing, China; 2002. p. 313−24.

[14] Luo J, Rohn J, Bayer M, Priess A. Modeling and experiments on energy loss in horizontal connecting pipe of vertical ground source heat pump system. Applied Thermal Engineering 2013;60:55−64.

[15] ASHRAE handbook. HVAC applications. Atlanta (GA): American Society of Heating, Refrigerating and Air−Conditioning Engineers; 2015.

[16]   ASHRAE. Commercial/institutional ground-source heat pump engineering manual. Atlanta (GA): American Society of Heating, Refrigerating and Air-Conditioning Engineers; 1995.

[17]   Rawlings RHD, Sykulski JR. Ground source heat pumps: a technology review. Building Services Engineering Research and Technology 1999;20(3):119–29.

[18]   Floridesa G, Kalogirou S. Ground heat exchanger—a review of systems, models and applications. Renewable Energy 2007;32(15):2461–78.

[19]   Philappacopoulus AJ, Berndt ML. Influence of rebounding in ground heat exchangers used with geothermal heat pumps. Geothermic 2001;30(5):527–45.

[20]   VIESSMANN. Heat pump systems—design guide. Romania. 2002.

[21]   Tinti F. Geotermia per la climatizzazione. Palermo (Italy): Dario Flaccovio Editore; 2008.

[22]   Sarbu I, Bura H. Thermal tests on borehole heat exchangers for ground-coupled heat pump systems. International Journal of Energy and Environment 2011;5(3):385–93.

[23]   Omer AM. Ground-source heat pumps systems and applications. Renewable and Sustainable Energy Reviews 2008;12(2):344–71.

[24]   Bernier M. Closed-loop ground-coupled heat pump systems. ASHRAE Journal 2006; 48(9):13–24.

[25]   Hellström G. Ground heat storage: thermal analyses of duct storage systems. Doctoral thesis. Sweden: Department of Mathematical Physics, University of Lund; 1991.

[26]   Kuang YH, Sumathy K, Wang RZ. Study on a direct-expansion solar-assisted heat pump water heating system. International Journal of Energy Research 2003;27:531–48.

[27]   Chaturvedi SK, Chiang YF, Roberts AS. Analysis of two-phase flow solar collectors with application to heat pumps. Journal of Solar Energy Engineering. ASME Transactions 1982;104:358–65.

[28]   Sarbu I, Sebarchievici C. Heat pumps. Timisoara. Romania: Polytechnic Publishing House; 2010 [in Romanian].

[29]   Jordan RC, Priester GB. Refrigeration and air conditioning. New York: Prentice-Hall, Inc; 1948.

[30]   Macía A, Bujedo LA, Magraner T, Chamorro CR. Influence parameters on the performance of an experimental solar-assisted ground-coupled absorption heat pump in cooling operation. Energy and Buildings 2013;66:282–8.

[31]   TRNSYS 17. A transient system simulation program user manual. Madison, WI: Solar Energy Laboratory, University of Wisconsin-Madison; 2012.

[32]   Monné C, Alonso S, Palacin F, Serra L. Monitoring and simulation of an existing solar powered absorption cooling system in Zaragoza (Spain). Applied Thermal Engineering 2011;31(1):28–35.

Index

'Note: Page numbers followed by "f" indicate figures, "t" indicate tables.'

Printed in the United States
By Bookmasters